Topology Optimization Theory for Laminar Flow

Yongbo Deng · Yihui Wu
Zhenyu Liu

Topology Optimization Theory for Laminar Flow

Applications in Inverse Design of Microfluidics

Springer

Yongbo Deng
Changchun Institute of Optics, Fine Mechanics and Physics (CIOMP)
Chinese Academy of Sciences
Changchun
China

Yihui Wu
Changchun Institute of Optics, Fine Mechanics and Physics (CIOMP)
Chinese Academy of Sciences
Changchun
China

Zhenyu Liu
Changchun Institute of Optics, Fine Mechanics and Physics (CIOMP)
Chinese Academy of Sciences
Changchun
China

ISBN 978-981-10-4686-5 ISBN 978-981-10-4687-2 (eBook)
https://doi.org/10.1007/978-981-10-4687-2

Library of Congress Control Number: 2017951416

Printed on acid-free paper

This Springer imprint is published by Springer Nature
The registered company is Springer Nature Singapore Pte Ltd.
The registered company address is: 152 Beach Road, #21-01/04 Gateway East, Singapore 189721, Singapore

To our parents and families.

Preface

Topology optimization can determine the topology, shape, and size of the structures, simultaneously. It has been regarded to be one of the most powerful approaches for the inverse design of structures in several scientific areas, including elasticity, fluid dynamics, thermodynamics, electromagnetics, etc. In fluid dynamics, one of the hot research areas is the microfluidics. And microfluidics has two typical characteristics, i.e., low Reynolds number and large surface–volume ratio. These two characteristics induces the low-efficient mass transfer and dominance of the surface tension in microfluidics. Because of these specialities, the usually used design approaches encounter the problem on synthetically considering the typical characteristics of microfluidics and manufacturability to achieve reasonable microfluidic performance. Therefore, the authors were inspired to utilize the topology optimization method and implement the inverse design of microfluidics based on the desired functional performance. Microfluidics usually has transient velocity distribution; and the actuation mechanism of microfluidics usually loads body forces or surface tensions on the microflows. However, the preliminary researches on topology optimization of flow problems mainly focused on the steady single-phase flows without body forces (including the researches of T. Borrvall & J. Petersson in 2003, A. Gersborg-Hansen, O. Sigmund & R.B. Haber in 2005, J.K. Guest & J.H. Proevost in 2006, etc). Therefore, the authors began their attempts on developing the topology optimization method for unsteady flows, flows with body forces and two-phase flows with surface tensions. In the authors' researches, the continuous adjoint method is systematically utilized to analyze the topology optimization problems; and this is can be one feature of this book. After the extension of topology optimization method, the authors have implemented several researches on the inverse design of typical microfluidic components, including micromixers, microvalves, and micropumps. There is still a plenty of room in the inverse design of microfluidics using topology optimization. The authors hope that this book can inspire more researchers in the relevant areas.

This book is a completely revised, updated, and expanded version of the authors' researches on topology optimization method for fluid dynamics. The authors knew each other since 2007, when Dr. Yongbo Deng was the Ph.D. student of Prof. Yihui

Wu and Prof. Zhenyu Liu. Before Yongbo Deng's Ph.D. defense, the authors collaborated together for pursuing the topology optimization of flow problems. During this period, they achieved the topology optimization for unsteady flows and flows with body forces; simultaneously, they have applied the developed topology optimization to implement the inverse design of the typical microfluidic components, including the passive micromixers, no-moving-part microvalves, and valveless micropumps. In the researches of passive micromixers, Prof. Zhenyu Liu's Master student Qingyong Gao played a part of the role on implementing the numerical computation; Dr. Yongshun Liu from Changchun Institute of Optics, Fine Mechanics and Physics (CIOMP) implemented the experimental verification of the derived layouts for the passive micromixers. In the researches of no-moving-part microvalves and valveless micropumps, Yongbo Deng and Zhenyu Liu began their academic collaboration with Prof. Jan Gerrit Korvink of Freiburg University, Germany (Prof. Korvink moved to Karlsruhe Institute of Technology in 2015). After his Ph.D. study, Yongbo Deng started his professional position in the State Key Laboratory of Applied Optics (SKLAO) of CIOMP, from July of 2012. Sequentially, he achieved his researches on topology optimization of two-phase flow with two immiscible fluids and combination of the topology optimization and optimal control methods. Simultaneously, he began his attempt on applying topology optimization in electroosmotics, which is an advantageous actuation mechanism for microfluidics. During his researches on electroosmotics, Yongbo Deng knew Prof. Shizhi Qian (Old Dominion University, Norfolk, USA) and began their relevant academic collaboration. Additionally, the authors' Ph.D. candidates Song Zhou, Jianhua Fan and Yuan Ji, and the authors' colleagues Dr. Teng Zhou, Prof. Ping Zhang and Dr. Junfeng Wu participated in the relevant researches.

The researches in this book were supported by the National Natural Science Foundation of China (Nos. 51405465, 50975272 and 11034007), the National High Technology Research and Development program of China (Nos. 2015AA042604, 2007AA042102 and 2012AA040503), Prof. Zhenyu Liu's Hundred Talent Project in Chinese Academy of Sciences (CAS), the Science and Technology Development Plan of Jilin Province and Changchun City(Nos. 20140201011GX, 20170312027ZG, 15SS12) and the SKLAO open funding in CIOMP. The authors are also grateful to the editors' and reviewers' kind attentions and valuable suggestions for their past research articles. Especially, the authors are grateful to Prof. Krister Svanberg (Royal Institute of Technology, Sweden) for his supplying the MATLAB codes of the method of moving asymptotes.

Changchun, China
July 2017

Yongbo Deng

Contents

Chapter 1
Introduction

Abstract In this chapter, the state of the art is introduced for the topology optimization of flow problems. The topology optimization method will be introduced for the steady and unsteady flows with body forces, respectively based on the density method and level set method. And the level set method is further used to extend the topology optimization into the area of two-phase flow with two immiscible fluids. It is highlighted that the continuous adjoint method is utilized to implement the analysis of the topology optimization, and this can effectively reduce the dependence on the numerical methods used to discretize the partial differential equations. And the developed topology optimization methods are applied to inversely design the microstructures for the hot microfluidics, and demonstrate the power of topology optimization.

1.1 Topology Optimization for Single-Phase Flow Problems

The layout optimization problems have attracted research interest of mathematicians and engineers for several decades [1, 2]. Usually, layout optimization can be categorized into three types, i.e. size optimization, shape optimization and topology optimization. Those three types can be easily distinguished as demonstrated in Fig. 1.1, where the compliance of the cantilever shown in Fig. 1.1 a is minimized respectively using the size optimization, shape optimization and topology optimization approaches. In size optimization, the structure is parameterized by the sizes and positions of geometrical characteristics, and the corresponding parameters are chosen to be the optimization variables; meanwhile, the shape optimization parameterizes the geometrical characteristics of structures using the spline interpolation [3, 4] etc., where the sample points are the optimization variables. After parameterization, the optimization variables are iteratively evolved in the preset feasible regions, using global optimization or gradient information-based optimization algorithm [5]. Size optimization and shape optimization are subjected to the inflexibility on changing the topology of the initial guesses of the structure, and hence have relatively strong dependence on the initial guesses of the optimization variables. Being different from the size and shape optimization, topology optimization is a more powerful tool for

Y. Deng et al., *Topology Optimization Theory for Laminar Flow*,
https://doi.org/10.1007/978-981-10-4687-2_1

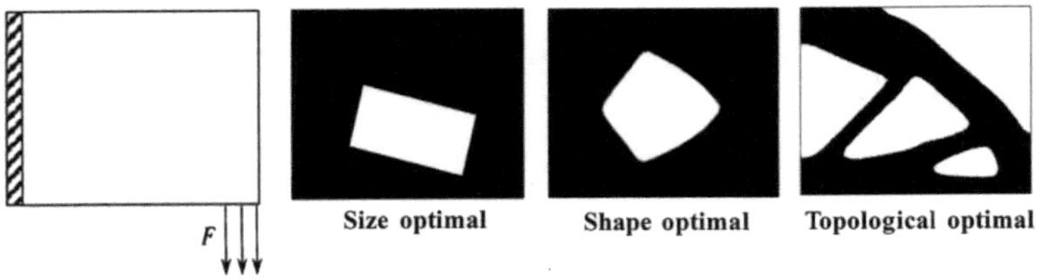

Fig. 1.1 Size optimization, shape optimization and topology optimization of a cantilever with minimizing compliance

optimizing the structures, with material distribution represented by the optimization variables. Because the material distribution is determined, topology optimization can fix the topology, shape and size of structures simultaneously. Therefore, topology optimization is more flexible and robust for implementing layout optimization.

Topology optimization is originated from the structural optimization problem in elasticity and compliance mechanisms [6–9]. And it has been extended to multiple physical problems, such as acoustics, electromagnetics, fluidics, optics and thermal dynamics [10–16]. Several approaches, such as the evolutionary techniques [17], the homogenization method [6, 18], the density method [19, 20] and the level set method [21–24], have been developed for the implementation of topology optimization. Especially, the density type method and level set method have been widespread utilized in the area of topology optimization. And the discussed topology optimization in this book are mainly based these two methods.

Topology optimization for fluid problems was pioneered by Steven, Li and Xie using the evolutionary techniques [17]. On the density method, Borrvall and Petersson performed the first attempt for the Stokes flow [11]. The density method was further researched for Stokes flows in [25, 26] and for Darcy-Stokes flows in [27, 28]. In the density method, an artificial friction force that is proportional to the fluid velocity was added to the Stokes equations in order to smoothly interpolate between the solid and fluid regions. This optimization model was extended to the Navier-Stokes flows with low and moderate Reynolds numbers in [26, 27, 29–31], and non-Newtonian flows in [32]. The density method-based topology optimization has been applied to design fluidic devices in [31, 33–35]. The density method-based topology optimization has the merits on rapid convergence and weak dependence on the initial guess. On the level set method, pioneered by Osher and Sethian [36], it accomplishes the change of topology by evolving and merging the zero contour of the level set function. This method can provide a general way to track the implicit interface between two phases, and it has been applied to fluidic shape and topology optimization [37, 38]. The major advantages of the level set method lies in expressing continuously moving interfaces and abstracting the material domains that correspond to the structural topology. It has been observed that the conventional level set method may be inadequate for the cases in which the initial shape of the structure has fewer holes than the optimal geometry [22], especially in two-dimensional cases. This difficulty can be overcome using topological sensitivity, which was introduced by

Sokolowski and Zochowski [39] for linear elasticity and that has been extended to several other linear and nonlinear physical problems [40–43]. The topological sensitivity has been researched for steady Stokes flows [44–47] and Navier-Stokes flows [48].

Topology optimization of fluidic flow channels has focused primarily on the steady flows without body forces [10, 11, 26, 27, 29, 30, 32, 38, 44]. In contrast, unsteady flows are widespread in reality, and the optimization of unsteady flows using the shape optimization method has been researched [3, 4, 49]. As such, it is desirable to extend topology optimization to the area of unsteady Navier-Stokes flows to reveal the dynamic effect of the unsteady flows on the optimal topology. Kreissl et al. implemented the topology optimization of unsteady flows [50]. In Kreissl's work, the optimization problem is analyzed using the stabilized SUPG and PSPG finite element formulation and the corresponding discretized adjoint equations. In Sect. 2.1, the density method-based topology optimization method is extended to unsteady incompressible Navier-Stokes flows at low and moderate Reynolds numbers based on work by Borrvall and Petersson [11] and Gersborg-Hansen et al. [29]. Being different with Kreissl's work, the topology optimization problem with more general objectives is analyzed using the continuous adjoint method; based on the Navier-Stokes equations and the derived continuous adjoint equations, the optimization can be implemented by choosing any stable spatial discretization method and adaptive temporal discretization method, such as the finite difference method, the finite element method or the finite volume method, although the numerical discretization of the Navier-stokes equations and the adjoint equations is implemented using the standard finite element method with Taylor-Hood elements. In Sect. 2.2, the topology optimization of unsteady Navier-Stokes flows is implemented using the level set method. This proposed optimization method can be used to optimize the unsteady flows with or without body forces and perform the optimization of unsteady flows in the engineering reality. By coupling the unsteady Navier-Stokes equations with the level set method, the optimization problem is constructed, and it is analyzed using the continuous adjoint method.

In the Navier-Stokes flows, external body forces that relate with the fluid inertia effect, such as the gravity, centrifugal force and Coriolis force, usually exist. It is necessary to develop a topology optimization method for the Navier-Stokes flows with body forces. However, the optimization model proposed by Borrvall and Petersson in [11] is hard to be extended directly to the flows with external forces. One of the reasons is that the two kinds of force terms exist in the Navier-Stokes equations, i.e. external force which has specific physical meaning, and the artificial friction force which is used to control the topology of fluid domains. Typically, the interpolation of artificial friction force must be modified so that the external body force can drive the motion of fluid and the artificial friction force can modify the topology of the fluid domain reasonably. A similar problem exists for the topology optimization of mechanisms using the density method. The standard interpolation function, such as the solid isotropic material with penalization (SIMP), has been used successfully for designing mechanisms with fixed loads. However, the SIMP can not be used directly to design mechanisms when the body forces (also called design-dependent loads), such as the self-weight of a structure or centrifugal force, are used as loads [51–53]. To overcome the above challenge, one can either modify the interpolation function for the design variable [54], or implement topology optimization using the level set method [22]. For both of these strategies, the key point is to choose a suitable way to deal with the design-dependent loads. Therefore, in view of the merits of the density method on efficient convergency, weak dependence of the initial guess and dealing with multiple constraints, the density method is used to implement the topology optimization of the steady and unsteady Navier-Stokes flows with body forces in Sect. 3.1; the continuous adjoint equations and adjoint sensitivity for the general objective are presented, and the numerical implementation of the optimization method is discussed with several numerical examples. In Sect. 3.2, the level set method is used to extend the topology optimization method to the Navier-Stokes flows with body forces; in this method, the shape and topological sensitivities are considered simultaneously; the solid and fluid domains are distinguished by constraining the fluid velocity to vanish in solid domain; the shape and topological sensitivities for two commonly used objectives, the dissipation power and the reverse of the kinetic energy, are considered separately, based on the continuous adjoint method. This topology optimization method can be used to perform the optimal design of the fluid devices with constant, nonuniform and solution-dependent forces.

1.2 Topology Optimization for Two-Phase Flow Problems

Along with the single phase flows without surface tension, the two-phase flow of immiscible fluids with surface tension plays a fundamental role in many natural and industrial phenomena. Examples can be found in the studies of capillarity, low-gravity fluid flow, hydrodynamic stability, surfactant behavior, cavitation, and droplet dynamics. Therefore, Chap. 4 of this book focuses on developing the topology optimization method for the two-phase flows of immiscible fluids, for the academic and engineering requirement.

At the interface of two immiscible fluids, the cohesive forces within the fluid and adhesive forces between the fluid and its surroundings give rise to the surface tension [55]. Due to the surface tension, the complexity on dealing with moving interfaces in two phase flows are encountered from the mathematical modeling and numerical algorithmic points of view. There are many approaches to characterize moving interfaces. The two main approaches are interface tracking and interface capturing methods. In interface tracking methods (e.g. volume-of-fluid [56], front-tracking [57], and immersed boundary [58, 59]), Lagrangian particles are used to track the interfaces, and these particles are advected by the velocity field, where the difficulties on capturing the topological changes of the moving interface are encountered. In the interface capturing methods (e.g. the level set method [60–64] and diffuse interface method [65–68]), the interface is implicitly captured by a contour of a particular scalar function, and the topological changes of the interface can be flexibly treated.

Therefore, the robust interface capturing methods are more popularly used in the simulation of two-phase flows, compared to the interface tracking methods.

In topology optimization of fluid flows, the solid phase is approximated using the fluid velocity penalization method, which is implemented by adding the artificial Darcy forces into the Navier-Stokes equations [11]. Based on the fluid velocity penalization method, the no-slip boundary condition is achieved at the implicit interface between the fluid and solid phases. In the simulation of two-phase flows, the stress singularity and infinite dissipation problem are encountered near the three-phase contact line of the two-phase flow, when the no-slip boundary condition is imposed on the walls of the two-phase flow directly [69, 70]. The singularity can be solved using the slip model [69] or diffuse interface method [70]. In the slip model, the slip boundary condition with the slip velocity proportional to the tangential stress is imposed on the walls. This is incompatible with the fluid velocity penalization method. On the other hand, the diffuse interface method solves the singularity and evolves the interface and three-phase contact line based on the molecular diffusion driven by chemical potential, with the no slip condition imposed on the wall. Therefore, the diffuse interface method is the reasonable choice to describe the two-phase flow in the corresponding topology optimization problem with no slip boundary condition at the fluid-solid interface with the solid phase approximated using the fluid velocity penalization method.

In the diffuse interface method for two-phase flow, the phase field is used to distinguish the two immiscible fluids, and the evolution of the phase field is governed by the Cahn-Hilliard equations, which is obtained based on the surface free energy minimum principle [71]. In topology optimization, the wall energy (i.e. the surface free energy at the fluid-solid interface) of the implicit solid-fluid boundary of the two-phase flow is one part of the total surface free energy. In order to consider the wall energy of the implicit boundary, the implicit boundary is required to have an exact mathematical expression on its normal vector and curvature. Therefore, the level set method, which can exactly satisfy this requirement [21, 22, 37, 38, 63, 64, 72, 73], is adopted to implement the topology optimization of two-phase flows.

It is noticed that the phase field is used to describe the two immiscible fluids in the dynamics of two-phase flow, and it is not the case in phase field method-based topology optimization [74]; instead, the level set method is utilized to implement the topology optimization of two-phase flow, where the solid phase is approximated by the fluid velocity penalization method.

1.3 Combination of Topology Optimization and Optimal Control

The optimal control method has numerous applications spanning many branches of engineering science, such as internal combustion engines, airbreathing engines, heat exchangers, control of heat rejection devices in computers, flow control problems in ink jet printers, manufacturing problems in microelectronic chips, control of pollutant transport, control of metal forming, control of chemical reactors, control of aerodynamic surfaces, and weather prediction [75]. For many years, optimal control has been an ad-hoc subject in engineering science despite its importance. Advances in high-performance computing of three dimensional time dependent problems and recent developments in nonlinear infinite dimensional systems theory and nonlinear partial differential equations have set the stage for optimal control theory. The subject has been under rapid development since the early works in [76–83]. The impact of this development to continuum mechanics has been essentially in the branch of solid mechanics. Several theorems relating the Navier-Stokes equations and the minimization of certain functionals were established in [84–88], thus starting the optimal control method in fluid mechanics.

In topology optimization, the optimal topology of materials is computed under the given control. At the same time, the optimal control is calculated under the fixed material layout in optimal control method. Naturally, one has the question that is how to find the optimal match of the material topology and control to present the reasonable designs of the material layout and control simultaneously. Therefore, Chap. 5 is focused on the combination of topology optimization and optimal control method to find the optimal match between the material topology and control. Based on the combination of topology optimization and optimal control method, the layout and control of a device can be optimized simultaneously, and the freedom of the engineering design can also be enlarged. Because the optimal control problem includes two types, i.e. the optimal distribution control and optimal boundary control, the combination discussed in this paper can be respectively categorized into the combination of topology optimization with optimal distribution control and optimal boundary control. The specific physical implication of control can be given according to different cases, e.g. physical implication of the distribution control in the heat transfer problem is heating source and that of the boundary control in the backstep flow control is blowing and sucking velocity.

In Chap. 5, the abstract form of the variational problem and the adjoint analysis are stated for the combination method; the numerical implementation on solving the corresponding variational problem are presented; and several test problems are presented to demonstrate and extend the combination method into the area of heat transferring and elasticity.

1.4 Topology Optimization-Based Inverse Design for Microfluidics

Miniaturization is a recent trend in analytical chemistry and life sciences. In the past two decades, miniaturization of fluid handling and fluid analysis has emerged in the interdisciplinary research field of microfluidics. Microfluidic applications cover micro arrays, DNA sequencing, sample preparation and analysis, cell separation and detection, and environmental monitoring. The use of microfluidics in these applications has attracted the interest of industry and academia, because of its potential and advantages, such as small amounts of sample and reagent, less time consumption, lower cost, and high throughput [89]. Miniaturization gives rise to the characteristics of microfluidics. One of these characteristics is the low Reynolds number, which means that microfluidic flow is laminar flow typically [90].

Lab on a chip is the carrier of the microfluidics. It is the generic term for the integration of the microdevices to carry out conventional analytical laboratory tests. Such devices offer significant benefits over traditional laboratory tests in terms of device size, sample/reagent usage, and can provide much faster results for chemical and biochemical analyses [91, 92]. Because of these advantages, lab on a chip devices are considered a promising option for the development of miniaturized devices for the environmental and defense monitoring, chemical synthesis and biomedical applications. In lab on a chip systems, various subcomponents such as pumps, mixers, reactors, and dilution chambers are integrated. Therefore, the study of fluid flow in microscale, i.e. microfluidics, has become central to the development of lab on a chip devices [93–95]. Because topology optimization has the merit on extended design space for structures compared with the conventional size and shape optimization, it can provide one powerful inverse design tool for microstructures in microfluidics. And several microfluidic components are inversely designed in Chap. 6, to demonstrate the power of the topology optimization for microfluidics.

1.4.1 Inverse Design of Micromixer

Micromixer is an important component in a microfluidic system. The applications of micromixing technologies can be categorized into: chemical applications, biological applications, and detection/analysis of chemical or biochemical content [96–98]. Mixing is fundamentally a diffusion process. At the boundary of an impurity in a fluid, the concentration gradient is large, and material flows until the gradient vanishes. However, diffusion alone is inefficient. Moving fluids can greatly enhance mixing through chaotic advection in which the region containing the impurity is strongly deformed; the length of its boundary grows exponentially, and diffusion becomes efficient [99]. Therefore, the two factors that influence the mixing performance of a micromixer are diffusion and chaotic advection [89, 100]. The relative importance of these two factors can be measured by the dimensionless number — Péclet number Pe. When the Péclet number satisfies $\mathrm{Pe} \gg 1$, the main mixing factor is the fluidic convection. Otherwise, the main factor is the diffusion. The convection intensity of the flow in a micromixer can be measured by the Reynolds number Re. When $\mathrm{Re} \gg 1$, convection dominates the flow; otherwise, viscosity dominates [89].

Two categories of micromixers are the active type and the passive type [89, 96, 101]. The active micromixers utilize external actuation, e.g. mechanical pulsation [102, 103] and electroosmotic flows [104–108], to create the chaotic advection of the flow. The passive micromixer implements the mixing of reagents for the steady flow. The chaotic advection of the steady flow that stretches and folds volumes of fluid over the cross section of the microchannel are induced by the layout of the microchannel. By the reasonable layout of the microchannel in a micromixer, the chaotic advection can be strengthened, and the mixing performance can be improved.

For passive micromixers, most of the designs are based on the trial and error method; the enhancement of the mixing performance of passive micromixers is based on the shape optimization method [109–116] and the mapping method [117–124]. Because the shape optimization method cannot change the topology of a micromixer, the extent to which it can improve the mixing efficiency is limited by the initial choice of the topology of the micromixer. The mapping method is used to optimize the micromixers with spatial periodic layouts, and this is convenient for the design of micromixers in practice. However, reasonable aperiodic design is more helpful to improve the mixing performance [125]. Compared to shape optimization, topology optimization can modify both the detailed topology and shape of microchannels simultaneously. In [35], Andreasen et al. performed the topology optimization of micromixers, where a constraint on hydrodynamic pressure drop across the micromixer is added to the optimization to regularize the topology optimization problem. Section 6.1 will focus on the layout design method of passive micromixers based on the topology optimization of fluidic flows to obtain the detailed topology and shape of passive micromixers, according to the desired mixing performance, where the pressure constraint is discarded to enlarge the feasible region of the optimization problem, improve the mixing performance of a micromixer in a larger extent, and reduce the CPU time costs in the adjoint analysis of the optimization problem.

For active micromixers, electroosmotic actuation is particularly suitable for microfluidic devices, because it can provides an attractive approach for manipulating liquids in microdevices, since microdevices operating on this principle do not require any moving parts. Electroosmosis has been investigated in several literatures, typically including the researches on electroosmotic micromixers [126, 127]. Electroosmosis is caused by the accumulation of a net electric charge on the solid surface that is in contact with an electrolyte solution [128]. As a result, charges accumulate in the thin liquid layer next to the solid surface. This thin layer is known as the Debye (or double) layer, and its thickness is typically in the magnitude of 10 nm [129]. Away from the solid surface, the electrolyte is neutral. This charge separation next to the solid wall causes either a positive or negative potential difference (i.e. Zeta potential) across the Debye layer. The magnitude of Zeta potential depends on the characteristics of the solid and the liquid. In the presence of an external electric field, the charges in the double layer are attracted to the oppositely charged electrode and drag the liquid along. Therefore, the electric field, through its action on the charges, creates a body force that induces fluid motion. Usually, the Debye layer is much smaller than the feature size of microfluidics, and the thin Debye layer can be approximated to be slip wall with velocity proportional to the tangential component of the external electric field. Therefore, the electrode patterns on the lateral walls of microchannel play the key role on the performance of electroosmotic microfluidics.

Several researches have been implemented for mathematically modeling electroosmotic microflows by analytical, numerical and experimental approaches [126, 130, 131]; shape optimization-based geometrical design of electroosmotic microchannel has been implemented [132, 133]; optimization of zeta potential distributions has been performed for minimal dispersion in an electroosmotic microchannel [134], to name the most relevant. In those researches, electrodes play the key role on generating the microfluidic motion, because they induces the external electric field with attracting charges. Therefore, reasonable electrode distribution is key for achieving the performance of the electroosmotic microflows. On the electrodes of electroosmosis, asymmetric polarization and nonplanar patterns have been adopted for electroosmotic actuation [135, 136]; staggered array of electrodes has been utilized to produce electroosmotic vortices for micromixing [137]; asymmetric electrode pair has been used in electroosmotic microconcentrator [138]; electric potential effect imposed on wall electrodes has been discussed in binary fluids [139]; facing rows of electrodes has been used for remediation of polluted soils by electrokinetic soil flushing [140], to name the most recent. Although those relevant results provided instructions for the control of electroosmosis, limits on determining the electrode patterns still exist in view of lacking generality, flexibility and efficiency because of the dependence on the designers' intuition. Section 6.2 focuses on the inverse design method for the electrode patterns of electroosmotic micromixers to overcome these limits. However, the inverse design of the electrode patterns in electroosmotic micromixer is the problem attributed to the case of the inverse determination of the surface topology. Therefore, in Sect. 6.2, it is proposed to carry out the inverse design of electrode patterns by implement the interpolation of electric insulation and electric potential on the specified walls of microchannels, and this can also be regarded to be the first attempt for the inverse determination of surface topology in the area of topology optimization.

1.4.2 Inverse Design of Microvalve

The absence of moving components in a microfluidic device can decrease the risk of wear, fatigue, or clogging when the working fluid contains cells or particles. Typical examples are the Tesla diode and Venturi diode [141]. Based on the diode effect, the micro Tesla diode has been used as mixer [89, 142], and the micro Venturi diode has been used as a micromixer [89], a component of a micropump [143–153], a micro dosing [154], droplets manipulation for the Newtonian fluid [155], and as micro memory elements [156] for the non-Newtonian fluid in microfluidics.

For micro Venturi diodes, several three dimensional layouts, such as conical [148], flat-walled layout [143, 144, 146] and pyramidal [147, 152], have been proposed. Among these choices, the flat-walled micro Venturi diode, which has rectangular channel profile, is a straight forward choice that can be fabricated using the standard MEMS techniques. When designing a fluidic resistance component, the layout of microfluidic channels needs to be optimized so that the manipulated liquid can flow through the specified channels under certain velocity or pressure distribution. In most cases, a micro Venturi diode is optimized by changing the diffuser angle [157] or the boundary shape for high Reynolds number cases [145, 149, 150, 158, 159]. However, the typical Reynolds number for the microfluidic devices is small due to small transverse length scale. In Sect. 6.3, the numerical optimization method is used to design flat-walled micro Venturi diodes for the case of steady incompressible Newtonian flow regimes, with Reynolds numbers less than 100. An optimization model is proposed to inversely design a micro Venturi diode. Numerical examples are presented to illustrate the optimization details in obtaining a design with good diode effect.

1.4.3 Inverse Design of Micropump

Micropump is an important class of devices in microfluidics [160]. In principle, there are two classes of micropump, one with check valves, the other valveless. Valveless micropumps need no interior moving parts, and so are potentially free of wear and clogging. A valveless micropump, in which fluid flow is driven by micro membranes, possesses wide application and, can be fabricated using standard Silicon processing. In most cases, a valveless micropump is mainly designed with nozzle-diffuser or peristaltic layout. In [31] it has been shown that the topologically optimized layout of moving-part-free fluidic resistance microvalves may be different than the commonly used Tesla valve in the steady flow case. Therefore, the application of optimization to the case of valveless micropumps for unsteady flow is a meaningful task.

The use of optimized fluidic channels can improve the performance of microfluidic devices drastically. Until now, the design of micropumps is mostly dependent on a designer's practical experience, and in the literature, merely the use of the parametric optimization method has been reported to date. Topology optimization of fluidic channel was first proposed for the stationary flow case [11]. Subsequently, the topology optimization method has been used to design micro reactors [161], and micro mixers [35]. In 2010, the topology optimization method was extended to unsteady flow problems [50, 162]. In Sect. 6.4, the topology optimization is used to inversely design the micropump layout.

References

1. O. Pironneau, On optimum profiles in Stokes flow. J. Fluid Mech. **59**(1), 117–128 (1973)
2. O. Pironneau, On optimum design in fluid mechanics. J. Fluid Mech. **64**(1), 97–110 (1974)
3. D.N. Srinath, S. Mittal, An adjoint method for shape optimization in unsteady viscous flows. J. Comput. Phys. **229**, 1994–2008 (2010)
4. C. Brandenburg, F. Lindemann, M. Ulbrich, S. Ulbrich, A continuous adjoint approach to shape optimization for Navier-Stokes flow. Int. Ser. Numer. Math. **158**, 35–56 (2009)
5. J. Nocedal, S.J. Wright, *Numerical Optimization* (Springer, 1998)
6. M.P. Bendsøe, N. Kikuchi, Generating optimal topologies in optimal design using a homogenization method. Comput. Methods Appl. Mech. Eng. **71**, 197–224 (1988)
7. O. Sigmund, A 99-line topology optimization code written in Matlab. Struct. Multidisc. Optim. **21**, 120–127 (2001)

8. O. Sigmund, On the design of compliant mechanisms using topology optimization. Mech. Struct. Mach. **25**, 495–526 (1997)
9. A. Saxena, Topology design of large displacement compliant mechanisms with multiple materials and multiple output ports. Struct. Multidisc. Optim. **30**, 477–490 (2005)
10. M. Bendsøe, O. Sigmund, *Topology Optimization-Theory, Methods and Applications* (Springer, 2003)
11. T. Borrvall, J. Petersson, Topology optimization of fluid in Stokes flow. Int. J. Numer. Meth. Fluids **41**, 77–107 (2003)
12. A. Gersborg-Hansen, M.P. Bendsøe, O. Sigmund, Topology optimization of heat conduction problems using the finite volume method. Struct. Multidisc. Optim. **31**, 251–259 (2006)
13. T. Nomura, K. Sato, K. Taguchi, T. Kashiwa, S. Nishiwaki, Structural topology optimization for the design of broadband dielectric resonator antennas using the finite difference time domain technique. Int. J. Numer. Meth. Eng. **71**, 1261–1296 (2007)
14. O. Sigmund, K.G. Hougaard, Geometric properties of optimal photonic crystals. Phys. Rev. Lett. **100**, 153904 (2008)
15. M.B. Duhring, J.S. Jensen, O. Sigmund, Acoustic design by topology optimization. J. Sound Vibr. **317**, 557–575 (2008)
16. W. Akl, A. El-Sabbagh, K. Al-Mitani, A. Baz, Topology optimization of a plate coupled with acoustic cavity. Int. J. Solids Struct. **46**, 2060–2074 (2008)
17. G.P. Steven, Q. Li, Y.M. Xie, Evolutionary topology and shape design for physical field problems. Comp. Methods Appl. Mech. Eng. Comput. Mech. **26**, 129–139 (2000)
18. G. Allaire, *Shape Optimization by the Homogenization Method* (Springer, New York, 2002)
19. G.I.N. Rozvany, Aims, scope, methods, history and unified terminology of computer-aided topology optimization in structural mechanics. Struct. Multidisc. Optim. **21**, 90–108 (2001)
20. M.P. Bendsøe, O. Sigmund, Material interpolations in topology optimization. Arch. Appl. Mech. **69**, 635–654 (1999)
21. M.Y. Wang, X. Wang, D. Guo, A level set method for structural optimization. Comput. Meth. Appl. Mech. Eng. **192**, 227–246 (2003)
22. G. Allaire, F. Jouve, A. Toader, Structural optimization using sensitivity analysis and a level-set method. J. Comput. Phys. **194**, 363–393 (2004)
23. Z. Liu, J.G. Korvink, Adaptive moving mesh level set method for structure optimization. Eng. Optim. **40**, 529–558 (2008)
24. X. Xing, P. Wei, M.Y. Wang, A finite element-based level set method for structural optimiztion. Int. J. Numer. Meth. Eng. **82**, 805–842 (2010)
25. P.H. Guillaume, K.S. Idris, Topological sensitivity and shape optimization for the Stokes equations. SIAM J. Cont. Optim. **43**, 1–31 (2004)
26. N. Aage, T.H. Poulsen, A. Gersborg-Hansen, O. Sigmund, Topology optimization of large scale stokes flow problems. Struct. Multidisc. Optim. **35**, 175–180 (2008)
27. J.K. Guest, J.H. Proevost, Topology optimization of creeping fluid flows using a Darcy-Stokes finite element. Int. J. Numer. Meth. Eng. **66**, 461–484 (2006)
28. N. Wiker, A. Klarbring, T. Borrvall, Topology optimization of regions of Darcy and Stokes flow. Int. J. Numer. Meth. Eng. **69**, 1374–1404 (2007)
29. A. Gersborg-Hansen, O. Sigmund, R.B. Haber, Topology optimization of channel flow problems. Struct. Multidisc. Optim. **29**, 1–12 (2005)
30. L.H. Olessen, F. Okkels, H. Bruus, A high-level programming-language implementation of topology optimization applied to steady-state Navier-Stokes flow. Int. J. Numer. Meth. Eng. **65**, 975–1001 (2006)
31. Y. Deng, Z. Liu, P. Zhang, Y. Wu, Jan G. Korvink, Optimization of no-moving-part fluidic resistance microvalves with low Reynolds number, in *IEEE MEMS Conference*, (HongKong, 2010) pp. 67–70
32. G. Pingen, K. Maute, Optimal design for non-Newtonian flows using a topology optimization approach. Comput. Math. Appl. **59**, 2340–2350 (2010)
33. F. Okkels, L.H. Olesen, H. Bruus, Application of topology optimization in the design of micro and nanofluidic systems. NSTI Nanotech. 575–578 (2005)

34. F. Okkels, H. Bruus, Scaling behavior of optimally structured catalytic microfluidic reactors. Phys. Rev. E **75**, 1–4 (2007)
35. C.S. Andreasen, A.R. Gersborg, O. Sigmund, Topology optimization of microfluidic mixers. Int. J. Numer. Meth. Fluids **61**, 498–513 (2008)
36. S. Osher, J.A. Sethian, Front propagating with curvature dependent speed: algorithms based on Hamilton-Jacobi formulations. J. Comput. Phys. **78**, 12–49 (1988)
37. X. Duan, Y. Ma, R. Zhang, Shape-topology optimization for Navier-Stokes problem using variational level set method. J. Comput. Appl. Math. **222**, 487–499 (2008)
38. S. Zhou, Q. Li, A variational level set method for the topology optimization of steady-state Navier-Stokes flow. J. Comput. Phys. **227**, 10178–10195 (2008)
39. J. Sokolowski, A. Zochowski, On the topological derivative in shape optimization. SIAM J. Control Optim. **37**, 1241–1272 (1999)
40. J. Sokolowski, A. Zochowski, Topological derivatives for elliptic problems. Inverse Prob. **15**, 123–134 (1999)
41. A.A. Novotny, R.A. Feijoo, E. Taroco, C. Padra, Topological sensitivity analysis. Comput. Meth. Appl. Mech. Engng **192**, 803–829 (2003)
42. M. Burger, B. Hackl, W. Ring, Incorporating topological derivatives into level set methods. J. Comput. Phys. **194**, 344–362 (2004)
43. S. Amstutz, Topological sensitivity analysis for some nonlinear PDE systems. J. Math. Pures Appl. **85**, 540–557 (2006)
44. V.J. Challis, J.K. Guest, Level set topology optimization of fluids in Stokes flow. Int. J. Numer. Meth. Eng. **79**, 1284–1308 (2009)
45. M. Abdelwahed, M. Hassine, Topological optimization method for a geometric control problem in Stokes flow. Appl. Numer. Math. **59**, 1823–1838 (2009)
46. H. Maatoug, Shape optimization for the Stokes equations using topological snesitivity analysis. ARIMA **5**, 216–229 (2006)
47. M.Y. Wang, Shape optimization with level set method incorporating topological derivatives, in *6th World Congresses of Structural and Multidisciplinary Optimization* (2005)
48. S. Amstutz, The topological asymptotic for the Navier-Stokes equations, ESAIM: control. Optim. Calc. Variat. **11**, 401–425 (2005)
49. M. Hinze, R. Pinnau, M. Ulbrich, S. Ulbrich, *Optimization with PDE constraints* (Springer, 2009)
50. S. Kreissl, G. Pingen, K. Maute, Topology optimization for unsteady flow. Int. J. Numer. Meth. Eng. (2011). doi:10.1002/nme.3151
51. B.C. Chen, N. Kikuchi, Topology optimization with design-dependent loads. Finite Elem. Anal. Design **37**, 57–70 (2001)
52. T. Gao, W.H. Zhang, J.H. Zhu, Topology optimization of heat conduction problem involving design-dependent heat load effect. Finite Elem. Anal. Design **44**, 805–813 (2008)
53. J.H. Zhu, P. Beckers, W.H. Zhang, On the multi-component layout design with inertial force. J. Comput. Appl. Math. **234**, 2222–2230 (2010)
54. T. Gao, W.H. Zhang, Topology optimization of multiphase material structures under design-dependent pressure loads. Int. J. Simul. Multidisci. Des. Optim. **3**, 297–306 (2009)
55. M.S. Victor, G.V. Manuel, J.R. Clayton, *Wetting and Spreading Dynamics* (CRC Press, 2007)
56. D. Gueyffier, J. Li, A. Nadim, R. Scardovelli, S. Zaleski, Volume-of-fluid interface tracking with smoothed surface stress methods for three-dimensional flows. J. Comput. Phys. **152**, 423–456 (1999)
57. J. Glimm, J.W. Grove, X.L. Li, K.M. Shyue, Q. Zhang, Y. Zeng, Three-dimensional front tracking. SIAM J. Sci. Comput. **19**, 703–727 (1998)
58. C.S. Peskin, D.M. McQueen, Modeling prosthetic heart valves for numerical analysis of blood flow in the heart. J. Comput. Phys. **37**, 113–132 (1980)
59. C.S. Peskin, The immersed boundary method. Acta Num. **11**, 1–39 (2002)
60. Y.C. Chang, T.Y. Hou, B. Merriman, S. Osher, A level set formulation of Eulerian interface capturing methods for incompressible fluid flows. J. Comput. Phys. **124**, 449–464 (1996)

61. M. Arienti, M. Sussman, An embedded level set method for sharp-interface multiphase simulations of Diesel injectors. Int. J. Multiphas. Flow **59**, 1–14 (2014)
62. R.F. Engberga, E.Y. Kenig, An investigation of the influence of initial deformation on fluid dynamics of toluene droplets in water. Int. J. Multiphas. Flow **76**, 144–157 (2015)
63. S. Osher, R.P. Fedkiw, *Level Set Methods and Dynamic Implicit Surfaces* (Springer, New York, 2002)
64. J.A. Sethian, P. Smereka, Level set methods for fluid interfaces. Annu. Rev. Fluid Mech. **35**, 341–372 (2003)
65. H.A.A. Amiri, A.A. Hamouda, Evaluation of level set and phase field methods in modeling two phase flow with viscosity contrast through dual-permeability porous medium. Int. J. Multiphas. Flow **52**, 22–34 (2013)
66. D.M. Anderson, G.B. McFadden, A.A. Wheeler, Diffuse-interface methods in fluid mechanics. Annu. Rev. Fluid Mech. **30**, 139–165 (1998)
67. D. Jacqmin, Calculation of two-phase Navier-Stokes flows using phase-field modeling. J. Comput. Phys. **155**, 96–127 (1999)
68. J.S. Kim, A continuous surface tension force formulation for diffuse-interface models. J. Comput. Phys. **204**, 784–804 (2005)
69. T. Qian, X.P. Wang, P. Sheng, Molecular hydrodynamics of the moving contact line in two-phase immiscible flows. Comm. Comput. Phys. **1**, 1–52 (2006)
70. D. Jacqmin, Contact-line dynamics of a diffuse fluid interface. J. Fluid Mech. **402**, 57–88 (2000)
71. J.W. Cahn, J. Hilliard, Free energy of a nonuniform system: I Interfacial free energy. J. Chem. Phys. **28**, 258–267 (1958)
72. J.A. Sethian, Level set methods and fast marching methods evolving interfaces, in *Computational Geometry, Fluid Mechanics, Computer Vision, and Materials Science* (Cambridge University Press, 1999)
73. Y. Deng, Z. Liu, Y. Wu, Optimization of unsteady incompressible Navier-Stokes flows using variational level set method. Int. J. Numer. Meth. Fluids **71**, 1475–1493 (2013)
74. A. Takezawa, S. Nishiwaki, M. Kitamura, Shape and topology optimization based on the phase field method and sensitivity analysis. Int. J. Comput. Phys. **229**, 2697–2718 (2010)
75. S.S. Sritharan, *Optimal Control Viscous Flow* (SIAM, Philadelphia, 1998)
76. J.L. Lions, *Optimal Control of Systems Governed by Partial Differential Equations* (Springer, New York, 1971)
77. J.L. Lions, *Some Aspects of the Optimal Control of Distributed Parameter Systems* (SIAM, Philadelphia, 1972)
78. J.L. Lions, *Some Methods in the Mathematical Analysis of Systems and Their Control* (Gordon and Breach, New York, 1981)
79. N.U. Ahmed, K.L. Teo, *Optimal Control of Distributed Parameter Systems* (North Holland, New York, 1981)
80. V. Barbu, G. Prato, *Da, Hamilton Jacobi Equations in Hilbert Spaces* (Pitman, Boston, 1983)
81. H.O. Fattorini, Time-optimal control of solutions of operational differential equation. SIAM J. Control **2**, 54–59 (1964)
82. A. Friedman, Optimal control of parabolic equations. J. Math. Anal. Appl. **18**, 479–491 (1967)
83. A.G. Butkovskii, *Theory of Optimal control of Distributed Parameter Systems* (Elsevier, 1969)
84. A.V. Fursikov, On some control problems and results related to the unique solution of mixed problems by the three-dimensional Navier-Stokes and Euler equations. Dokl. Akad. Nauk SSSR **252**, 1066–1070 (1980)
85. A.V. Fursikov, Control problems and results on the unique solution of mixed problems by the three-dimensional Navier-Stokes and Euler equations. Math. Sbornik **115**, 281–306 (1981)
86. A.V. Fursikov, Properties of solutions of certain extremum problems related to the Navier-Stokes and Euler equations. Math. Sbornik **117**, 323–349 (1981)
87. A. Jameson, Computational aerodynamics for aircraft design. Math. Sbornik **245**, 361–371 (1989)

88. R. Glowinski, A.J. Kearsley, T.W. Pan, J. Periaux, Numerical simulation and optimal shape for viscous flow by a fictitious domain method. Int. J. Num. Meth. Fluids **20**, 695–711 (1995)
89. N.T. Nguyen, Z. Wu, Micromixers - a review. J. Micromech. Microeng. **15**, 1–16 (2005)
90. G. Karniadakis, A. Beskok, N. Aluru, *Microflows and Nanoflows: Fundamentals and Simulation* (Springer, 2005)
91. P.S. Dittrich, A. Manz, Lab-on-a-chip: microfluidics in drug discovery. Nat. Rev. Drug Disc. **5**, 210 (2006)
92. A. Manz, J.C.T. Eijkel, Miniaturization and chip technology. What can we expect? Pure Appl. Chem. **73**, 1555 (2001)
93. H.A. Stone, A.D. Stroock, A. Ajdari, Annu. Rev. Fluid Mech. **36**, 381 (2004)
94. T.M. Squires, S.R. Quake, Microfluidics: fluid physics on the nanoliter scale. Rev. Mod. Phys. **77**, 977 (2005)
95. L. OuYang, C. Wang, F. Du, T. Zheng, H. Liang, Paper electrochromatographic separations of multi-component metal complexes on a microfluidic paper-based device with a simplified photolithography. RSC Adv. **4**, 1093–1101 (2014)
96. G.S. Jeong, S. Chung, C.B. Kim, S.H. Lee, Applications of micromixing technology. Analyst **135**, 460–473 (2010)
97. A.D. Stroock, S.K.W. Dertinger, A. Ajdari, I. Mezić, H.A. Stone, G.M. Whitesides, Chaotic mixer for microchannels. Science **295**, 647 (2002)
98. D. Li, *Encyclopedia of Microfluidics and Nanofluidics* (Springer, 2007)
99. D.H. Kelley, N.T. Ouellette, Separating stretching from folding in fluid mixing. Nature Physics (2011). doi:10.1038/NPHYS1941
100. N.T. Nguyen, *Micromixer: Fundamentals, Design and Fabrication* (William Andrew, 2008)
101. C.Y. Lee, C.L. Chang, Y.N. Wang, L.M. Fu, Microfluidic mixing: a review. Int. J. Mol. Sci. **12**, 3263–3287 (2011)
102. I. Glasgow, N. Aubry, Lab Chip **3**, 114 (2003)
103. C.H. Lin, C.H. Tsai, C.W. Pan, L.M. Fu, Rapid circular microfluidic mixer utilizing unbalanced driving force. Biomed. Microdevic. **9**, 43–50 (2007)
104. M.Z. Huang, R.J. Yang, C.H. Tai, C.H. Tsai, L.M. Fu, Application of electrokinetic instability flow for enhanced micromixing in cross-shaped microchannel. Biomed. Microdevic. **8**, 309–315 (2006)
105. C.H. Wu, R.J. Yang, Improving the mixing performance of side channel type micromixers using an optimal voltage control model. Biomed. Microdevic. **8**, 119–131 (2006)
106. W. Ng, S. Goh, Y. Lam, C. Yang, I. Rodriguez, Lab Chip **9**, 802 (2009)
107. C.H. Lin, L.M. Fu, Y.S. Chien, Anal. Chem. **76**, 5265 (2004)
108. H. Song, Z. Cai, H. Noh, D. Bennett, Lab Chip **10**, 734 (2010)
109. C.A. Cortes-Quiroza, A. Azarbadegana, M. Zangeneha, A. Goto, Analysis and multi-criteria design optimization of geometric characteristics of grooved micromixer. Chem. Eng. J. **160**, 852–864 (2010)
110. R. Choudhary, T. Bhakat, R.K. Singh, A. Ghubade, S. Mandal, A. Ghosh, A. Rammohan, A. Sharma, S. Bhattacharya, Bilayer staggered herringbone micro-mixers with symmetric and asymmetric geometries. Microfluid. Nanofluid. **10**, 271–286 (2011)
111. M.A. Ansari, K.Y. Kim, Evaluation of surrogate models for optimization of herringbone groove micromixer. J. Mechanic. Sci. Technol. **22**, 387–396 (2008)
112. N.S. Lynn, D.S. Dandy, Geometrical optimization of helical flow in grooved micromixers. Lab Chip **7**, 580–587 (2007)
113. C.A. Cortes-Quiroz, M. Zangeneh, A. Goto, On multi-objective optimization of geometry of staggered herringbone micromixer. Microfluid. Nanofluid. **7**, 29–43 (2009)
114. H. Song, X.Z. Yin, D.J. Bennett, Optimization analysis of the staggered herringbone micromixer based on the slip-driven method. Chem. Eng. Res. Design **86**, 883–891 (2008)
115. M.A. Ansari, K.Y. Kim, Shape optimization of a micromixer with staggered herringbone groove. Chem. Eng. Sci **62**, 6687–6695 (2007)
116. S. Hossain, A. Husain, K.Y. Kim, Shape optimization of a micromixer with staggered-herringbone grooves patterned on opposite walls. Chem. Eng. J. **162**, 730–737 (2010)

117. H.Y. Lee, J. Voldman, Optimizing micromixer design for enhancing dielectrophoretic microconcentrator performance. Anal. Chem. **79**, 1833–1839 (2007)
118. T.G. Kang, M.K. Singh, P.D. Anderson, H.E.H. Meijer, A chaotic serpentine mixer efficient in the creeping flow regime: from design concept to optimization. Microfluid. Nanofluid. **7**, 783–794 (2009)
119. P. B H. Jr, D. R Mott, F. S Ligler, J. P Golden, C. R Kaplan, E. S Oran, A combinatorial approach to microfluidic mixing. J. Micromech. Microeng. **18**, 115019(7pp) (2008)
120. E. Schall, Y. L. Guer, S. Gibout, An anisotropic unstructured meshes mapping method (AUM3) to optimize mixing in laminar periodic flows. *Proc. PSFVIP-4* (June 3–5, Chamonix, France, 2003)
121. M.K. Singh, T.G. Kang, P.D. Anderson, H.E.H. Meijer, A.N. Hrymak, Analysis and optimization of low-pressure drop static mixers. AIChE J. **55**, 2208–2216 (2009)
122. M.K. Singh, T.G. Kang, H.E.H. Meijer, P.D. Anderson, The mapping method as a toolbox to analyze, design, and optimize micromixers. Microfluid. Nanofluid. (2007). doi:10.1007/s10404-007-0251-7
123. D.R. Mott, P.B.H. Jr, K.S. Obenschain, E.S. Oran, The numerical toolbox: an approach for modeling and optimizing microfluidic components. Mech. Res. Commun. **36**, 104–109 (2009)
124. D.R. Mott, P.B. Howell Jr., J.P. Golden, C.R. Kaplan, F.S. Ligler, E.S. Oran, Toolbox for the design of optimized microfluidic components. Lab Chip **6**, 540–549 (2006)
125. T.G. Kang, M.K. Singh, T.H. Kwon, P.D. Anderson, Chaotic mixing using periodic and aperiodic sequences of mixing protocols in a micromixer. Microfluid Nanofluid **4**, 589–599 (2008)
126. S. Qian, H.H. Bau, A Chaotic Electroosmotic Stirrer. Anal. Chem. **74**, 3616–3625 (2002)
127. J.B. Zhang, G.W. He, F. Liu, electroosmotic flow and mixing in heterogeneous microchannels. Phys. Rev. E **73**, 056305 (2006)
128. M. Ammam, Electrophoretic deposition under modulated electric fields: a review. RSC Advanc. **2**, 7633–7646 (2012)
129. R.J. Hunter, *Zeta Potential in Colloid Science: Principles and Applications* (Academic, New York, 1981)
130. T.J. Johnson, D. Ross, L.E. Locascio, Rapid microfluidic mixing. Anal. Chem. **74**, 45–51 (2002)
131. C.C. Chang, R.J. Yang, Electrokinetic mixing in microfluidic systems. Microfluid Nanofluid **3**, 501–525 (2007)
132. M. Jain, A. Yeung, K. Nandakumar, Induced charge electro osmotic mixer: Obstacle shape optimization. Biomicrofluidics **3**, 022413 (2009)
133. T.J. Johnson, L.E. Locascio, Characterization and optimization of slanted well designs for microfluidic mixing under electroosmotic flow. Lab Chip **2**, 135–140 (2002)
134. H.S. Woo, B.J. Yoon, I.S. Kang, Optimization of zeta potential distributions for minimal dispersion in an electroosmotic microchannel. Int. J. Heat Mass Trans. **51**, 4551–4562 (2008)
135. J. Wu, Ac electro-osmotic micropump by asymmetric electrode polarization. J. Appl. Phys. **103**, 024907 (2008)
136. J.P. Urbanski, T. Thorsen, J.A. Levitan, M.Z. Bazant, Fast ac electro-osmotic micropumps with nonplanar electrodes. Appl. Phys. Leet. **89**, 143508 (2006)
137. K. Matsubara, T. Narumi, Microfluidic mixing using unsteady electroosmotic vortices produced by a staggered array of electrodes. Chem. Eng. J. **288**, 638–647 (2016)
138. J.H. Chen, Y.C. Lee, W.H. Hsieh, AC electroosmotic microconcentrator using a face-to-face, asymmetric electrode pair with expanded sections in the bottom electrode. Microfluid Nanofluid **20**, 72 (2016)
139. A.K. Nayak, A. Haque, A. Banerjee, B. Weigand, Flow mixing and electric potential effect of binary fluids in micro/nano channels, Colloids and Surfaces A: Physicochem. Eng. Aspects **512**, 145–157 (2017)
140. C. Risco, R. López-Vizcaíno, C. Sáez, A. Yustres, P. Cañizares, V. Navarro, M.A. Rodrigo, Remediation of soils polluted with 2,4-D by electrokinetic soil flushing with facing rows of electrodes: a case study in a pilot plant. Chem. Eng. J. **285**, 128–136 (2016)

141. K.W. Oh, C.H. Ahn, A review of microvalves. J. Micromech. Microeng. **16**, 13–39 (2006)
142. C.C. Hong, J.W. Choi, C.H. Ahn, A novel in-plane microfluidic mixer with modified tesla structures. Lab chip **4**, 109–11 (2004)
143. A. Olsson, G. Stemme, E. Stemme, A valve-less planar fluid pump with two pump chambers. Sensors Actuators A **46–47**, 549–556 (1995)
144. A. Olsson, P. Enoksson, G. Stemme, E. Stemme, A valve-less planar pump isotropically etched in silicon. J. Micromech. Microeng. **6**, 87–91 (1996a)
145. A. Olsson, G. Stemme, E. Stemme, Diffuser-element design investigation for valve-less pumps. Sensors Actuators A **57**, 137–143 (1996b)
146. A. Olsson, P. Enoksson, G. Stemme, E. Stemme, Micromachined flatwalled valveless diffuser pumps. J. Micromech. Microeng. **6**, 161–166 (1997)
147. A. Olsson, G. Stemme, E. Stemme, Numerical and experimental studies of flatwalled diffuser elements for valve-less micropumps. Sensors Actuators A **84**, 165–175 (2000)
148. E. Stemme, G. Stemme, A valveless diffuser/nozzle based fluid pump. Sensors Actuators A **39**, 159–167 (1993)
149. Y.C. Wang, J.C. Hsu, P.C. Kuo, Loss characteristics and flow rectification property of diffuser valves for micropump applications. Int. J. Heat Mass Trans. **52**, 328–336 (2009)
150. C. Yamahata, C. Lotto, E. Al-Assaf, M.A.M. Gijs, A PMMA valveless micropump using electromagnetic actuation. Microfluid Nanofluid **1**, 197–207 (2005)
151. N.T. Nguyen, X. Huang, T.K. Chuan, MEMS-micropumps: a review. Trans. ASME J. Fluids Eng. **124**, 384–392 (2002)
152. T. Gerlach, Microdiffusers as dynamic passive valves for micropump applications. Sensors Actuators A **69**, 181–191 (1998)
153. A. Groisman, S.R. Quake, A microfluidic rectifier: anisotropic flow resistance at low Reynolds numbers. Phys. Rev. Lett. **92**, 094501 (2004)
154. M. Carmona, S. Marco, J. Samitier, M.C. Acero, J.A. Plaza, J. Esteve, Microvalve analysis: wall shear and diffuser effects, in *International Conference on Modeling and Simulation of Microsystems*, pp. 554–557 (1999)
155. J. Liu, Y.F. Yap, N.T. Nguyen, Behavior of microdroplets in diffuser/nozzle structures. Microfluid. Nanofluid. **6**, 835–846 (2009)
156. A. Groisman, M. Enzelberger, S.R. Quake, Microfluidic memory and control devices. Science **300**, 955–958 (2003)
157. C.T. Wang, P.C. Chang, C.C. Huang, Optimization of diffuser valves. J. Mar. Sci. Tech. **16**, 134–138 (2008)
158. P.W. Runstadler, F.X. Dolan, R.C. Dean, *Diffuser Data Book* (Creare, New Hampshire, 1975)
159. F.M. White, *Fluid Mech.*, 4th edn. (WCB/McGraw-Hill, Singapore, 1999)
160. D.J. Laser, J.G. Santiago, A review of micropumps. J. Micromech. Microeng. **14**, 35–64 (2004)
161. F. Okkels, H. Bruus, Design of micro-fluidic bio-reactors using topology optimization, (ECCOMAS CFD, 2006)
162. Y. Deng, Y. Wu, Z. Liu, Topology optimization of unsteady incompressible Navier-Stokes flows. J. Comput. Phys. **230**, 6688–6708 (2011)

Chapter 2
Topology Optimization for Unsteady Flows

This chapter presents the topology optimization of unsteady incompressible Navier-Stokes flows, where the density method and level set method are respectively used to implement the implicitly expression of the unsteady flows. In the density method, the optimization problem is formulated by adding the artificial Darcy frictional force into the incompressible Navier-Stokes equations; the continuous adjoint method is used to derive the adjoint derivative; and the method of moving asymptotes is used to evolve the design variable. In the level set method, the fluid velocity is constrained to be zero in the implicitly expressed solid domain; the variational level set method is used to derive the optimization sensitivity; and the level set function is evolved by solving the Hamilton-Jacobian equation with the upwind finite difference method. Furthermore, numerical examples are provided to demonstrate the feasibility and necessity of the topology optimization method for unsteady Navier-Stokes flows, where the dynamic effect and Reynolds number are investigated for the optimal topology.

2.1 Density Method-Based Topology Optimization for Unsteady Flows

In this section, the density method-based topology optimization for unsteady flows is introduced. The key ideas behind this work is on the implicit expression of fluidic flow using artificial Darcy force and the continuous adjoint method-based analysis of the topology optimization problem.

Y. Deng et al., *Topology Optimization Theory for Laminar Flow*,
https://doi.org/10.1007/978-981-10-4687-2_2

2.1.1 Density Method-Based Optimization Problem

The dynamics of the incompressible fluidic flows can be expressed by the incompressible Navier-Stokes equations as [1]

$$\begin{aligned} \rho \frac{\partial \mathbf{u}}{\partial t} - \eta \nabla \cdot \left(\nabla \mathbf{u} + \nabla \mathbf{u}^{\mathrm{T}} \right) + \rho \left(\mathbf{u} \cdot \nabla \right) \mathbf{u} + \nabla p = \mathbf{f}, \ \text{ in } Q \\ -\nabla \cdot \mathbf{u} = 0, \ \text{ in } Q \end{aligned} \tag{2.1}$$

where $\mathbf{u}$ is the fluidic velocity; p is the fluidic pressure; ρ is the fluidic density; η is the fluidic viscosity; $\mathbf{f}$ is the body force loaded on the fluid; and t is the time. $Q = (0, T) \times \Omega$ is the time-space, where $(0, T)$ is the computational time interval and Ω is the computational domain. To solve the transient eq. 2.1, the initial condition needs to be enforced

$$\mathbf{u}(0, \mathbf{x}) = \mathbf{u}_0(\mathbf{x}), \ \text{ in } \Omega \tag{2.2}$$

where $\mathbf{u}_0(\mathbf{x})$ satisfies the incompressible condition $\nabla \cdot \mathbf{u}_0 = 0$. The commonly used boundary conditions for incompressible Navier-Stokes equations include the Dirichlet and Neumann type boundary conditions

$$\mathbf{u} = \mathbf{u}_D(t, \mathbf{x}), \ \text{ on } \Sigma_D \tag{2.3}$$

$$\left[-p\mathbf{I} + \eta \left(\nabla \mathbf{u} + \nabla \mathbf{u}^{\mathrm{T}} \right) \right] \mathbf{n} = \mathbf{g}(t, \mathbf{x}), \ \text{ on } \Sigma_N \tag{2.4}$$

where $\mathbf{u}_D$ and $\mathbf{g}$ are the specified velocity and stress distribution on the boundaries Γ_D and Γ_N; $\mathbf{n}$ is the outward unitary normal on $\partial\Omega$; $\Sigma_D = (0, T) \times \Gamma_D$ and $\Sigma_N = (0, T) \times \Gamma_N$ are the time-space boundaries. Specifically, the no-slip boundary is a particular Dirichlet type boundary condition with $\mathbf{u}_D = \mathbf{0}$, and the open-boundary on the outlet can be expressed by the Neumann type boundary condition with $\mathbf{g} = \mathbf{0}$. In density method-based topology optimization for the Navier-Stokes flow, the body force is expressed to be [2, 3]

$$\mathbf{f} = -\alpha \mathbf{u} \tag{2.5}$$

where α is the impermeability of an artificial porous medium. Its value depends on the optimization design variable γ, and the functional relation is termed *material interpolation* in topology optimization [2, 3]

$$\alpha(\gamma) = \alpha_{min} + (\alpha_{max} - \alpha_{min}) \frac{q(1-\gamma)}{q+\gamma} \tag{2.6}$$

where α_{min} and α_{max} are the minimal and maximal values of α, respectively; and q is a real and positive parameter used to tune the convexity of the interpolation function in Eq. 2.6. The value of γ can vary between zero and one, where $\gamma = 0$ corresponds to an artificial solid phase and $\gamma = 1$ corresponds to the fluid. Usually, α_{min} is chosen to be 0; α_{max} is chosen to be a finite but large number to simultaneously ensure the numerical stability of the optimization procedure and approximate the solid phase with negligible permeability [3, 4]. It is valuable to note that the design variable γ is time independent, because the layout of fluidic domain is kept unchanged during the solving of the transient Navier-Stokes equations.

Then, the topology optimization problem for the unsteady incompressible Navier-Stokes flow can be formulated to be

$$
\text{Find } \gamma \in [0, 1] \text{ to minimize } J(\mathbf{u}, p; \gamma), \text{ subject to}
$$
$$
\begin{cases}
\rho \dfrac{\partial \mathbf{u}}{\partial t} - \eta \nabla \cdot \left(\nabla \mathbf{u} + \nabla \mathbf{u}^{\mathrm{T}} \right) + \rho (\mathbf{u} \cdot \nabla) \mathbf{u} + \nabla p = -\alpha \mathbf{u}, \text{ in } Q \\
- \nabla \cdot \mathbf{u} = 0, \text{ in } Q \\
\mathbf{u}(0, \mathbf{x}) = \mathbf{u}_0(\mathbf{x}), \text{ in } \Omega \\
\mathbf{u}(t, \mathbf{x}) = \mathbf{u}_D(t, \mathbf{x}), \text{ on } \Sigma_D \\
\left[-p\mathbf{I} + \eta \left(\nabla \mathbf{u} + \nabla \mathbf{u}^{\mathrm{T}} \right) \right] \mathbf{n} = \mathbf{g}(t, \mathbf{x}), \text{ on } \Sigma_N \\
\int_\Omega \gamma \, d\Omega \le V_r \cdot V_0
\end{cases}
\tag{2.7}
$$

where $V_0 = \int_\Omega 1 \, \mathrm{d}\Omega$ is the volume of the whole design domain; $V_r \in (0, 1)$ is the upper bound for the volume fraction of the fluid phase. One general optimization objective, which includes both the domain and boundary integrations about the fluidic velocity and pressure, is chosen to be

$$
J(\mathbf{u}, p; \gamma) = \int_0^T \int_\Omega \beta_1 A(\mathbf{u}, \nabla \mathbf{u}, p; \gamma) \, \mathrm{d}\Omega \mathrm{d}t + \int_0^T \int_{\partial\Omega} \beta_2 B(\mathbf{u}, p; \gamma) \, \mathrm{d}\Gamma \mathrm{d}t
\tag{2.8}
$$

where β_1 and β_2 are space-independent parameters.

2.1.2 *Continuous Adjoint Analysis*

According to the adjoint method for the Navier-Stokes equations in [5–7], the adjoint analysis of the topology optimization problem for unsteady Navier-Stokes flows is implemented as follows. Without considering the inequality constraint on the volume of fluid phase at first, the topology optimization problem into Eq. 2.7 can be rewritten in the following abstract form

$$
\min \; J(\mathbf{u}, p; \gamma), \quad \text{s.t. } e(\mathbf{u}, p; \gamma) = 0, \; \gamma \in \mathscr{K}
\tag{2.9}
$$

where $\mathscr{K}$ is the feasible space of the design variable γ, and $e(\cdot)$ is the weak operator of the Navier-Stokes equations. According to the Karush-Kuhn-Tucker conditions for partial differential equation constrained optimization problems [5, 8], the optimization problem in Eq. 2.9 can be solved with the solution of the following abstract equations

$$e(\mathbf{u}, p; \gamma) = 0 \tag{2.10}$$

$$\begin{pmatrix} (e_{\mathbf{u}}(\mathbf{u}, p; \gamma))^* & \mathbf{0} \\ \mathbf{0} & (e_p(\mathbf{u}, p; \gamma))^* \end{pmatrix} \begin{pmatrix} \mathbf{u}_a \\ p_a \end{pmatrix} = \begin{pmatrix} -J_{\mathbf{u}}(\mathbf{u}, p; \gamma) \\ -J_p(\mathbf{u}, p; \gamma) \end{pmatrix} \tag{2.11}$$

$$\left(e_\gamma(\mathbf{u}, p; \gamma)\right)^* (\mathbf{u}_a, p_a) + J_\gamma(\mathbf{u}, p; \gamma) = 0 \tag{2.12}$$

where $\mathbf{u}_a$ and p_a are the adjoint variables for the fluid velocity $\mathbf{u}$ and pressure p, respectively; $(\cdot)^*$ is the adjoint of the corresponding operator. Based on Eq. 2.11, the adjoint equations of the Navier-Stokes equations can be written as follows (see Appendix 2.3.1 for more details)

$$\begin{aligned}
-\rho\frac{\partial \mathbf{u}_a}{\partial t} - \eta\nabla\cdot\left(\nabla\mathbf{u}_a + \nabla\mathbf{u}_a^{\mathrm{T}}\right) &- \rho(\mathbf{u}\cdot\nabla)\mathbf{u}_a + \rho(\nabla\mathbf{u})\cdot\mathbf{u}_a + \nabla p_a \\
&= -\beta_1\left(\frac{\partial A}{\partial \mathbf{u}} - \nabla\cdot\frac{\partial A}{\partial \nabla\mathbf{u}}\right) + \frac{\partial \mathbf{f}}{\partial \mathbf{u}}\mathbf{u}_a, \text{ in } Q \\
-\nabla\cdot\mathbf{u}_a &= -\beta_1\frac{\partial A}{\partial p} + \frac{\partial \mathbf{f}}{\partial p}\cdot\mathbf{u}_a, \text{ in } Q \\
\mathbf{u}_a(T, \mathbf{x}) &= \mathbf{0}, \text{ in } \Omega \\
\mathbf{u}_a &= -\frac{\partial B}{\partial p}\mathbf{n}, \text{ on } \Sigma_D \\
\left[-p_a\mathbf{I} + \eta\left(\nabla\mathbf{u}_a + \nabla\mathbf{u}_a^{\mathrm{T}}\right)\right]\mathbf{n} &= -\rho(\mathbf{u}\cdot\mathbf{n})\mathbf{u}_a - \beta_1\frac{\partial A}{\partial \nabla\mathbf{u}}\mathbf{n} - \beta_2\frac{\partial B}{\partial \mathbf{u}}, \text{ on } \Sigma_N
\end{aligned} \tag{2.13}$$

The transient adjoint Eq. 2.13 are terminal value problems, where the value of $\boldsymbol{\mu}$ at the terminal time T is specified and the transient solver need to be implemented from time T to 0. Based on Eq. 2.12, the adjoint derivatives of the optimization objective in Eq. 2.8 can be expressed to be (see Appendix 2.3.2 for more details)

$$\begin{aligned}
\left.\frac{D\hat{J}}{D\gamma}\right|_{\Omega} &= \int_0^T \beta_1\left(\frac{\partial A}{\partial \gamma} + \frac{\partial \alpha}{\partial \gamma}\mathbf{u}\cdot\mathbf{u}_a\right)\mathrm{d}t, \text{ in } \Omega \\
\left.\frac{D\hat{J}}{D\gamma}\right|_{\partial\Omega} &= \int_0^T \beta_2\frac{\partial B}{\partial \gamma}\mathrm{d}t, \text{ on } \partial\Omega
\end{aligned} \tag{2.14}$$

2.1.3 Numerical Implementation

The topology optimization for unsteady Navier-Stokes flows is implemented using the gradient based iterative approach. The procedure for an iterative optimization includes the following steps (Fig. 2.1)

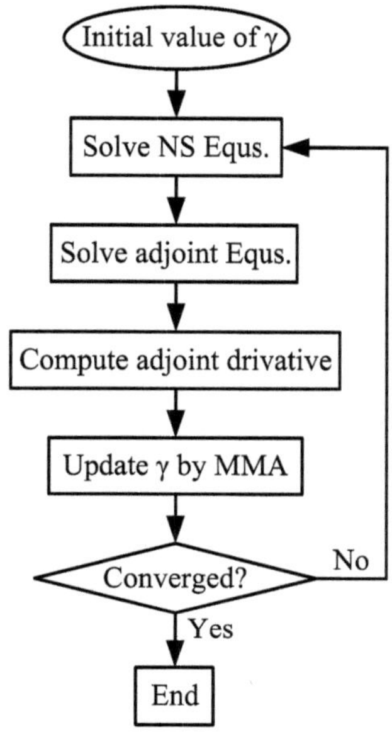

Fig. 2.1 The flowchart of the iterative optimization

- the Navier-Stokes equations are solved with the given value of the design variable;
- the adjoint equations are solved based on the numerical solution of the Navier-Stokes equations;
- the adjoint derivatives of the objective function are computed by Eq. 2.14, and the adjoint derivatives of the design constraint are computed similarly;
- the design variable is updated by the method of moving asymptotes (MMA) [9].

The above steps are implemented iteratively until the stopping criteria are satisfied. In the iterative procedure, the transient Navier-Stokes equations and the corresponding adjoint equations are solved by the mixed finite element method using the finite element package *Comsol Multiphysics* [10], where all the numerical implementation is based on the package's basic module: *COMSOL Multiphysics→PDE Modes→PDE, General Form*.

The PDE modes of *Comsol Multiphysics* can solve partial differential equations of the form

$$\begin{aligned} \mathbf{d}_a \frac{\partial \mathbf{u}}{\partial t} + \nabla \cdot \boldsymbol{\Gamma} &= \mathbf{F}, \text{ in } \Omega \\ -\mathbf{n} \cdot \boldsymbol{\Gamma} = \mathbf{G} + \left(\frac{\partial \mathbf{R}}{\partial \mathbf{u}} \right)^{\mathrm{T}} \boldsymbol{\lambda}, \ \mathbf{R} &= \mathbf{0}, \ \text{ on } \Sigma \end{aligned} \tag{2.15}$$

where $\mathbf{d}_a$ and $\boldsymbol{\Gamma}$ are tensors; $\mathbf{F}$, $\mathbf{G}$ and $\mathbf{R}$ are vectors; $\mathbf{n}$ is the unitary outward normal; $\Sigma = \Sigma_D \cup \Sigma_N$. For the 2D case, the transient Navier-Stokes equations can be solved by setting

$$\mathbf{d}_a = \begin{pmatrix} \rho & 0 \\ 0 & \rho \\ 0 & 0 \end{pmatrix}; \quad \boldsymbol{\Gamma} = \begin{pmatrix} -2\eta \frac{\partial u_1}{\partial x} + p & -\eta \left(\frac{\partial u_1}{\partial y} + \frac{\partial u_2}{\partial x} \right) \\ -\eta \left(\frac{\partial u_1}{\partial y} + \frac{\partial u_2}{\partial x} \right) & -2\eta \frac{\partial u_2}{\partial y} + p \\ 0 & 0 \end{pmatrix};$$
$$\mathbf{F} = \begin{pmatrix} -\alpha u_1 - \rho \left(u_1 \frac{\partial u_1}{\partial x} + u_2 \frac{\partial u_1}{\partial y} \right) \\ -\alpha u_2 - \rho \left(u_1 \frac{\partial u_2}{\partial x} + u_2 \frac{\partial u_2}{\partial y} \right) \\ -\left(\frac{\partial u_1}{\partial x} + \frac{\partial u_2}{\partial y} \right) \end{pmatrix}; \mathbf{R} = \begin{pmatrix} u_1 - u_{1D} \\ u_2 - u_{2D} \\ 0 \end{pmatrix} \text{ on } \Sigma_D; \tag{2.16}$$
$$\mathbf{G} = \begin{pmatrix} 0 \\ 0 \\ 0 \end{pmatrix} \text{ on } \Sigma_D; \mathbf{R} = \begin{pmatrix} 0 \\ 0 \\ 0 \end{pmatrix} \text{ on } \Sigma_N; \; \mathbf{G} = \begin{pmatrix} g_1 \\ g_2 \\ 0 \end{pmatrix} \text{ on } \Sigma_N$$

where $\mathbf{u} = (u_1, u_2)$, $\mathbf{u}_D = (u_{1D}, u_{2D})$ and $\mathbf{g} = (g_1, g_2)$. Similarly, the adjoint equations can be solved by setting

$$\mathbf{d}_a = \begin{pmatrix} -\rho & 0 \\ 0 & -\rho \\ 0 & 0 \end{pmatrix};$$
$$\boldsymbol{\Gamma} = \begin{pmatrix} -2\eta \frac{\partial u_{a1}}{\partial x} - \beta_1 \frac{\partial A}{\partial u_{a1x}} + p_a & -\eta \left(\frac{\partial u_{a1}}{\partial y} + \frac{\partial u_{a2}}{\partial x} \right) - \beta_1 \frac{\partial A}{\partial u_{a1y}} \\ -\eta \left(\frac{\partial u_{a1}}{\partial y} + \frac{\partial u_{a2}}{\partial x} \right) - \beta_1 \frac{\partial A}{\partial u_{a2x}} & -2\eta \frac{\partial u_{a2}}{\partial y} - \beta_1 \frac{\partial A}{\partial u_{a2y}} + p_a \\ 0 & 0 \end{pmatrix};$$
$$\mathbf{F} = \begin{pmatrix} -\alpha u_{a1} - \beta_1 \frac{\partial A}{\partial u_1} + \rho \left(u_1 \frac{\partial u_{a1}}{\partial x} + u_2 \frac{\partial u_{a1}}{\partial y} - u_{a1} \frac{\partial u_1}{\partial x} - u_{a2} \frac{\partial u_1}{\partial y} \right) \\ -\alpha u_{a2} - \beta_1 \frac{\partial A}{\partial u_2} + \rho \left(u_1 \frac{\partial u_{a2}}{\partial x} + u_2 \frac{\partial u_{a2}}{\partial y} - u_{a1} \frac{\partial u_2}{\partial x} - u_{a2} \frac{\partial u_2}{\partial y} \right) \\ \beta_1 \frac{\partial A}{\partial p} - \left(\frac{\partial u_{a1}}{\partial x} + \frac{\partial u_{a2}}{\partial y} \right) \end{pmatrix}; \tag{2.17}$$
$$\mathbf{R} = \begin{pmatrix} u_{a1} + n_1 \frac{\partial B}{\partial p} \\ u_{a2} + n_2 \frac{\partial B}{\partial p} \\ 0 \end{pmatrix} \text{ on } \Sigma_D; \; \mathbf{G} = \begin{pmatrix} 0 \\ 0 \\ 0 \end{pmatrix} \text{ on } \Sigma_D; \mathbf{R} = \begin{pmatrix} 0 \\ 0 \\ 0 \end{pmatrix} \text{ on } \Sigma_N;$$
$$\mathbf{G} = \begin{pmatrix} -\rho u_{a1} (u_1 n_1 + u_2 n_2) - \beta_1 \left(n_1 \frac{\partial A}{\partial u_{1x}} + n_2 \frac{\partial A}{\partial u_{1y}} \right) - \beta_2 \frac{\partial B}{\partial u_1} \\ -\rho u_{a2} (u_1 n_1 + u_2 n_2) - \beta_1 \left(n_1 \frac{\partial A}{\partial u_{2x}} + n_2 \frac{\partial A}{\partial u_{2y}} \right) - \beta_2 \frac{\partial B}{\partial u_2} \\ 0 \end{pmatrix} \text{ on } \Sigma_N$$

where $\mathbf{u}_a = (u_{a1}, u_{a2})$ and $\mathbf{n} = (n_1, n_2)$. The temporal integrations are numerically implemented by solving a scalar general form equation (Eq. 2.18) in *Comsol*

$$\begin{aligned} d_a \frac{\partial u}{\partial t} + \nabla \cdot \Gamma = F, \text{ in } Q \\ -\mathbf{n} \cdot \Gamma = G + \left(\frac{\partial R}{\partial u} \right)^{\mathrm{T}} \boldsymbol{\lambda}, \; R = 0, \text{ on } \Sigma \end{aligned} \tag{2.18}$$

By setting

$$d_a = 1, \ \Gamma = \mathbf{0}, \ G = 0, \ R = 0, \ F = \beta_1 \left(\frac{\partial A}{\partial \gamma} + \frac{\partial \alpha}{\partial \gamma} \mathbf{u} \cdot \mathbf{u}_a \right) \tag{2.19}$$

and solving Eq. 2.18 for u, the adjoint derivative $\frac{D\hat{J}}{D\gamma}\big|_\Omega$ in Eq. 2.14 can be obtained as $u|_{t=T}$. Changing the value of F in Eq. 2.19 to be $\beta_2 \frac{\partial B}{\partial \gamma}$, the adjoint derivative $\frac{D\hat{J}}{D\gamma}\big|_{\partial\Omega}$ can be solved as $u|_{\partial\Omega, t=T}$. The spatial integrations in the iterative procedure can be performed by the inner function of *Comsol*. For the 3D case, the settings for Eqs. 2.15 and 2.18 are similar to the 2D case. During the optimization procedure, the Navier-Stokes equations and the adjoint equations are solved using Taylor-Hood elements [11], which interpolate the fluidic velocity quadratically and the pressure linearly. The design variable is interpolated linearly based on the corner nodes of the elements (Figs. 2.2 and 2.3). The transient equations are solved using the backward differentiation formula method in the time dependent solver of *Comsol* [12].

The stopping criteria are specified as the change of values of the objective between two consecutive iterations and the residual of the volume constraint satisfying

$$\begin{aligned} & |J_k - J_{k-1}| / |J_0| < 1 \times 10^{-6} \\ & \left| \int_\Omega \gamma \, d\Omega / V_0 - V_r \right| / V_r < 1 \times 10^{-3} \end{aligned} \tag{2.20}$$

One can note that it is convenient to implement transient optimization problems, when the continuous adjoint method is adopted.

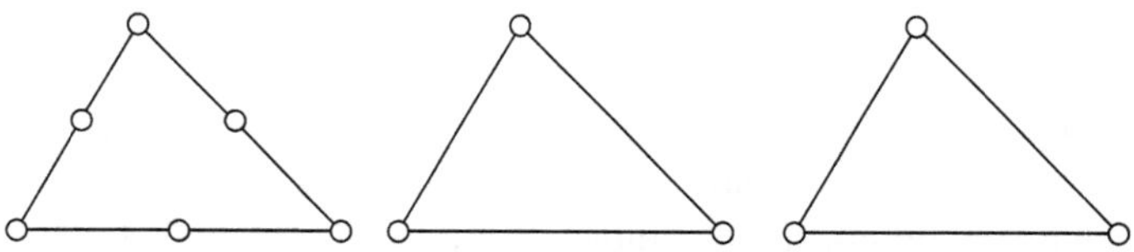

Fig. 2.2 The finite element nodes used to express the velocity, pressure and design variable on a triangular element

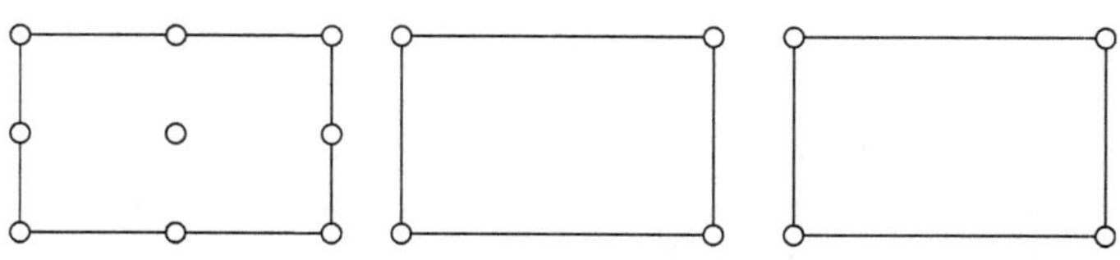

Fig. 2.3 The finite element nodes used to express the velocity, pressure and design variable on a rectangular element

2.1.4 Numerical Examples

In this section, several numerical examples are presented to demonstrate the capability of the topology optimization method for unsteady Navier-Stokes flows. The density and viscosity of the fluid are set to be unitary value, if there is no specification. The values of α_{max} and q are chosen based on numerical experiments. More details are available in [4]. The Reynolds number is calculated as

$$\mathrm{Re} = \rho U_{max} L / \eta \tag{2.21}$$

where L is the width of the inlet, and U_{max} is the maximal value of the velocity on the inlet. The initial value condition of the transient Navier-Stokes equations is $\mathbf{u}_0 = \mathbf{0}$. The objective function for the topology optimization problem in Eq. 2.7 is chosen to be the energy dissipation inside the design domain and the pressure on the inlet

$$\begin{aligned} J(\mathbf{u}, p; \gamma) = &\int_0^T \int_\Omega \beta_1 \left[\frac{\eta}{2} \left(\nabla \mathbf{u} + \nabla \mathbf{u}^{\mathrm{T}} \right) : \left(\nabla \mathbf{u} + \nabla \mathbf{u}^{\mathrm{T}} \right) + \alpha \mathbf{u}^2 \right] \mathrm{d}\Omega \mathrm{d}t \\ &+ \int_0^T \int_{\Gamma_i} \beta_2 p \, \mathrm{d}\Gamma \mathrm{d}t \end{aligned} \tag{2.22}$$

where Γ_i is the inlet boundary. The parameters β_1 and β_2 depend on specified examples and can be tuned by the designer based on design necessity or numerical experiments.

2.1.4.1 Double Pipe

One double pipe is used to investigate the feasibility of the proposed optimization method. The design domain is shown in Fig. 2.4 and is discretized by 60×60 rectangular elements. The optimization parameters are listed in Table 2.1. The Neumann boundary condition in Eq. 2.4 is loaded on the outlets Γ_{o1} and Γ_{o2} by setting $\mathbf{g} = \mathbf{0}$. When the transient velocity in Eq. 2.23

$$\begin{aligned} \mathbf{u}_{in1} &= -144(y - 4/6)(5/6 - y)\cos(t)\mathbf{n}, \ t \in [0, 2\pi] \\ \mathbf{u}_{in2} &= -144(y - 1/6)(2/6 - y)\sin(t)\mathbf{n}, \ t \in [0, 2\pi] \end{aligned} \tag{2.23}$$

is imposed on the inlets, the optimized result is derived as shown in Fig. 2.5. Snapshots of the optimization procedure are shown in Fig. 2.6. The convergent histories of the objective values and the volume of the fluidic channel is shown in Fig. 2.7. Snapshots of the streamline of the unsteady flow at specified time are shown in Fig. 2.8.

By solving the topology optimization of the double pipe for the steady case, instead of the transient case, one can obtain the optimized channels in Fig. 2.9a, with the injecting velocity in Eq. 2.24 imposed on the inlets

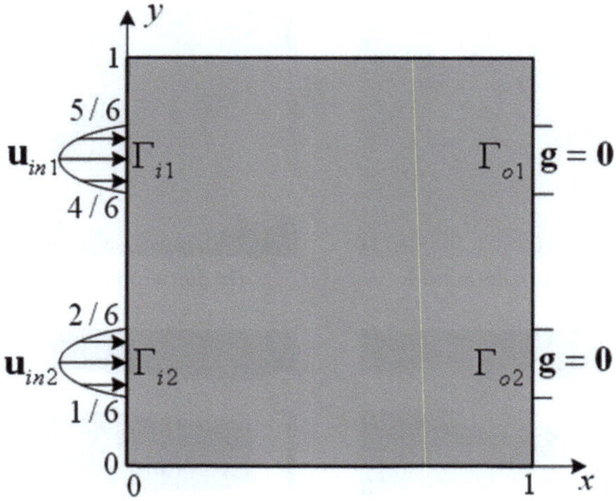

Fig. 2.4 Design domain of the double pipe. The values $\mathbf{u}_{in1}$ and $\mathbf{u}_{in2}$ are the velocity distribution on the inlets

Table 2.1 Parameter settings in the topology optimization of the double pipe

β_1	β_2	V_r	α_{min}	α_{max}	q
1	0	1/3	0	1×10^4	1

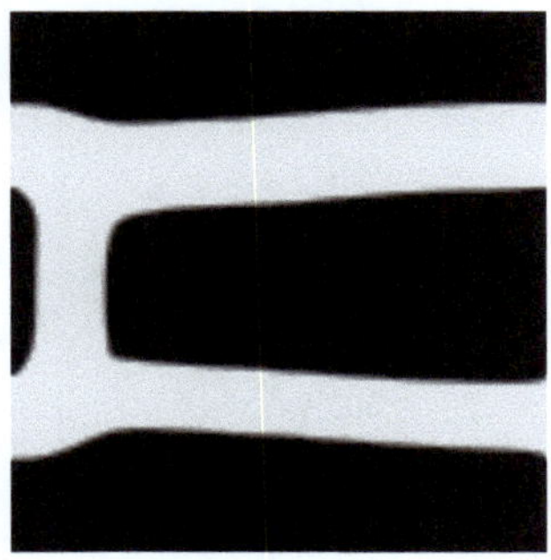

Fig. 2.5 Optimal design of the double pipe for unsteady Navier-Stokes flows corresponding to the inlet velocity in Eq. 2.23

$$\begin{aligned}\mathbf{u}_{in1} &= -144(y - 4/6)(5/6 - y)\mathbf{n} \\ \mathbf{u}_{in2} &= -144(y - 1/6)(2/6 - y)\mathbf{n}\end{aligned} \tag{2.24}$$

The optimized result in Fig. 2.9a agrees with the numerical results derived by Borrvall and Petersson in [2] for steady flows. By changing $\mathbf{u}_{in2}$ to be a velocity of suction and maintaining $\mathbf{u}_{in1}$ a velocity of injection (Eq. 2.25),

$$\begin{aligned}\mathbf{u}_{in1} &= -144(y - 4/6)(5/6 - y)\mathbf{n} \\ \mathbf{u}_{in2} &= 144(y - 1/6)(2/6 - y)\mathbf{n}\end{aligned} \tag{2.25}$$

a bend pipe is obtained as shown in Fig. 2.9b, where the fluid flows from Γ_{i1} to Γ_{i2} directly. When both $\mathbf{u}_{in1}$ and $\mathbf{u}_{in2}$ are velocities of suction (Eq. 2.26),

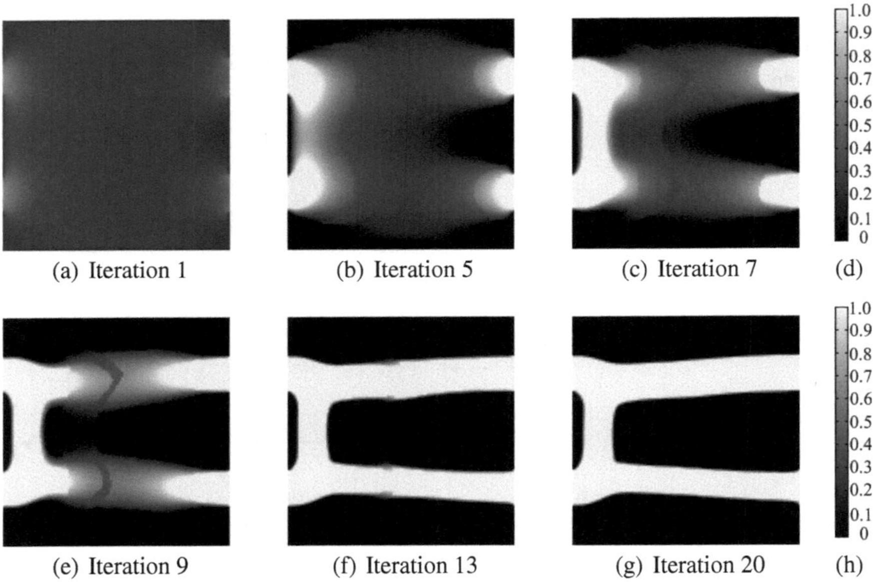

Fig. 2.6 Snapshots of optimization procedure for double pipe example in Fig. 2.5

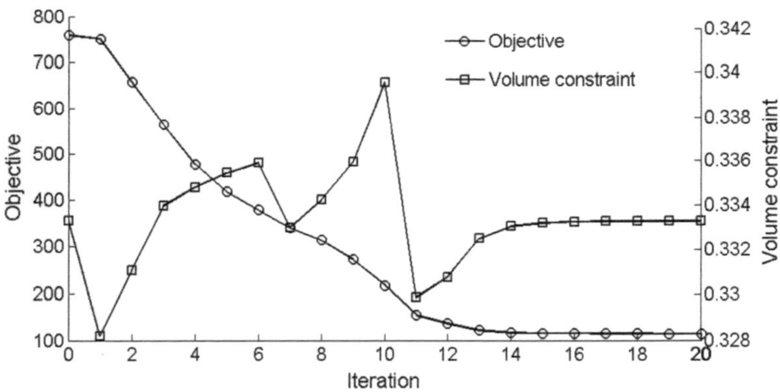

Fig. 2.7 Convergent history of the objective and volume constraint for the optimal design shown in Fig. 2.5

$$\begin{aligned} \mathbf{u}_{in1} &= 144(y - 4/6)(5/6 - y)\mathbf{n} \\ \mathbf{u}_{in2} &= 144(y - 1/6)(2/6 - y)\mathbf{n} \end{aligned} \tag{2.26}$$

the optimized channel is shown in Fig. 2.9c. Figure 2.9a–c show that the optimal design is a double pipe, when $\mathbf{u}_{in1}$ and $\mathbf{u}_{in2}$ are both the velocity of injection or suction; and the optimal design is a bend channel, when $\mathbf{u}_{in1}$ and $\mathbf{u}_{in2}$ are the velocity of suction and injection, respectively. According to Eq. 2.23, $\mathbf{u}_{in1}$ and $\mathbf{u}_{in2}$ are the velocity of injection as $t \in (0, \pi/2)$; the velocity of suction and injection as $t \in (\pi/2, \pi)$; the

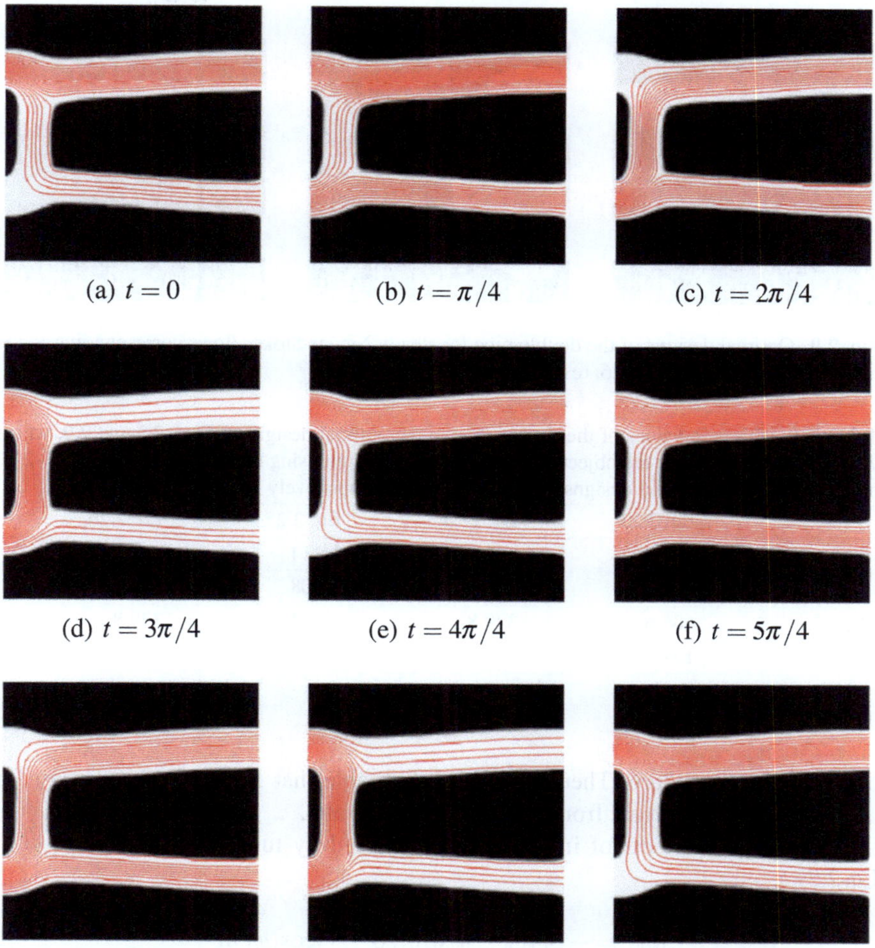

(a) $t = 0$ (b) $t = \pi/4$ (c) $t = 2\pi/4$

(d) $t = 3\pi/4$ (e) $t = 4\pi/4$ (f) $t = 5\pi/4$

(g) $t = 6\pi/4$ (h) $t = 7\pi/4$ (i) $t = 8\pi/4$

Fig. 2.8 Streamline snapshots of the unsteady double pipe flow corresponding to the optimal design in Fig. 2.5

velocity of suction as $t \in (\pi, 3\pi/2)$; and the velocity of injection and suction as $t \in (3\pi/2, 2\pi)$. Therefore, the fluid could be transported between Γ_{i1} and Γ_{i2} as $t \in (\pi/2, \pi) \cup (3\pi/2, 2\pi)$ and could flow in parallel between the inlets and outlets as $t \in (0, \pi/2) \cup (\pi, 3\pi/2)$. This analysis is consistent with the streamline distribution in Fig. 2.8. By respectively imposing the velocity of inlet in Eqs. 2.23–2.26 for the optimized results in Figs. 2.5 and 2.9, the values of the objective in Eq. 2.22 shown in Table 2.2 are obtained. By cross comparison the data in Table 2.2 and the optimized channels in Figs. 2.5 and 2.9, one can conclude that the optimal topology of a channel is valid only corresponding to the specific inlet velocity that is specified during the

(a) (b) (c)

Fig. 2.9 Optimal designs of the double pipe for steady Navier-Stokes flows corresponding to the inlet velocity in Eqs. 2.24–2.26, respectively

Table 2.2 Objective values of the double pipe in the optimal designs of Figs. 2.5 and 2.9, where J_U, J_{Sa}, J_{Sb} and J_{Sc} are the objective values obtained by imposing the velocity in Eqs. 2.23–2.26 on the inlets of the optimal designs in Figs. 2.5 and 2.9, respectively

	Eq. 2.23	Eq. 2.24	Eq. 2.25	Eq. 2.26
J_U	113.8	56.65	16.14	56.65
J_{Sa}	133.6	48.60	37.08	48.73
J_{Sb}	606.6	379.8	6.04	379.8
J_{Sc}	134.8	48.78	48.73	48.71

optimization procedure. Therefore, it is reasonable that the optimal topology for unsteady flow is different from its steady counterparts.

The dynamic effects of inflow can be adjusted by tuning the parameter ω in Eq. 2.27

$$\begin{aligned} \mathbf{u}_{in1} &= -144(y - 4/6)(5/6 - y)\cos(\omega t)\mathbf{n} \\ \mathbf{u}_{in2} &= -144(y - 1/6)(2/6 - y)\sin(\omega t)\mathbf{n} \end{aligned} \tag{2.27}$$

A larger value of ω corresponds to stronger oscillation of inflow. Figure 2.10 shows the optimized double pipes for different values of ω. These results illustrate the necessity of the optimization for unsteady flow.

2.1.4.2 Three-Terminal Device

Fluidic channels with periodic dynamic input on the inlet have been widely used in fluidic devices [13]. In this example, a three-terminal device with periodic transient velocity on the inlet Γ_i, given as the equation

$$\mathbf{u}_{in} = -1 \times 10^4(y - 1)(1.2 - y)\sin(t)\mathbf{n} \tag{2.28}$$

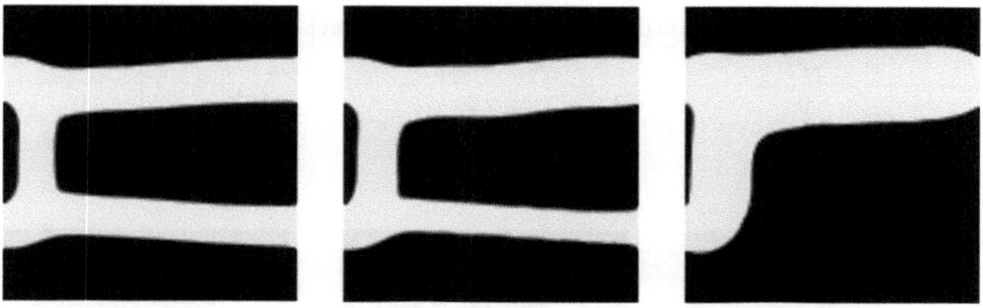

(a) $\omega = 1/2$ (b) $\omega = 2$ (c) $\omega = 8$

Fig. 2.10 Optimal designs of the double pipe corresponding to different values of ω in Eq. 2.27

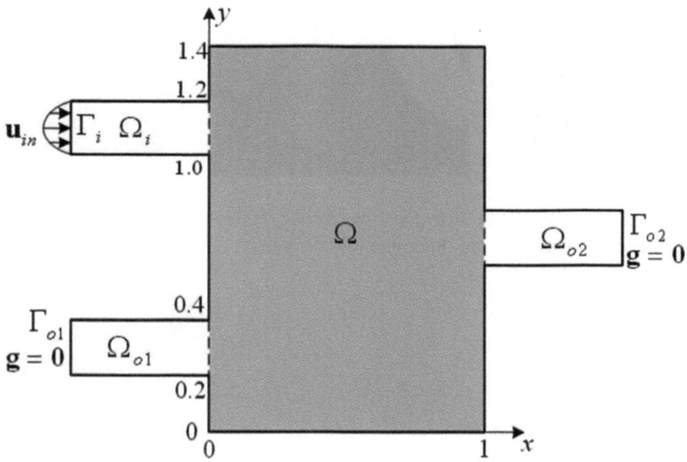

Fig. 2.11 Design domain of the three-terminal device. The value $\mathbf{u}_{in}$ is the velocity distribution on the inlet

is optimized. According to Eq. 2.21, the Reynolds number is 100 in this example. The design domain Ω is shown in Fig. 2.11, where the inlet duct is Ω_i and the outlet ducts are Ω_{o1} and Ω_{o2}. The design domain is discretized by 100×140 rectangular elements. The Neumann boundary condition in Eq. 2.4 is loaded on the outlets Γ_{o1} and Γ_{o2} by setting $\mathbf{g} = \mathbf{0}$. The other boundaries are set as no-slip boundaries. Three time intervals $[0, \pi]$, $[\pi, 2\pi]$ and $[0, 2\pi]$ which correspond to pure injection flow, pure suction flow and periodic injection-suction flow, are considered separately. The optimization parameter values are shown in Table 2.3 and the optimal topologies of the device are shown in Fig. 2.12a–c. Figure 2.13 shows the optimized device where the steady velocity condition is loaded on the inlet Γ_i as

$$\mathbf{u}_{in} = -1 \times 10^4 (y-1)(1.2-y)\mathbf{n} \tag{2.29}$$

Table 2.3 Parameter settings in the topology optimization of three-terminal device

β_1	β_2	V_r	α_{min}	α_{max}	q
1	0	0.3	0	1×10^4	1

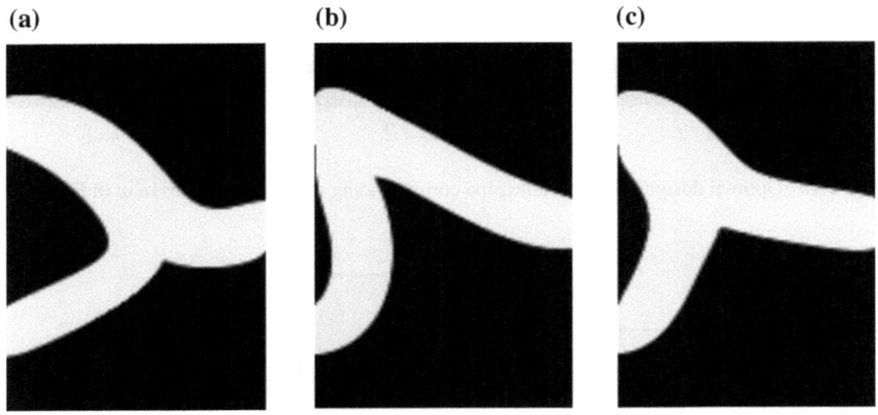

Fig. 2.12 Optimal designs of the three-terminal device. Here, **a**, **b** and **c** are optimal designs for unsteady flows corresponding to the time intervals $[0, \pi]$, $[\pi, 2\pi]$ and $[0, 2\pi]$, respectively

Fig. 2.13 Optimal design of the three-terminal device for steady flow

From Fig. 2.12, one can see that the suction flow occurring as $t \in (\pi, 2\pi)$ is a key point in the difference of the shape of the device compared to the steady flow case. By imposing the velocity boundary in Eqs. 2.28 and 2.29 on the inlets of the optimal designs in Figs. 2.12 and 2.13 respectively, the values of objective listed in Table 2.4 can be obtained. The cross comparison of the values of energy dissipation confirms further that the dynamic effect of the unsteady flow can influence the detailed shape of the device. This example illustrates that the optimal design of a unsteady flow is influenced by the dynamic effect induced by the different choice of time intervals.

Table 2.4 Objective values of the optimal designs in Figs. 2.12 and 2.13. J_{Ta}, J_{Tb}, J_{Tc} and J_S are objective values obtained by imposing the velocity in Eqs. 2.28 and 2.29 on the inlets of optimal designs in Figs. 2.12 and 2.13

	Steady	$t \in [0, \pi]$	$t \in [\pi, 2\pi]$	$t \in [0, 2\pi]$
J_{Ta}	1.064×10^5	1.298×10^5	1.305×10^5	2.605×10^5
J_{Tb}	1.347×10^5	1.579×10^5	1.258×10^5	2.841×10^5
J_{Tc}	1.138×10^5	1.302×10^5	1.271×10^5	2.574×10^5
J_S	1.062×10^5	1.299×10^5	1.300×10^5	2.602×10^5

2.1.4.3 Bend Channel

One example of a bend channel with different Reynolds numbers under steady flow has been discussed by Gersborg-Hansen et al. in [3]. The similar example is discussed here for unsteady flows. The computational domain includes the inlet duct Ω_i, the outlet duct Ω_o and the design domain Ω (Fig. 2.14), which are respectively discretized by 40×20, 40×20 and 100×100 rectangular elements. The transient velocity is set to be

$$\mathbf{u}_{in} = -4U_{max}(y - 3.5)(4.5 - y)t\mathbf{n}, \ t \in [0, 1] \tag{2.30}$$

where U_{max} is specified to be 1, 50 and 300, respectively corresponding to the Reynolds numbers of 1, 50 and 300. The optimization parameters are listed in Table 2.5. The optimized channels with different Reynolds numbers and the corresponding objective values are shown in Fig. 2.15 and Table 2.6. The results in Fig. 2.15 show that the bend channel has sharp corners for flow with low Reynolds

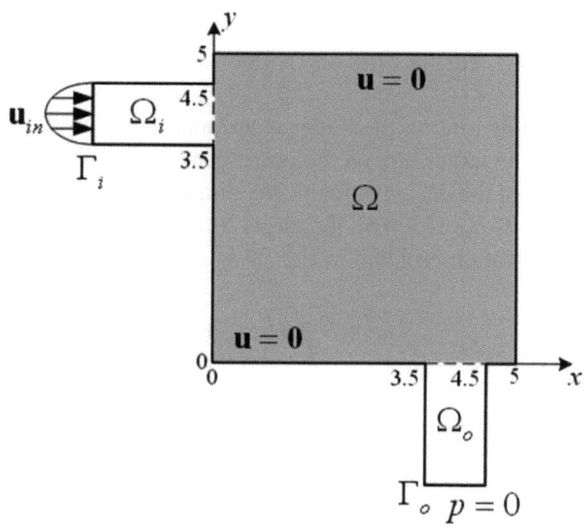

Fig. 2.14 Design domain of the bend channel. The value $\mathbf{u}_{in}$ is the velocity distribution on the inlet

Table 2.5 Parameter settings in the topology optimization of the bend channel

β_1	β_2	V_r	α_{min}	α_{max}	q
1	0.1	0.25	0	1×10^4	1

(a) Re=1 (b) Re=50 (c) Re=300

Fig. 2.15 Optimal designs of the bend channel corresponding to different Reynolds numbers

Table 2.6 Objective values of the optimal designs in Fig. 2.15

	Re = 1	Re = 50	Re = 300
J	2.094×10^1	5.722×10^4	2.745×10^6

number, and the corners become rounder along with the increase of the Reynolds number. Therefore, the optimized channel is relatively straight for the flow with low Reynolds number, and it becomes bending for the flow with large enough Reynolds number. The results agree with those obtained by Gersborg-Hansen et al. in [3] for steady flows.

2.1.4.4 Target Flux on Outlet

The flow rate on the outlet is an important output performance for fluid flow [13, 14]. The target flux on the outlet can be produced by adding a flux constraint at the outlet of the design domain. According to the flow rate constraint for optimization of steady flows [15, 16], the target flux for a unsteady flow is added into the topology optimization problem in Eq. 2.7 as an inequality constraint

$$\left(\frac{\int_{T_1}^{T_2} \int_{\Gamma_o} \mathbf{u} \cdot \mathbf{n} \, \mathrm{d}\Gamma \mathrm{d}t}{Q_{tar}} - 1 \right)^2 \leq \varepsilon \tag{2.31}$$

where $[T_1, T_2] \subseteq [0, T]$ is the time interval; Γ_o is the outlet boundary; Q_{tar} is the target flux on the corresponding outlet; and ε is the allowable tolerance. The adjoint analysis for the flux constraint follows the same procedure as that for the objective in

Sect. 2.1.2. Additionally, the objective or certain design constraints may be defined on the partial time interval $[T_1, T_2] \subseteq [0, T]$, while the others are still defined on the whole time interval $[0, T]$. Therefore, the adjoint analysis of the corresponding expression is implemented on $[0, T_2]$ instead of $[0, T]$, although the unsteady flow problem is defined on the whole time interval $[0, T]$. Sequentially, the corresponding adjoint equations and adjoint derivatives are transient equations and integrations on $[0, T_2]$ instead of $[0, T]$ (see Sect. 2.3.2.2 for more details).

First, a flux distribution structure is optimized. The design domain is shown in Fig. 2.16a, and is discretized by 100×100 rectangular elements. The velocity loaded on the inlet is (Fig. 2.17c)

$$\begin{aligned}\mathbf{u}_{in} = &- [4(y-3.5)(4.5-y)(t \le 0.5)\\ &+ 4(y-3.5)(4.5-y)(18t-8)(0.5 < t \le 1)\\ &+ 40(y-3.5)(4.5-y)(1 < t \le 2)\\ &+ 4(y-3.5)(4.5-y)(-18t+46)(2 < t \le 2.5)\\ &+ 4(y-3.5)(4.5-y)(t > 2.5)]\mathbf{n},\ t \in [0, 3]\end{aligned} \tag{2.32}$$

The optimization parameter values are listed in Table 2.7. The flux constraint expressed in Eq. 2.31 is imposed on the outlet Γ_{o1} to constrain the flux distribution between the two outlets Γ_{o1} and Γ_{o2}. The parameters in Eq. 2.31 are chosen to be 1×10^{-4} for ε and $\frac{1}{3}\int_0^3 \int_{\Gamma_i} -\mathbf{u}_{in} \cdot \mathbf{n}\, \mathrm{d}\Gamma \mathrm{d}t$ for Q_{tar}. When $[T_1, T_2]$ is set equal to $[0, T]$, the design constraint is

$$\left(\frac{\int_0^3 \int_{\Gamma_{o1}} \mathbf{u} \cdot \mathbf{n}\, \mathrm{d}\Gamma \mathrm{d}t}{\frac{1}{3}\int_0^3 \int_{\Gamma_i} -\mathbf{u}_{in} \cdot \mathbf{n}\, \mathrm{d}\Gamma \mathrm{d}t} - 1 \right)^2 \le 1 \times 10^{-4} \tag{2.33}$$

Then the objective and the design constraint are both defined on the whole time interval. The optimized flux distribution device is shown in Fig. 2.17a. And the corresponding objective value is 1.740×10^3. When the time interval $[T_1, T_2]$ is set to be $[1, 2]$, the design constraint is modified to be

$$\left(\frac{\int_1^2 \int_{\Gamma_{o1}} \mathbf{u} \cdot \mathbf{n}\, \mathrm{d}\Gamma \mathrm{d}t}{\frac{1}{3}\int_0^3 \int_{\Gamma_i} -\mathbf{u}_{in} \cdot \mathbf{n}\, \mathrm{d}\Gamma \mathrm{d}t} - 1 \right)^2 \le 1 \times 10^{-4} \tag{2.34}$$

and the objective is still defined on the whole time interval. According to the derivation of adjoint sensitivity in Appendix 2.3.2.2, the adjoint equations and derivative corresponding to the flux constraint are defined on the time interval $[0, 2]$. By keeping the other parameters unchanged, the optimization problem is solved. The optimized result is derived as shown in Fig. 2.17b. The corresponding objective value is 1.257×10^3. The outflow rate on the outlet Γ_{o1} corresponding to the above two optimal designs is shown in Fig. 2.17c. Because the fluid considered is incompressible, the left fluid must flow out from the outlet Γ_{o2}. Therefore, the optimized structures

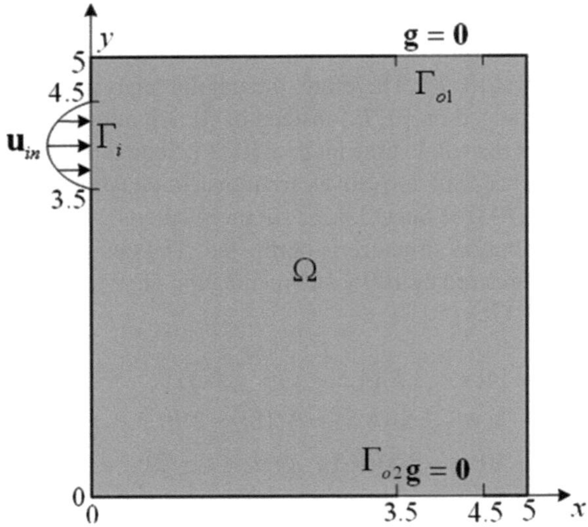

Fig. 2.16 Design domain of the flux distribution structure

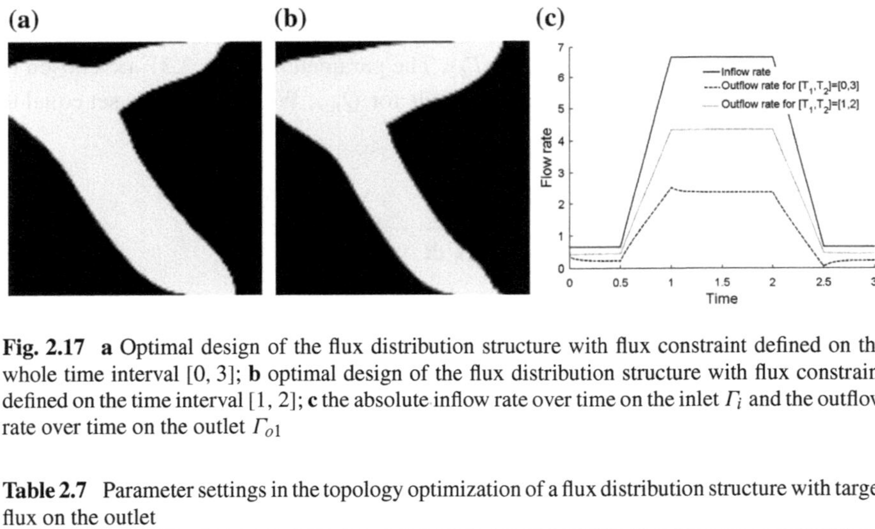

Fig. 2.17 **a** Optimal design of the flux distribution structure with flux constraint defined on the whole time interval [0, 3]; **b** optimal design of the flux distribution structure with flux constraint defined on the time interval [1, 2]; **c** the absolute inflow rate over time on the inlet Γ_i and the outflow rate over time on the outlet Γ_{o1}

Table 2.7 Parameter settings in the topology optimization of a flux distribution structure with target flux on the outlet

β_1	β_2	V_r	α_{min}	α_{max}	q
1	0	0.4	0	1×10^3	1

in Fig. 2.17a, b have two branches. The flux constraints in Eqs. 2.33 and 2.34 mean that the flux at the outlet Γ_{o1} of the optimal design in Fig. 2.17a is lower than that in Fig. 2.17b. Therefore, the thick branch connected to Γ_{o1} in Fig. 2.17b is helpful for decreasing the velocity gradient and dissipation of the flow.

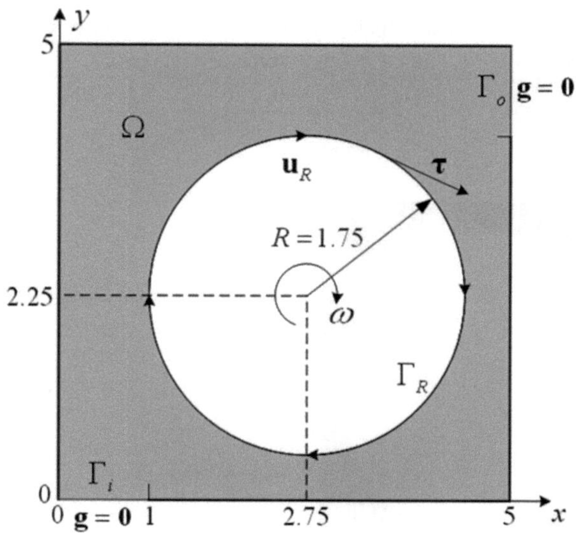

Fig. 2.18 Design domain of the roller-type pump

Second, a roller-type pump, which pumps the liquid using a rotating roller, is optimized. Figure 2.18 shows the design domain, where the rotating roller drives the liquid flowing from the inlet Γ_i to the outlet Γ_o. The fluidic velocity on the surface of the roller is equal to the rotational velocity of the roller surface,

$$\mathbf{u}_r = \omega(t) R \boldsymbol{\tau}, \ t \in [0, 2] \tag{2.35}$$

where $\omega(t)$ is the transient angular speed of the roller; R is the radius of the roller; and $\boldsymbol{\tau}$ is the unitary tangential vector of the roller surface. The optimization parameters are listed in Table 2.8. By respectively setting Q_{tar} and ε to be 20 and 1×10^{-4}, the flux constraint shown in Eq. 2.31 is imposed on the boundary Γ_o as

$$\left(\frac{\int_0^2 \int_{\Gamma_o} \mathbf{u} \cdot \mathbf{n} \, \mathrm{d}\Gamma \mathrm{d}t}{20} - 1 \right)^2 \leq 1 \times 10^{-4} \tag{2.36}$$

The design domain is discretized by 171,519 triangular elements. After the transient angular speed is set to $\omega(t) = 400 \min(t, 1)/7$ (Fig. 2.19b), the optimized roller-type pump is derived as shown in Fig. 2.19a. The corresponding objective value is 5.866×10^5. The outflow rate over time is shown in Fig. 2.19b. The streamline distribution of the flow in the optimized pump is shown in Fig. 2.20. Figure 2.20 shows that the streamlines near the surface of the pump's roller is closed, while the others start at the inlet and end at the outlet. Therefore, a net flux between the inlet and outlet of the pump is produced by the rotating roller.

Table 2.8 Parameter settings in the topology optimization of the roller-type pump

β_1	β_2	V_r	α_{min}	α_{max}	q
1	0	0.4	0	1×10^3	1

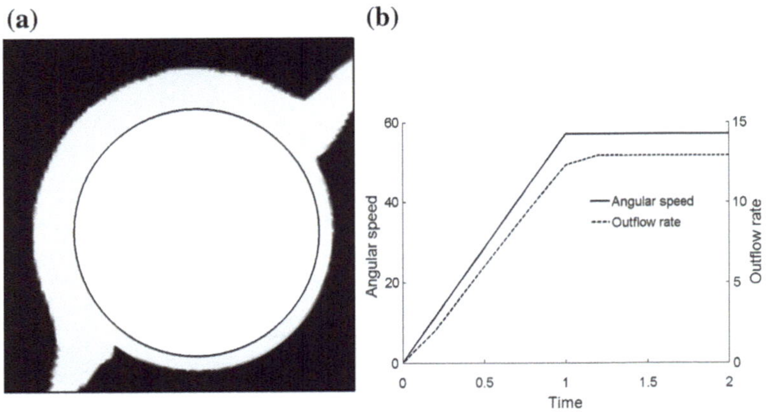

Fig. 2.19 **a** Optimal design of the roller-type pump; **b** the angular velocity over time and the outflow rate over time on the outlet Γ_o

Fig. 2.20 Streamlines of the flow in the optimal design of the roller-type pump at $t = 0.8$

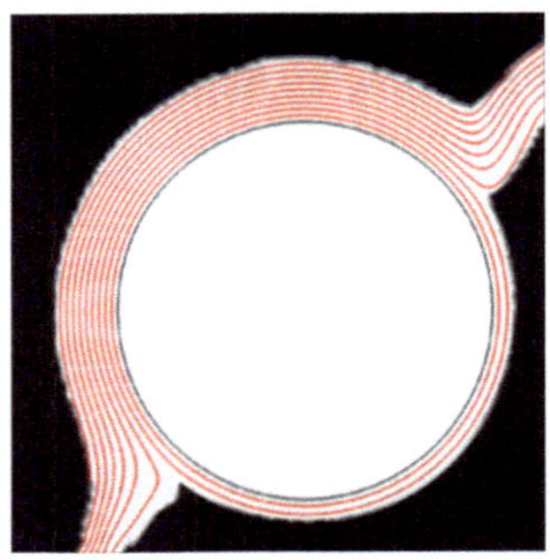

2.1.4.5 Star-Shaped Microchannel Chip

To demonstrate the applications in microfluidics, a popular technique in biochemistry and bioengineering [17, 18], this example involves the design of an infuser, a device that feeds a reactor or a piece of analysis equipment with a specific amount of fluid. Flushing the fluidic channel is an important step to maintain consistent performance and enhance the efficiency of an infuser. The fluid flowing through the infuser is water, with density 1×10^3 kg/m^3 and viscosity 1×10^{-3} Pa s. The design domain is in Fig. 2.21a and it is discretized by 160,866 triangular elements. The transient velocity loaded on the inlets of the Star-shaped microchannel is

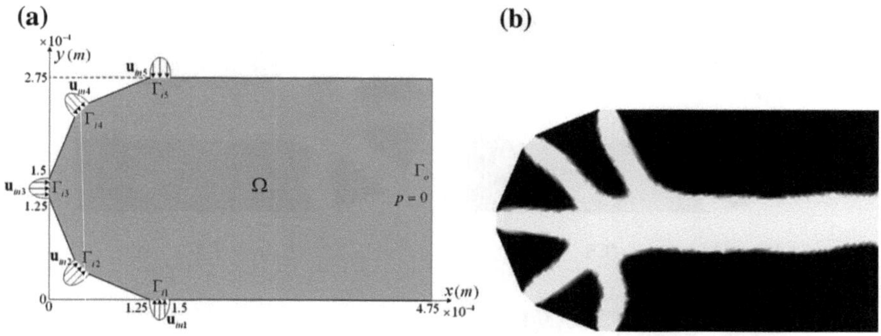

Fig. 2.21 **a** Design domain of the star-shaped microchannel chip; **b** optimal design of the star-shaped microchannel chip

$$\begin{aligned}
\mathbf{u}_{in1} &= -U\left(1+2\sin(\pi t)\right)\mathbf{n} \\
\mathbf{u}_{in2} &= -U\left(1+2\sin(\pi t+\pi/4)\right)\mathbf{n} \\
\mathbf{u}_{in3} &= -U\left(1+2\sin(\pi t+\pi/2)\right)\mathbf{n} \\
\mathbf{u}_{in4} &= -U\left(1+2\sin(\pi t+3\pi/4)\right)\mathbf{n} \\
\mathbf{u}_{in5} &= -U\left(1+2\sin(\pi t+\pi)\right)\mathbf{n}
\end{aligned} \tag{2.37}$$

where $U = 4 \times 10^{-2} s\,(1-s)$ is a parabolic distribution on the inlets of the design domain, and s is the parametric coordinate on the corresponding inlet. The energy dissipation and pressure distribution at the inlet (Eq. 2.22) are set to be the objective. The time interval is chosen to be [0, 2]. The optimization parameters are listed in Table 2.9. The optimized infuser has star-shaped channels (Fig. 2.21b). The corresponding objective value is 795.6. Snapshots for the distribution of the fluidic velocity are shown in Fig. 2.22. It shows that the fluid pulses through the star-shaped chip periodically. This periodic pulsing is helpful for the effective flushing of leftovers in microfluidic chips.

2.1.5 Summary

This topology optimization method can used to design the unsteady incompressible Navier-Stokes flows at low and moderate Reynolds numbers. It makes topology optimization more useful for practical engineering designs, such as the flow with unsteady

Table 2.9 Parameter settings in the topology optimization of a star-shaped microchannel chip

β_1	β_2	V_r	α_{min}	α_{max}	q
1	0.1	0.35	0	1×10^6	1

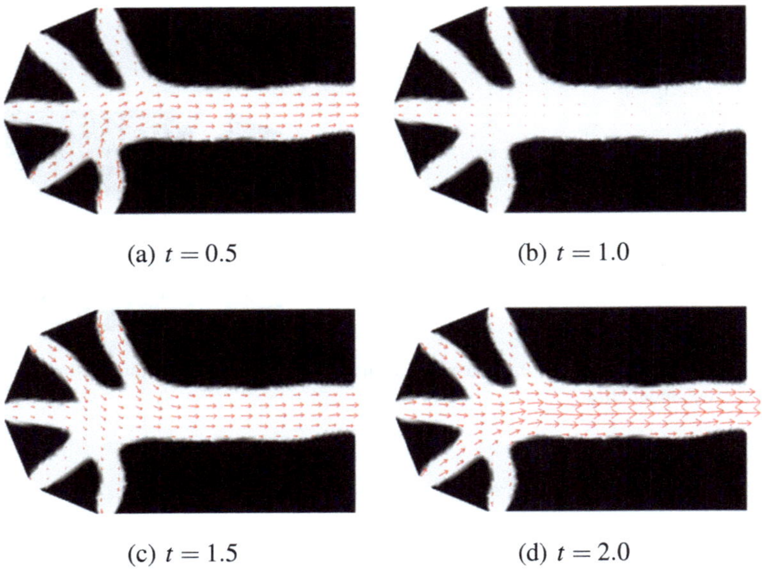

Fig. 2.22 Snapshots of the distribution of fluidic velocity

state or optimization focusing on the dynamic effect of fluid. The topology optimization problem of the unsteady Navier-Stokes flow is analyzed using continuous adjoint method. Then the numerical optimization procedure can be implemented by user-available numerical computational methods, such as finite difference method, finite element method or finite volume method. The numerical examples demonstrated that the optimized design of the unsteady Navier-Stokes flow is influenced by the dynamic effect, the Reynolds number and the constraints on the flux of fluid.

2.2 Level Set Method-Based Topology Optimization for Unsteady Flows

This section introduces the level set method-based topology optimization for unsteady flows. Being different from the density method, the level set method distinguish the fluid and solid phases by the sign of the level set function defined on the design domain. The key ideas behind this work is on constrain the fluidic velocity to be zero in the implicitly expressed solid phase, and the analysis of the topology optimization problem using variational level set method.

2.2.1 Level Set Method-Based Optimization Problem

In order to express the solid-liquid interface implicitly, the unsteady incompressible Navier-Stokes equations are coupled with the level set function, expressed by a signed distance function ϕ defined on the optimization domain $\Omega \subset \mathbb{R}^d \rightarrow \mathbb{R}$ ($d = 2$ or 3 is the space dimension). The solid and fluid domains is distinguished as

$$\begin{cases} \phi(\mathbf{x}) > 0, & \forall \mathbf{x} \in \Omega_s \\ \phi(\mathbf{x}) < 0, & \forall \mathbf{x} \in \Omega_l \\ \phi(\mathbf{x}) = 0, & \forall \mathbf{x} \in \Gamma \end{cases} \tag{2.38}$$

where Ω_s, Ω_l and Γ are the solid region, fluid region and implicit boundary, respectively. Ω_s and Ω_l satisfy $\Omega_s \cup \Omega_l = \Omega$. Then, the unsteady incompressible Navier-Stokes equations are modified by constraining the fluid velocity to be zero in the solid phase Ω_s

$$\begin{aligned} \rho \frac{\partial \mathbf{u}}{\partial t} - \eta \Delta \mathbf{u} + \rho (\mathbf{u} \cdot \nabla) \mathbf{u} + \nabla p = \mathbf{0}, \ \text{in } Q \\ -\nabla \cdot \mathbf{u} = 0, \ \text{in } Q \\ H(\phi) \mathbf{u} = \mathbf{0}, \ \text{in } Q \end{aligned} \tag{2.39}$$

where $\mathbf{u}$ is the fluid velocity; p is the pressure; ρ is the fluid density; η is the fluid viscosity; Q is the computational domain $(0, T) \times \Omega$; t is the time and $(0, T)$ is the time interval; $H(\phi)$ is the Heaviside function:

$$H(\phi) = \begin{cases} 1, & \phi \geq 0 \\ 0, & \phi < 0 \end{cases} \tag{2.40}$$

and the derivative of $H(\phi)$ is the Dirac function:

$$\tau(\phi) = \begin{cases} +\infty, \ \phi = 0 \\ 0, \ \phi \neq 0 \end{cases} \tag{2.41}$$

The implicit no slip boundary, lying between the solid and fluid domains, is represented by the zero level set of the level set function ϕ. Compared with the method in [19], the above coupling method avoids abstracting the fluid domain and changing the computational domain. The initial condition of Navier-Stokes equation 2.39 is

$$\mathbf{u}(0, \mathbf{x}) = \mathbf{u}_0(\mathbf{x}), \ \text{in } \Omega \tag{2.42}$$

where $\mathbf{u}_0(\mathbf{x})$ satisfies $\nabla \cdot \mathbf{u}_0 = 0$. The boundary conditions contains the Dirichlet type

$$\mathbf{u}(t, \mathbf{x}) = \mathbf{u}_D(t, \mathbf{x}), \ \text{on } \Sigma_D \tag{2.43}$$

and the Neumann type

$$(-p\mathbf{I}+\eta\nabla\mathbf{u})\cdot\mathbf{n}=\mathbf{g},\quad \text{on } \Sigma_N \tag{2.44}$$

where $\mathbf{u}_D$ and $\mathbf{g}$ are the known velocity and stress distribution on Γ_D and Γ_N, respectively; $\mathbf{n}$ is the outward unitary normal on $\partial\Omega$; Σ_D and Σ_N represent $(0,T)\times\Gamma_D$ and $(0,T)\times\Gamma_N$, respectively. Particularly, the no slip boundary condition is Dirichlet type, where $\mathbf{u}_D$ is set to be $\mathbf{0}$; the open boundary condition is Neumann type, where $\mathbf{g}$ is set to be $\mathbf{0}$.

Based on the above description, the level set method-based topology optimization problem for the unsteady Navier-Stokes flow is formulated to be

$$\text{Find } \phi,\ \text{to minimize } J(\mathbf{u},p;\phi),\ \text{subject to}$$
$$\begin{cases} \rho\dfrac{\partial\mathbf{u}}{\partial t}-\eta\Delta\mathbf{u}+\rho(\mathbf{u}\cdot\nabla)\mathbf{u}+\nabla p=\mathbf{0}, & \text{in } Q\\ -\nabla\cdot\mathbf{u}=0, & \text{in } Q\\ H(\phi)\,\mathbf{u}=\mathbf{0}, & \text{in } Q\\ \mathbf{u}(0,\mathbf{x})=\mathbf{u}_0(\mathbf{x}), & \text{in } \Omega\\ \mathbf{u}(t,\mathbf{x})=\mathbf{u}_D(t,\mathbf{x}), & \text{on } \Sigma_D\\ (-p\mathbf{I}+\eta\nabla\mathbf{u})\,\mathbf{n}=\mathbf{g}, & \text{on } \Sigma_N\\ \displaystyle\int_\Omega H(-\phi)\,\mathrm{d}\Omega=V_rV_0 \end{cases} \tag{2.45}$$

where $V_r\in(0,1)$ is the volume fraction; $V_0=\int_\Omega 1\,\mathrm{d}\Omega$ is the volume of the optimization domain Ω. The objective is set to be the general form

$$\begin{aligned} J(\mathbf{u},p;\phi)=&\int_0^T\int_\Omega \beta_1H(-\phi)\,A(\mathbf{u},\nabla\mathbf{u},p)\,\mathrm{d}\Omega\mathrm{d}t+\int_0^T\int_\Gamma \beta_2B(\mathbf{u},\nabla\mathbf{u},p)\,\mathrm{d}\Gamma\mathrm{d}t\\ &+\int_0^T\int_{\partial\Omega}\beta_3C(\mathbf{u},p)\,\mathrm{d}\Gamma\mathrm{d}t \end{aligned} \tag{2.46}$$

which can be transformed into

$$\begin{aligned} J(\mathbf{u},p;\phi)=&\int_0^T\int_\Omega \left[\beta_1H(-\phi)\,A(\mathbf{u},\nabla\mathbf{u},p)+\beta_2\tau(\phi)\,\|\nabla\phi\|\,B(\mathbf{u},\nabla\mathbf{u},p)\right]\mathrm{d}\Omega\mathrm{d}t\\ &+\int_0^T\int_{\partial\Omega}\beta_3C(\mathbf{u},p)\,\mathrm{d}\Gamma\mathrm{d}t \end{aligned} \tag{2.47}$$

where $A(\mathbf{u},\nabla\mathbf{u},p)$ is a functional defined on Q; $B(\mathbf{u},\nabla\mathbf{u},p)$ is a functional defined on the implicit boundary $(0,T)\times\Gamma$; $C(\mathbf{u},p)$ is a functional defined on $(0,T)\times\partial\Omega$; β_1, β_2 and β_3 are space-independent coefficients. The definition of the general objective in Eq. 2.46 makes the optimization with the objectives defined on the design domain, the boundary of the design domain or the implicit boundary (i.e. zero level set) be feasible using this optimization method for unsteady flows. The particular objectives, e.g. the dissipation of the flow, the kinetic energy of the fluid and the pressure drop between the inlet and outlet, can be included in the general formulation in Eq. 2.46.

2.2.2 *Sensitivity Analysis*

The sensitivity analysis of the optimization problem is implemented using the Lagrangian multiplier based adjoint approach [5–7], and the volume constraint is treated using the Lagrangian multiplier based quadratic penalty method [20]. Then the adjoint equations of the Navier-Stokes equations in the optimization problem (Eq. 2.45) are derived to be (see Appendix 2.4.1 for more details)

$$\begin{aligned} -\rho \frac{\partial \mathbf{u}_a}{\partial t} - \eta \Delta \mathbf{u}_a - \rho \left(\mathbf{u} \cdot \nabla\right) \mathbf{u}_a + \rho \left(\nabla \mathbf{u}\right) \cdot \mathbf{u}_a + \nabla p_a &= -\left(\frac{\partial \tilde{A}}{\partial \mathbf{u}} + \nabla \cdot \frac{\partial \tilde{A}}{\partial \nabla \mathbf{u}}\right), \text{ in } Q \\ -\nabla \cdot \mathbf{u}_a &= -\frac{\partial \tilde{A}}{\partial p}, \text{ in } Q \\ \mathbf{u}_a \left(T, \mathbf{x}\right) &= \mathbf{0}, \text{ in } \Omega \\ \mathbf{u}_a &= -\beta_3 \frac{\partial C}{\partial p} \mathbf{n}, \text{ on } \Sigma_D \\ \left(-p_a \mathbf{I} + \eta \nabla \mathbf{u}_a\right) \mathbf{n} &= -\rho \left(\mathbf{u} \cdot \mathbf{n}\right) \mathbf{u}_a - \frac{\partial \tilde{A}}{\partial \nabla \mathbf{u}} \mathbf{n} \\ &\quad - \beta_3 \frac{\partial C}{\partial \mathbf{u}}, \text{ on } \Sigma_N \end{aligned} \tag{2.48}$$

where $\tilde{A} = \beta_1 H(-\phi) A + \beta_2 \tau(\phi) \|\nabla \phi\| B$; $\mathbf{u}_a$ and p_a are the adjoint variables of the fluid velocity $\mathbf{u}$ and pressure p, respectively. According to [21], the evolution of the level set function is achieved by solving the Hamilton-Jacobian equation

$$\frac{\partial \phi}{\partial \theta} + v_n \left\|\nabla \phi\right\| = 0 \tag{2.49}$$

where θ is the level set evolving time, and it is different from the time in Eq. 2.39; v_n is the normal velocity loaded on the level set function ϕ. Therefore, the variational of ϕ is $\delta \phi = -v_n \left\|\nabla \phi\right\| \delta \theta$, where $\delta \theta$ is the evolution of time, and it must be positive [21]. In order to ensure the descendent of the objective, the shape sensitivity should satisfy $\delta \hat{J} < 0$, where $\hat{J}$ is the augmented Lagragian objective functional and $\delta \hat{J}$ is (see Appendix 2.4.2 for more details on the derivation of $\delta \hat{J}$)

$$\delta \hat{J} = -\int_{\Omega}\left[\int_0^T (\beta_1 A + \beta_2 \nabla B \cdot \mathbf{n}_\Gamma + \beta_2 B\kappa - \mathbf{u}\cdot\mathbf{u}_a)\,\mathrm{d}t - \lambda + \Lambda\left(\int_{\Omega} H(-\phi)\,\mathrm{d}\Omega - V_r V_0\right)\right]\tau(\phi)\delta\phi\,\mathrm{d}\Omega \tag{2.50}$$

where λ and Λ are the Lagrangian multiplier and the quadratic penalty parameter, respectively. Because $\tau(\phi)$, $\|\nabla\phi\|$ and $\delta\theta$ are all nonnegative, the normal velocity used to evolve the level set function is set to be

$$v_n = -\int_0^T [\beta_1 A + \beta_2 (\nabla B \cdot \mathbf{n}_\Gamma + B\kappa) - \mathbf{u}\cdot\mathbf{u}_a]\,\mathrm{d}t + \lambda - \Lambda\left(\int_{\Omega} H(-\phi)\,\mathrm{d}\Omega - V_r V_0\right) \tag{2.51}$$

where $\mathbf{n}_\Gamma = \nabla\phi/\|\nabla\phi\|$, $\kappa = \nabla\cdot\mathbf{n}_\Gamma$ and $\mathbf{u}_a$ is derived by solving the adjoint Eq. 2.48.

2.2.3 Numerical Implementation

The flowchart of the optimization procedure includes the following steps (Fig. 2.23):

- specify the initial distribution of the level set function ϕ, the initial values of the Lagrangian multiplier λ and the penalty parameter Λ;
- compute the velocity $\mathbf{u}$ and the pressure p by solving the unsteady incompressible Navier-Stokes eq. 2.39, and the corresponding adjoint variables $\mathbf{u}_a$ and p_a are computed by solving the adjoint Eq. 2.48;
- compute the normal velocity v_n in Eq. 2.51;
- evolve the level set function ϕ by solving the Hamilton-Jacobin eq. 2.49;
- reinitialize the level set function ϕ after several iterations;
- stop the iterative optimization, when the change of the objective value and the volume fraction are less than the user-specified tolerance in five consecutive iterations or the maximum number of iterations is reached.

The level set function is evolved on a grid mesh with ghost elements (the combination of the dash and solid mesh in Fig. 2.24). The mesh for solving the Navier-Stokes equations is a set of elements embedded in the grid mesh for evolving the level set function (solid mesh in Fig. 2.24). The Navier-Stokes equations and the adjoint equations are solved using the finite element package *COMSOL Multiphysics* [10]. One of the advantages of the *COMSOL Multiphysics* package is that one can input the user-defined partial differential equations (PDEs) using the so-called *General PDE* format.

During the optimization procedure, the Navier-Stokes equations and the adjoint equations are solved by the stable Taylor-Hood Q2–Q1 elements [11, 22], which interpolate the fluid velocity quadratically and the fluid pressure linearly (Fig. 2.25a,

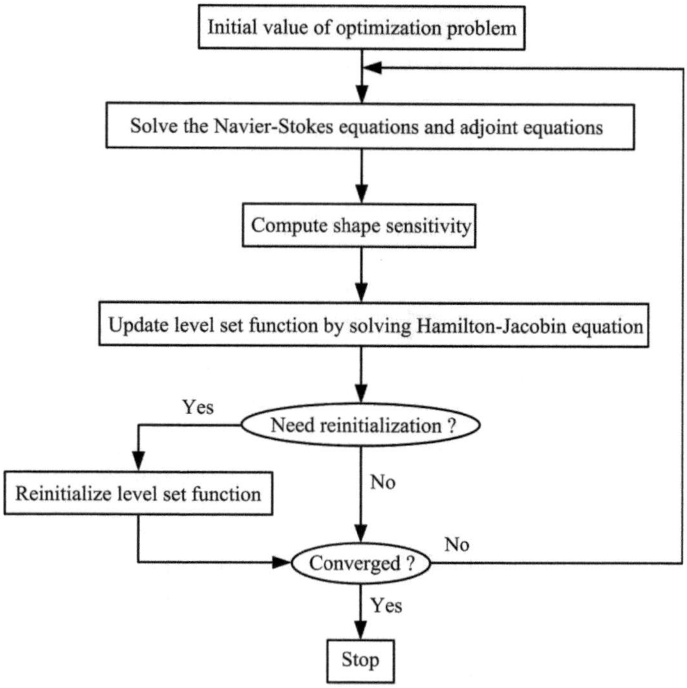

Fig. 2.23 Flowchart of the optimization procedure

b). The level set function is interpolated by the linear Q1 elements (Fig. 2.25c). The unsteady Navier-Stokes equations are solved using the fifth order backward differentiation formula method in the transient solver. The Lagrangian multiplier λ and penalty parameter Λ are updated as [20, 23]

$$\begin{aligned} \lambda_k &= \lambda_{k-1} - \Lambda_{k-1}\left(\int_\Omega H\left(-\phi_{k-1}\right)\,\mathrm{d}\Omega - V_r V_0\right) \\ \Lambda_k &= \frac{1}{\alpha_\Lambda}\Lambda_{k-1}, \quad \alpha_\Lambda \in (0,1) \end{aligned} \tag{2.52}$$

where α_Λ is chosen to be 0.9, and the initial values λ_0 and Λ_0 are chosen based on numerical experiments.

In order to evolve the level set function, the Hamilton-Jacobin equation is solved by the upwind finite difference method. The time step for the finite difference scheme is chosen based on the CFL stability condition [24]

$$\Delta\theta \le \beta_{CFL}\frac{h_E}{\max\{|v_n|\}}, \quad \beta_{CFL} \in (0,1) \tag{2.53}$$

where h_E is the size of the elements and β_{CFL} is chosen to be 0.1.

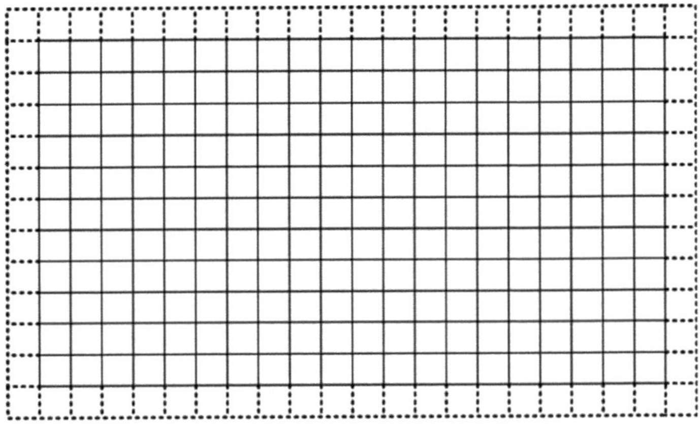

Fig. 2.24 Schematic for the grid mesh used to solve the Navier-Stokes equations (*solid mesh*) and evolve the level set function (the combination of the *solid* and *dash meshes*)

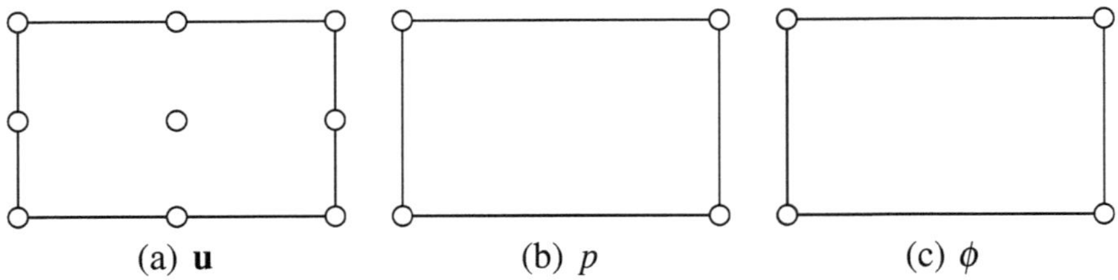

Fig. 2.25 Finite element nodes used to express the velocity, pressure and level set function on a rectangular element

The evolved level set function is reinitialized after several iterations. The reinitialization of the level set function is performed by computing the Euclidean distance transform of the binary value that corresponds to the level set function [25].

The Heaviside function and the Dirac function need to be smoothed in the numerical implementation. The smoothed Heaviside function $H(\phi, h)$ and the smoothed Dirac function $\tau(\phi, h)$ are

$$
\begin{aligned}
H(\phi, h) &= \begin{cases} 1, \ \phi \geq h \\ \dfrac{1}{2} + \dfrac{15\phi}{16h} - \dfrac{5\phi^3}{8h^3} + \dfrac{3\phi^5}{16h^5}, \ |\phi| < h \\ 0, \ \phi \leq -h \end{cases} \\
\tau(\phi, h) &= \begin{cases} \dfrac{15}{16h}\left(1 - \dfrac{\phi^2}{h^2}\right)^2, \ |\phi| < h \\ 0, \ |\phi| \geq h \end{cases}
\end{aligned}
\tag{2.54}
$$

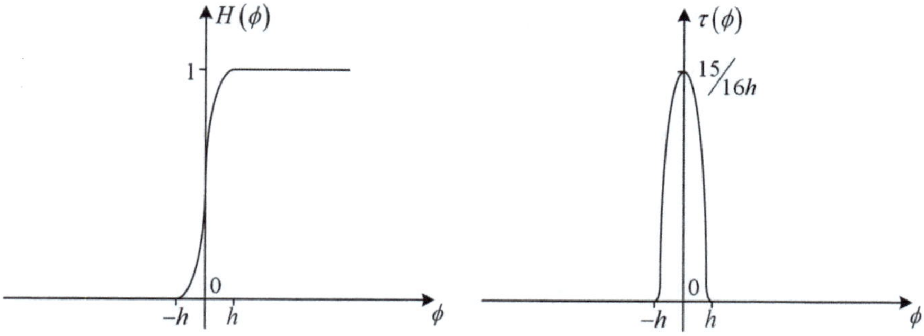

(a) Heaviside function (b) Dirac function

Fig. 2.26 The smoothed Heaviside and Dirac functions, where $h = 5 \times 10^{-5}$

where h is the support size and is chosen to be 5×10^{-5} in the numerical examples (Fig. 2.26).

2.2.4 *Numerical Examples*

Several numerical examples are presented for the optimization of unsteady Navier-Stokes flows, where the density and viscosity of fluid are set to be unity, the initial condition of Navier-Stokes equations is $\mathbf{u}_0 = \mathbf{0}$, the objective in Eq. 2.46 is fixed by setting

$$A = \frac{1}{2}\eta \left(\nabla \mathbf{u} + \nabla \mathbf{u}^{\mathrm{T}}\right) : \left(\nabla \mathbf{u} + \nabla \mathbf{u}^{\mathrm{T}}\right); \; B = 1; \; C = \begin{cases} p, & \text{on } \Gamma_i \\ 0, & \text{on } \partial\Omega \backslash \Gamma_i \end{cases} \tag{2.55}$$

where A is the dissipation of the flow [4, 26]; B is set to be 1 to limit the perimeter of the zero level set and ensure smoothness of the boundaries in the optimized result; and C is set to be p at the inlet Γ_i to minimize the pressure drop of flows in the case where the pressure is zero at the outlet.

2.2.4.1 Straight Channel

A straight channel, which has been investigated for the unsteady flow in [27], is optimized firstly to show the robustness of the optimization method. The optimization domain is shown in Fig. 2.27a, which is discretized by 100×100 rectangular elements. The initial distribution of the level set function and the corresponding zero level set are shown in Fig. 2.27b, c, respectively. The velocity imposed on the inlet Γ_i of the channel is

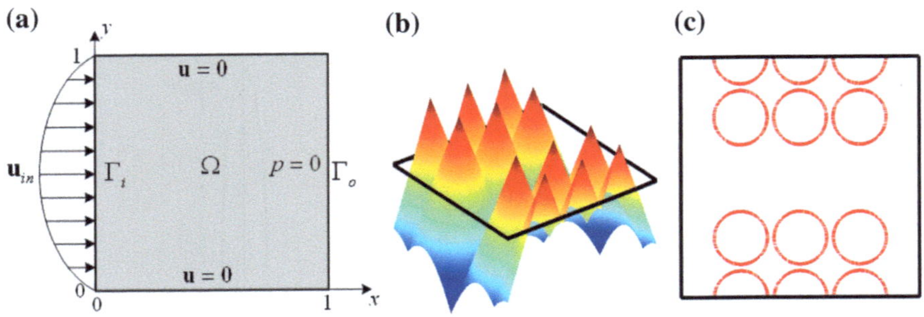

Fig. 2.27 **a** Design domain of the straight channel; **b** initial distribution of the level set function; **c** initial zero level set. $\mathbf{u}_{in}$ is the velocity distribution loaded on the inlet of the domain

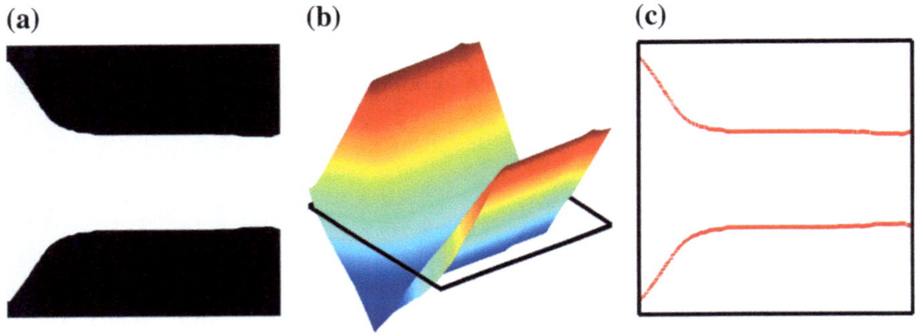

Fig. 2.28 **a** Optimized straight channel for the unsteady incompressible Navier-Stokes flow; **b** distribution of the level set function corresponding to the optimized straight channel; **c** zero level set corresponding to the optimized straight channel

$$\mathbf{u}_{in} = \left(4y\left(1-y\right)t, 0\right), \quad t \in [0, 1] \tag{2.56}$$

At the outlet Γ_o, the Neumann boundary condition in Eq. 2.44 is imposed by setting $p = 0$ and $\eta \nabla \mathbf{u} \cdot \mathbf{n} = \mathbf{0}$. The other boundaries are set to be no slip type. The volume fraction V_r is set as 0.4. The parameters β_1, β_2 and β_3 in the Eq. 2.46 are set as 1, 0.1 and 0.1, respectively. By respectively choosing λ_0 and Λ_0 to be -2.25×10^2 and 1×10^{10}, the optimization problem is solved.

The optimized topology, the corresponding distribution of the level set function and zero level set are derived as shown in Fig. 2.28. Snapshots for the evolution of the zero level set are shown in Fig. 2.29. The convergent histories of the objective value and volume fraction are shown in Fig. 2.30. When the initial distribution of the level set function is chosen as shown in Fig. 2.31a, the optimized straight channel and corresponding zero level set are obtained as shown in Fig. 2.31b. And the snapshots for the evolution of the zero level set are shown in Fig. 2.32.

From Figs. 2.28 and 2.31b, one can see that the obtained results have the same topology with the example of the straight channel in [27]. And the results in Figs. 2.28 and 2.31b illustrate that the proposed method has relatively weak dependence on

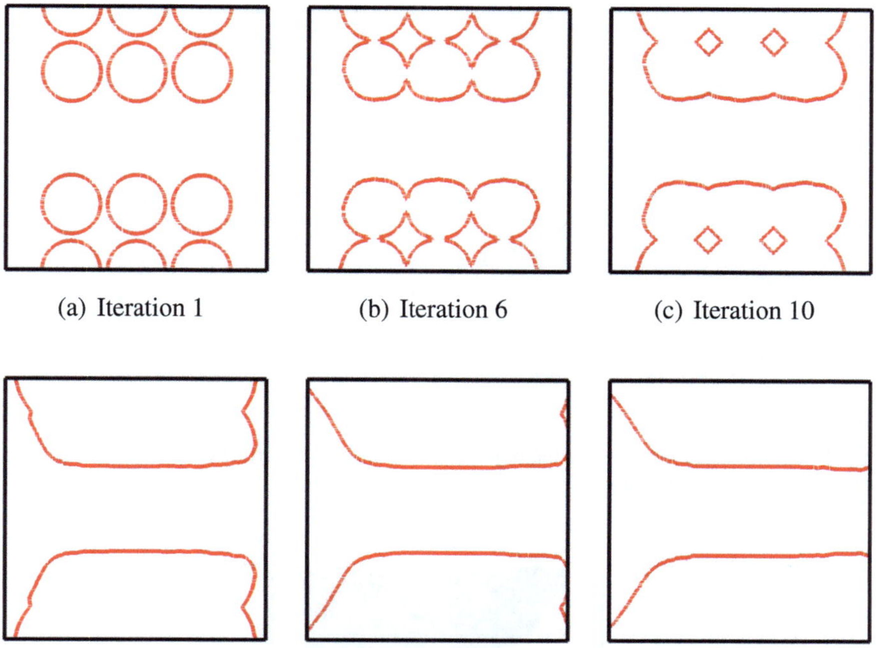

(a) Iteration 1 (b) Iteration 6 (c) Iteration 10

(d) Iteration 16 (e) Iteration 30 (f) Iteration 60

Fig. 2.29 Snapshots for the evolution of the zero level set corresponding to the result shown in Fig. 2.28

the choice of the initial distribution of the level set function. This example on the optimization of the straight channel for the unsteady flow demonstrates that this method can not only change the shape but also the topology of the fluid domain, and the topology optimization of the unsteady Navier-Stokes flows can be achieved using the variational level set method.

2.2.4.2 A Double Pipe

A double pipe is designed in this numerical example. The design domain is shown in Fig. 2.33a, which is discretized by 120 × 120 rectangular elements. The initial distribution of the level set function and the corresponding zero level set are shown in Figs. 2.33b, c, respectively. The parameters β_1, β_2 and β_3 in Eq. 2.46 are set to be 1, 0.1 and 0, respectively. The volume fraction V_r is set as 1/3. The Neumann boundary condition in Eq. 2.44 is imposed on the outlets Γ_{o1} and Γ_{o2} by setting $\mathbf{g} = \mathbf{0}$. By solving the optimization problem with λ_0 and Λ_0 respectively chosen to be -2.75×10^2 and 1×10^{10}, the optimized design of the double pipe is obtained as shown in Fig. 2.34a, where the velocity at the inlets is

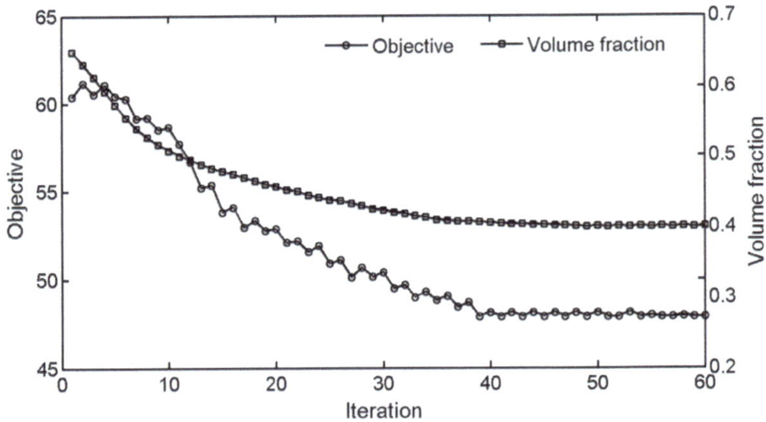

Fig. 2.30 Convergent histories of the objective value and volume fraction corresponding to the optimized straight channel in Fig. 2.28

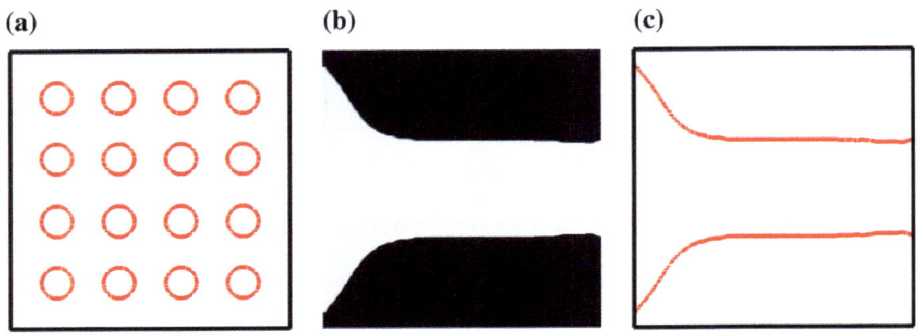

Fig. 2.31 **a** Initial distribution of the level set function; **b** optimized straight channel and the corresponding zero level set

$$\begin{aligned} \mathbf{u}_{in1} &= -144(y-4/6)(5/6-y)\mathbf{n} \\ \mathbf{u}_{in2} &= -144(y-1/6)(2/6-y)\mathbf{n} \end{aligned} \tag{2.57}$$

This optimized design agrees with the numerical example of a double pipe carried out by Borrvall and Petersson [2] and Challis and Guest in [19] for the steady Stokes flows. When the following time dependent boundary conditions

$$\begin{aligned} \mathbf{u}_{in1} &= -144\,(y-4/6)\,(5/6-y)\sin(t)\,\mathbf{n} \\ \mathbf{u}_{in2} &= -144\,(y-1/6)\,(2/6-y)\cos(t)\,\mathbf{n} \end{aligned} \tag{2.58}$$

which have the same maximum of the velocity as the corresponding steady case, are imposed on the inlets of the design domain, the optimized design is obtained as shown in Fig. 2.34b, where the time interval is $[0, \pi]$.

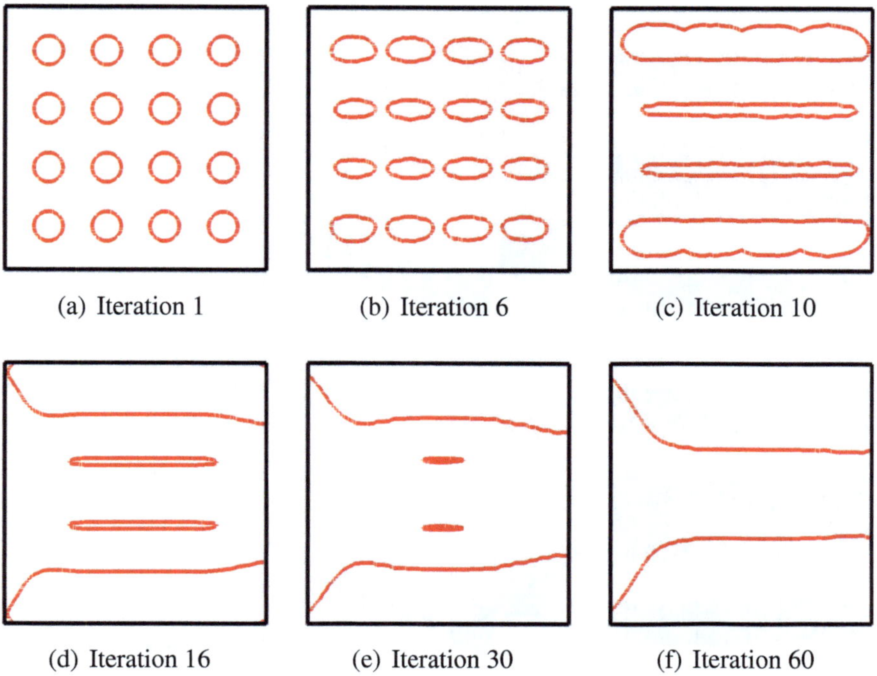

Fig. 2.32 Snapshots for the evolution of the zero level set corresponding to the result shown in Fig. 2.31b

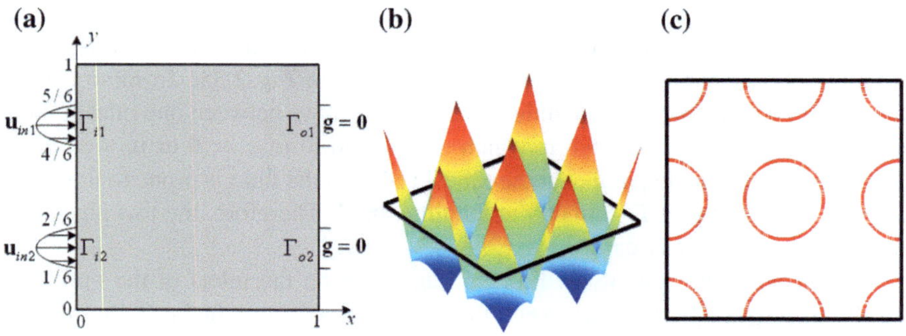

Fig. 2.33 **a** Design domain of the double pipe, where $\mathbf{u}_{in1}$ and $\mathbf{u}_{in2}$ are the velocity distribution at the inlets; **b** initial distribution of the level set function; **c** initial zero level set

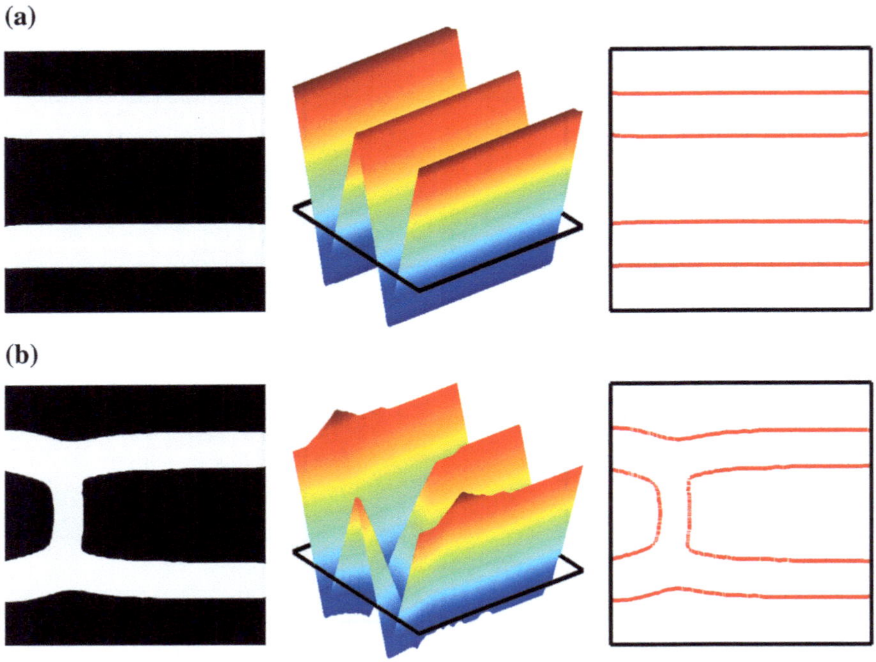

Fig. 2.34 The optimized shape, the level set function and the corresponding zero level set for a double pipe corresponding to the steady flow (**a**) and unsteady flow (**b**)

In the obtained optimized double pipe for the unsteady flow, the streamline distribution corresponding to different time is shown in Fig. 2.35. To minimize the dissipation of the flow in the double pipe, the fluid flows between one inlet and two outlets at the time 0, $\pi/2$ and π, which correspond to $\mathbf{u}_{in1} = \mathbf{0}$ or $\mathbf{u}_{in2} = \mathbf{0}$; the fluid is prone to flow in parallel at the time $3\pi/4$; and the fluid is prone to flow from the port Γ_{i1} to the port Γ_{i2} directly at the time $3\pi/4$. Therefore, the two pipes joint together in the unsteady case.

By imposing the velocity in Eqs. 2.57 and 2.58 on the inlets of the optimized results in Figs. 2.34a, b respectively, the values of objective in Eq. 2.46 are computed (Table 2.10). The cross comparison of the objective values in Table 2.10 and the results in Fig. 2.34 demonstrates that the optimized topologies are different corresponding to the steady and unsteady flows in the double pipe.

Being different from steady flows, unsteady flows have dynamic effect. Figure 2.34 shows that the dynamic effect of the unsteady flow influences the optimized topology of the flow. This further demonstrates the significance on extending the level set method-based topology optimization to the unsteady Navier-Stokes flows.

When the initial distribution of the level set function is chosen as shown in Fig. 2.36a, the optimized double pipe and corresponding zero level set are obtained as shown in Fig. 2.36b. The comparison of the results in Figs. 2.34b and 2.36b

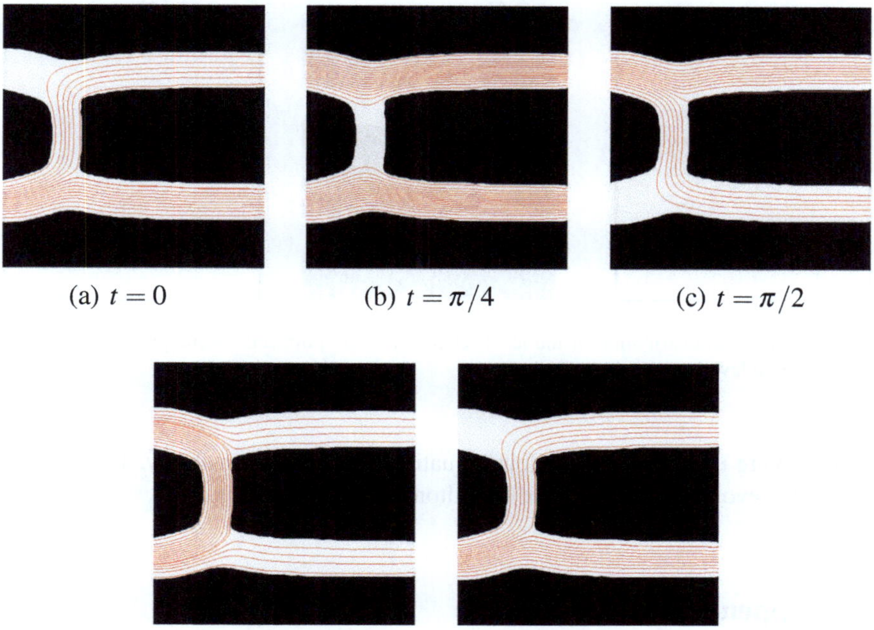

(a) $t = 0$ (b) $t = \pi/4$ (c) $t = \pi/2$

(d) $t = 3\pi/4$ (e) $t = \pi$

Fig. 2.35 Streamline distribution in the optimized double pipe for the unsteady flow

Table 2.10 Objective values obtained by imposing the velocity in Eqs. 2.57 and 2.58 at the inlets of the optimized results in Fig. 2.34

	Steady	Unsteady
Figure 2.34a	54.77	88.72
Figure 2.34b	59.93	85.49

demonstrates that the proposed method has weak dependence on the initial distribution of the level set function, furthermore.

2.2.5 *Summary*

The level set method-based topology optimization of the unsteady Navier-Stokes flow can be achieved using the variational level set method. The solid-fluid interface and the zero velocity condition in the solid domain has been expressed implicitly, by coupling the level set method with the unsteady Navier-Stokes equations. The topology optimization problem is analyzed using the Lagrangian-based adjoint approach,

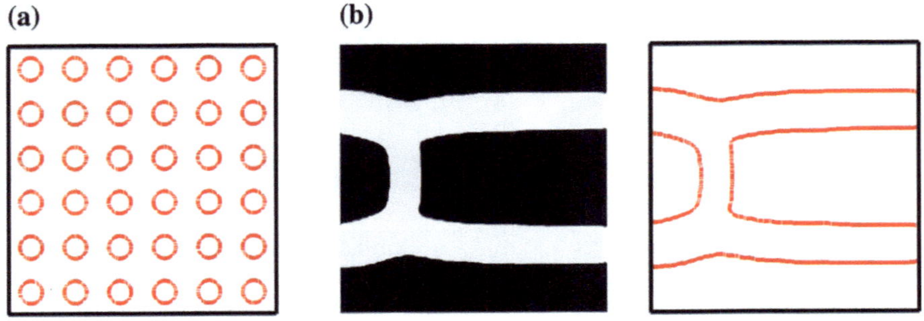

Fig. 2.36 **a** Initial distribution of the level set function; **b** optimized double pipe and the corresponding zero level set

with deriving the continuous adjoint equations and shape sensitivity. The level set function is evolved by solving the Hamilton-Jacobian equation.

2.3 Appendix for Section 2.1

2.3.1 *Adjoint Equations in Density Method-Based Topology Optimization*

To analyze the topology optimization problem in Eq. 2.7, the functional spaces for variables are chosen to be

$$\begin{aligned}&\mathbf{u} \in \mathscr{U}_\Omega^d := \left(\mathscr{L}^2\left((0,T);\mathscr{H}(\Omega)\right)\right)^d;\ p \in \mathscr{P}_\Omega := \mathscr{L}^2\left((0,T);\mathscr{L}^2(\Omega)\right)\\ &\mathbf{f} \in \mathscr{F}_\Omega^{d*} := \left(\mathscr{L}^2\left((0,T);\mathscr{H}^*(\Omega)\right)\right)^d;\ \ \mathbf{u}_0 \in \left(\mathscr{H}_0^1(\Omega)\right)^d;\ \gamma \in \mathscr{L}^1(\Omega)\end{aligned} \tag{2.59}$$

where d is the spatial dimension; $\mathscr{L}^1(\Omega)$ and $\mathscr{L}^2(\Omega)$ are the first-order and second-order integrable Lebesgue spaces respectively; $\mathscr{H}^*(\Omega)$ is the dual space of the first order Hilbert space $\mathscr{H}(\Omega)$; and $\left(\mathscr{H}_0^1(\Omega)\right)^d = \{\mathbf{u} \in (\mathscr{H}(\Omega))^d \mid \nabla\cdot\mathbf{u} = 0\}$. According to the Rietz representation theorem and the Hölder inequality [28], it is known that the Bochner space V_Ω is reflexive, i.e., $\mathscr{V}_\Omega^* = \mathscr{V}_\Omega$. Based on the adjoint analysis method, the sensitivity analysis of the optimization problem in Eq. 2.7 can be derived as follows: based on [29]

$$\left\langle u^*, \int_0^T f(t)\,\mathrm{d}t\right\rangle_{\mathscr{X}^*,\mathscr{X}} = \int_0^T \langle u^*, f(t)\rangle_{\mathscr{X}^*,\mathscr{X}}\,\mathrm{d}t \tag{2.60}$$

where $\mathscr{X}$ is a Bochner space, and its dual is $\mathscr{X}^*$, $u^* \in \mathscr{X}^*$, $f : (0,T) \to \mathscr{X}$ is Bochner integrable, then the integration on time and on space can change sequence.

Therefore,

$$
\begin{aligned}
e(\mathbf{u}, p; \gamma) = & \int_0^T \int_\Omega \left[\rho \frac{\partial \mathbf{u}}{\partial t} - \eta \nabla \cdot \left(\nabla \mathbf{u} + \nabla \mathbf{u}^{\mathrm{T}}\right) + \rho (\mathbf{u} \cdot \nabla) \mathbf{u} + \nabla p - \mathbf{f}\right] \cdot \mathbf{v} \, \mathrm{d}\Omega \mathrm{d}t \\
& - \int_0^T \int_\Omega q \nabla \cdot \mathbf{u} \, \mathrm{d}\Omega \mathrm{d}t + \int_0^T \int_{\Gamma_D} \mathbf{u} \cdot \mathbf{v} \, \mathrm{d}\Gamma \mathrm{d}t \\
& + \int_0^T \int_{\Gamma_N} \left[-p\mathbf{I} + \eta \left(\nabla \mathbf{u} + \nabla \mathbf{u}^{\mathrm{T}}\right)\right] \mathbf{n} \cdot \mathbf{v} \, \mathrm{d}\Gamma \mathrm{d}t + \int_\Omega (\mathbf{u} \cdot \mathbf{v}) \big|_{t=0} \mathrm{d}\Omega
\end{aligned}
\tag{2.61}
$$

for $\forall \mathbf{v} \in \mathscr{U}_\Omega^d$ and $\forall q \in \mathscr{P}_\Omega$. Based on Eq. 2.60, the following transformation of Eq. 2.61 is derived by partial integration on time:

$$
\begin{aligned}
e(\mathbf{u}, p; \gamma) = & \rho \int_\Omega \int_0^T \left[\frac{\partial (\mathbf{u} \cdot \mathbf{v})}{\partial t} - \mathbf{u} \cdot \frac{\partial \mathbf{v}}{\partial t}\right] \mathrm{d}t \mathrm{d}\Omega \\
& + \eta \int_0^T \int_\Omega \left[\left(\nabla \mathbf{u} + \nabla \mathbf{u}^{\mathrm{T}}\right) : \nabla \mathbf{v} - \nabla \cdot \left(\left(\nabla \mathbf{u} + \nabla \mathbf{u}^{\mathrm{T}}\right) \cdot \mathbf{v}\right)\right] \mathrm{d}\Omega \mathrm{d}t \\
& + \int_0^T \int_\Omega \rho (\mathbf{u} \cdot \nabla) \mathbf{u} \cdot \mathbf{v} \, \mathrm{d}\Omega \mathrm{d}t + \int_0^T \int_\Omega \left[\nabla \cdot (p\mathbf{v}) - p \nabla \cdot \mathbf{v}\right] \mathrm{d}\Omega \mathrm{d}t \\
& - \int_0^T \int_\Omega \mathbf{f} \cdot \mathbf{v} \, \mathrm{d}\Omega \mathrm{d}t - \int_0^T \int_\Omega q \nabla \cdot \mathbf{u} \, \mathrm{d}\Omega \mathrm{d}t \\
& + \int_0^T \int_{\Gamma_N} \left[-p\mathbf{I} + \eta \left(\nabla \mathbf{u} + \nabla \mathbf{u}^{\mathrm{T}}\right)\right] \mathbf{n} \cdot \mathbf{v} \, \mathrm{d}\Gamma \mathrm{d}t + \int_\Omega (\mathbf{u} \cdot \mathbf{v}) \big|_{t=0} \mathrm{d}\Omega
\end{aligned}
\tag{2.62}
$$

Based on Gauss theory [5], Eq. 2.62 can be transformed into

$$
\begin{aligned}
e(\mathbf{u}, p; \gamma) = & \rho \int_\Omega \left[(\mathbf{u} \cdot \mathbf{v}) \big|_{t=T} - (\mathbf{u} \cdot \mathbf{v}) \big|_{t=0}\right] \mathrm{d}\Omega - \rho \int_0^T \int_\Omega \mathbf{u} \cdot \frac{\partial \mathbf{v}}{\partial t} \mathrm{d}\Omega \mathrm{d}t \\
& + \eta \int_0^T \int_\Omega \left(\nabla \mathbf{u} + \nabla \mathbf{u}^{\mathrm{T}}\right) : \nabla \mathbf{v} \, \mathrm{d}\Omega \mathrm{d}t + \int_0^T \int_\Omega \rho (\mathbf{u} \cdot \nabla) \mathbf{u} \cdot \mathbf{v} \, \mathrm{d}\Omega \mathrm{d}t \\
& - \int_0^T \int_\Omega p \nabla \cdot \mathbf{v} \, \mathrm{d}\Omega \mathrm{d}t - \int_0^T \int_\Omega \mathbf{f} \cdot \mathbf{v} \, \mathrm{d}\Omega \mathrm{d}t \\
& + \int_0^T \int_{\partial\Omega} \left[p\mathbf{I} - \eta \left(\nabla \mathbf{u} + \nabla \mathbf{u}^{\mathrm{T}}\right)\right] \mathbf{n} \cdot \mathbf{v} \, \mathrm{d}\Gamma \mathrm{d}t - \int_0^T \int_\Omega q \nabla \cdot \mathbf{u} \, \mathrm{d}\Omega \mathrm{d}t \\
& + \int_0^T \int_{\Gamma_D} \mathbf{u} \cdot \mathbf{v} \, \mathrm{d}\Gamma \mathrm{d}t + \int_0^T \int_{\Gamma_N} \left[-p\mathbf{I} + \eta \left(\nabla \mathbf{u} + \nabla \mathbf{u}^{\mathrm{T}}\right)\right] \mathbf{n} \cdot \mathbf{v} \, \mathrm{d}\Gamma \mathrm{d}t \\
& + \int_\Omega (\mathbf{u} \cdot \mathbf{v}) \big|_{t=0} \mathrm{d}\Omega
\end{aligned}
\tag{2.63}
$$

By inserting the boundary conditions into Eq. 2.63, the reduced weak operator of transient Navier-Stokes equations is obtained as

$$
\begin{aligned}
e(\mathbf{u}, p; \gamma) = {} & \rho \int_\Omega \left[(\mathbf{u} \cdot \mathbf{v}) \big|_{t=T} - \mathbf{u}_0 \cdot \mathbf{v} \big|_{t=0} \right] \mathrm{d}\Omega - \rho \int_0^T \int_\Omega \mathbf{u} \cdot \frac{\partial \mathbf{v}}{\partial t} \mathrm{d}\Omega \mathrm{d}t \\
& + \eta \int_0^T \int_\Omega \left(\nabla \mathbf{u} + \nabla \mathbf{u}^\mathrm{T} \right) : \nabla \mathbf{v} \, \mathrm{d}\Omega \mathrm{d}t + \int_0^T \int_\Omega \rho (\mathbf{u} \cdot \nabla) \mathbf{u} \cdot \mathbf{v} \, \mathrm{d}\Omega \mathrm{d}t \\
& - \int_0^T \int_\Omega p \nabla \cdot \mathbf{v} \, \mathrm{d}\Omega \mathrm{d}t - \int_0^T \int_\Omega \mathbf{f} \cdot \mathbf{v} \, \mathrm{d}\Omega \mathrm{d}t - \int_0^T \int_\Omega q \nabla \cdot \mathbf{u} \, \mathrm{d}\Omega \mathrm{d}t \\
& + \int_0^T \int_{\Gamma_D} \left[p\mathbf{I} - \eta \left(\nabla \mathbf{u} + \nabla \mathbf{u}^\mathrm{T} \right) \mathbf{n} \cdot \mathbf{v} \right] \Big|_{\mathbf{u}=\mathbf{u}_D} \mathrm{d}\Gamma \mathrm{d}t + \int_0^T \int_{\Gamma_D} \mathbf{u}_D \cdot \mathbf{v} \, \mathrm{d}\Gamma \mathrm{d}t
\end{aligned}
\tag{2.64}
$$

According to the definition of the Gâteaux derivative [5], the Gâteaux derivatives of Eq. 2.92 in the direction $(\mathbf{w}, r) \in \mathscr{U}_\Omega^d \times \mathscr{P}_\Omega$ are

$$
\begin{aligned}
& \langle e_\mathbf{u}(\mathbf{u}, p; \gamma), \mathbf{w} \rangle_{\mathscr{U}_\Omega^{d*}, \mathscr{U}_\Omega^d} \\
= {} & \lim_{l \to 0^+} \frac{e(\mathbf{u} + l\mathbf{w}, p) - e(\mathbf{u}, p)}{l} \\
= {} & \rho \int_\Omega (\mathbf{w} \cdot \mathbf{v}) \big|_{t=T} \mathrm{d}\Omega - \rho \int_0^T \int_\Omega \mathbf{w} \cdot \frac{\partial \mathbf{v}}{\partial t} \mathrm{d}t \mathrm{d}\Omega + \eta \int_0^T \int_\Omega \left(\nabla \mathbf{w} + \nabla \mathbf{w}^\mathrm{T} \right) : \nabla \mathbf{v} \, \mathrm{d}\Omega \mathrm{d}t \\
& + \int_0^T \int_\Omega \rho (\mathbf{w} \cdot \nabla) \mathbf{u} \cdot \mathbf{v} \, \mathrm{d}\Omega \mathrm{d}t + \int_0^T \int_\Omega \rho (\mathbf{u} \cdot \nabla) \mathbf{w} \cdot \mathbf{v} \, \mathrm{d}\Omega \mathrm{d}t \\
& - \int_0^T \int_\Omega \frac{\partial \mathbf{f}}{\partial \mathbf{u}} \mathbf{w} \cdot \mathbf{v} \, \mathrm{d}\Omega \mathrm{d}t - \int_0^T \int_\Omega q \nabla \cdot \mathbf{w} \, \mathrm{d}\Omega \mathrm{d}t
\end{aligned}
\tag{2.65}
$$

and

$$
\begin{aligned}
& \left\langle e_p(\mathbf{u}, p; \gamma), r \right\rangle_{\mathscr{P}_\Omega^*, \mathscr{P}_\Omega} \\
= {} & \lim_{l \to 0^+} \frac{e(\mathbf{u}, p + lr) - e(\mathbf{u}, p)}{l} \\
= {} & - \int_0^T \int_\Omega r \nabla \cdot \mathbf{v} \, \mathrm{d}\Omega \mathrm{d}t - \int_0^T \int_\Omega \frac{\partial \mathbf{f}}{\partial p} r \cdot \mathbf{v} \, \mathrm{d}\Omega \mathrm{d}t + \int_0^T \int_{\Gamma_D} r \mathbf{n} \cdot \mathbf{v} \, \mathrm{d}\Gamma \mathrm{d}t
\end{aligned}
\tag{2.66}
$$

Based on the linear property of operators $\langle e_\mathbf{u}(\mathbf{u}, p; \gamma), \mathbf{w} \rangle_{\mathscr{U}_\Omega^{d*}, \mathscr{U}_\Omega^d}$ and $\left\langle e_p(\mathbf{u}, p; \gamma), r \right\rangle_{\mathscr{P}_\Omega^*, \mathscr{P}_\Omega}$, they are bounded based on the Cauchy-Schwarz inequality and the Poincaré's inequality. Therefore, Eqs. 2.65 and 2.66 are the Gâteaux derivatives of $e(\mathbf{u}, p)$. The dual operator of a linear operator is defined as [5]

$$
\left\langle D^* u, v \right\rangle_{\mathscr{X}^*, \mathscr{X}} = \langle u, Dv \rangle_{\mathscr{Y}^*, \mathscr{Y}}, \quad \forall u \in \mathscr{Y}^*, \ v \in \mathscr{X} \tag{2.67}
$$

where $\mathscr{X}$ and $\mathscr{Y}$ are Banach spaces, $D \in \mathscr{L}(\mathscr{X}, \mathscr{Y})$ and $D^* \in \mathscr{L}(\mathscr{Y}^*, \mathscr{X}^*)$. By rewriting $e_\mathbf{u}(\mathbf{u}, p; \gamma)$ and $e_p(\mathbf{u}, p; \gamma)$ in the formulation of linear operators, one can derive

$$\begin{aligned} \langle e_{\mathbf{u}} (\mathbf{u}, p; \gamma), \mathbf{w} \rangle_{\mathscr{U}_\Omega^{d*}, \mathscr{U}_\Omega^d} &= \langle A_1 \mathbf{v} + A_2 q, \mathbf{w} \rangle_{\mathscr{U}_\Omega^{d*}, \mathscr{U}_\Omega^d} \\ \left\langle e_p (\mathbf{u}, p; \gamma), r \right\rangle_{\mathscr{P}_\Omega^*, \mathscr{P}_\Omega} &= \langle A_3 \mathbf{v}, r \rangle_{\mathscr{P}_\Omega^*, \mathscr{P}_\Omega} \end{aligned} \tag{2.68}$$

and

$$\begin{aligned} \left\langle e_{\mathbf{u}}^* (\mathbf{u}_a, p_a; \gamma), \mathbf{w} \right\rangle_{\mathscr{U}_\Omega^{d*}, \mathscr{U}_\Omega^d} &= \langle A_1 \mathbf{u}_a + A_2 p_a, \mathbf{w} \rangle_{\mathscr{U}_\Omega^{d*}, \mathscr{U}_\Omega^d} \\ \left\langle e_p^* (\mathbf{u}_a, p_a; \gamma), r \right\rangle_{\mathscr{P}_\Omega^*, \mathscr{P}_\Omega} &= \langle A_3 \mathbf{u}_a, r \rangle_{\mathscr{P}_\Omega^*, \mathscr{P}_\Omega} \end{aligned} \tag{2.69}$$

where $\mathbf{u}_a \in \mathscr{U}_\Omega^d$ and $p_a \in \mathscr{P}_\Omega$, A_1, A_2 and A_3 are linear operators. According to Eqs. 2.11 and 2.69, the weak form of the adjoint equations for the topology optimization problem of unsteady Navier-Stokes flows in Eq. 2.7 can be derived as

$$\begin{aligned} &\rho \int_\Omega (\mathbf{w} \cdot \mathbf{u}_a) \big|_{t=T} \,\mathrm{d}\Omega - \rho \int_0^T \int_\Omega \mathbf{w} \cdot \frac{\partial \mathbf{u}_a}{\partial t} \,\mathrm{d}\Omega \mathrm{d}t + \eta \int_0^T \int_\Omega \left(\nabla \mathbf{w} + \nabla \mathbf{w}^\mathrm{T} \right) : \nabla \mathbf{u}_a \,\mathrm{d}\Omega \mathrm{d}t \\ &+ \int_0^T \int_\Omega \rho \left(\mathbf{w} \cdot \nabla \right) \mathbf{u} \cdot \mathbf{u}_a \,\mathrm{d}\Omega \mathrm{d}t + \int_0^T \int_\Omega \rho (\mathbf{u} \cdot \nabla) \mathbf{w} \cdot \mathbf{u}_a \,\mathrm{d}\Omega \mathrm{d}t - \int_0^T \int_\Omega p_a \nabla \cdot \mathbf{w} \,\mathrm{d}\Omega \mathrm{d}t \\ &- \int_0^T \int_\Omega \frac{\partial \mathbf{f}}{\partial \mathbf{u}} \mathbf{w} \cdot \mathbf{u}_a \,\mathrm{d}\Omega \mathrm{d}t = - \left\langle J_{\mathbf{u}} (\mathbf{u}, \nabla \mathbf{u}, p; \gamma), \mathbf{w} \right\rangle_{\mathscr{U}_\Omega^{d*}, \mathscr{U}_\Omega^d} \end{aligned} \tag{2.70}$$

and

$$\begin{aligned} &- \int_0^T \int_\Omega r \nabla \cdot \mathbf{u}_a \,\mathrm{d}\Omega \mathrm{d}t - \int_0^T \int_\Omega \frac{\partial \mathbf{f}}{\partial p} r \cdot \mathbf{u}_a \,\mathrm{d}\Omega \mathrm{d}t + \int_0^T \int_{\Gamma_D} r \mathbf{n} \cdot \mathbf{u}_a \,\mathrm{d}\Gamma \mathrm{d}t \\ &= - \left\langle J_p (\mathbf{u}, \nabla \mathbf{u}, p; \gamma), r \right\rangle_{\mathscr{P}_\Omega^*, \mathscr{P}_\Omega} \end{aligned} \tag{2.71}$$

Additionally, there are

$$\begin{aligned} &\int_0^T \int_\Omega \left(\nabla \mathbf{w} + \nabla \mathbf{w}^\mathrm{T} \right) : \nabla \mathbf{u}_a \,\mathrm{d}\Omega \mathrm{d}t \\ &= - \int_0^T \int_\Omega \nabla \cdot \left(\nabla \mathbf{u}_a + \nabla \mathbf{u}_a^\mathrm{T} \right) \cdot \mathbf{w} \,\mathrm{d}\Omega \mathrm{d}t + \int_0^T \int_{\Gamma_N} \left(\nabla \mathbf{u}_a + \nabla \mathbf{u}_a^\mathrm{T} \right) \mathbf{n} \cdot \mathbf{w} \,\mathrm{d}\Gamma \mathrm{d}t \end{aligned} \tag{2.72}$$

$$\int_0^T \int_\Omega p_a \nabla \cdot \mathbf{w} \,\mathrm{d}\Omega \mathrm{d}t = - \int_0^T \int_\Omega \nabla p_a \cdot \mathbf{w} \,\mathrm{d}\Omega \mathrm{d}t + \int_0^T \int_{\Gamma_N} p_a \mathbf{n} \cdot \mathbf{w} \,\mathrm{d}\Gamma \mathrm{d}t \tag{2.73}$$

and

$$\int_0^T \int_\Omega (\mathbf{u} \cdot \nabla) \mathbf{w} \cdot \mathbf{u}_a \, \mathrm{d}\Omega \mathrm{d}t = -\int_0^T \int_\Omega (\mathbf{u} \cdot \nabla) \mathbf{u}_a \cdot \mathbf{w} \, \mathrm{d}\Omega \mathrm{d}t + \int_0^T \int_{\Gamma_N} (\mathbf{u} \cdot \mathbf{n}) \mathbf{u}_a \cdot \mathbf{w} \, \mathrm{d}\Gamma \mathrm{d}t \tag{2.74}$$

Because $\mathbf{u}$ is known on Σ_D and p can be expressed by $\nabla \mathbf{u}$ on Σ_N, according to Eq. 2.4

$$B(\mathbf{u}, p; \gamma) = \begin{cases} B(p; \gamma), & \text{on } \Sigma_D \\ B(\mathbf{u}, \nabla \mathbf{u}; \gamma), & \text{on } \Sigma_N \end{cases} \tag{2.75}$$

By inserting Eqs. 2.72–2.75 into Eqs. 2.70 and 2.71, one can obtain

$$\begin{aligned} &\langle J_{\mathbf{u}}(\mathbf{u}, p; \gamma), \mathbf{w} \rangle_{\mathscr{U}_\Omega^{d*}, \mathscr{U}_\Omega^d} \\ &= \int_0^T \int_\Omega \beta_1 \left[\frac{\partial A}{\partial \mathbf{u}} \cdot \mathbf{w} - \left(\nabla \cdot \frac{\partial A}{\partial \nabla \mathbf{u}} \right) \cdot \mathbf{w} \right] \mathrm{d}\Omega \mathrm{d}t \\ &+ \int_0^T \int_{\Gamma_N} \beta_1 \frac{\partial A}{\partial \nabla \mathbf{u}} \mathbf{n} \cdot \mathbf{w} \, \mathrm{d}\Gamma \mathrm{d}t + \int_0^T \int_{\Gamma_N} \beta_2 \frac{\partial B}{\partial \mathbf{u}} \cdot \mathbf{w} \, \mathrm{d}\Gamma \mathrm{d}t \end{aligned} \tag{2.76}$$

and

$$\langle J_p(\mathbf{u}, p; \gamma), r \rangle_{\mathscr{P}_\Omega^*, \mathscr{P}_\Omega} = \int_0^T \int_\Omega \beta_1 \frac{\partial A}{\partial p} r \, \mathrm{d}\Omega \mathrm{d}t + \int_0^T \int_{\Gamma_D} \beta_2 \frac{\partial B}{\partial p} r \, \mathrm{d}\Gamma \mathrm{d}t \tag{2.77}$$

Therefore, the adjoint equations of the Navier-Stokes equations can be written as

$$\begin{aligned} -\rho \frac{\partial \mathbf{u}_a}{\partial t} - \eta \nabla \cdot \left(\nabla \mathbf{u}_a + \nabla \mathbf{u}_a^{\mathrm{T}} \right) - \rho (\mathbf{u} \cdot \nabla) \mathbf{u}_a + \rho (\nabla \mathbf{u}) \cdot \mathbf{u}_a + \nabla p_a & \\ = -\beta_1 \left(\frac{\partial A}{\partial \mathbf{u}} - \nabla \cdot \frac{\partial A}{\partial \nabla \mathbf{u}} \right) + \frac{\partial \mathbf{f}}{\partial \mathbf{u}} \mathbf{u}_a, & \text{ in } Q \\ -\nabla \cdot \mathbf{u}_a = -\beta_1 \frac{\partial A}{\partial p} + \frac{\partial \mathbf{f}}{\partial p} \cdot \mathbf{u}_a, & \text{ in } Q \\ \mathbf{u}_a(T, \mathbf{x}) = \mathbf{0}, & \text{ in } \Omega \\ \mathbf{u}_a = -\frac{\partial B}{\partial p} \mathbf{n}, & \text{ on } \Sigma_D \\ \left[-p_a \mathbf{I} + \eta \left(\nabla \mathbf{u}_a + \nabla \mathbf{u}_a^{\mathrm{T}} \right) \right] \mathbf{n} = -\rho (\mathbf{u} \cdot \mathbf{n}) \mathbf{u}_a - \beta_1 \frac{\partial A}{\partial \nabla \mathbf{u}} \mathbf{n} - \beta_2 \frac{\partial B}{\partial \mathbf{u}}, & \text{ on } \Sigma_N \end{aligned} \tag{2.78}$$

2.3.2 *Adjoint Sensitivity of Density Method-Based Optimization Problem*

2.3.2.1 For Expression Defined on [0, *T*]

Based on the similar analysis of the dual operator of $e_\gamma(\mathbf{u}, p; \gamma)$, the adjoint derivatives of the optimization problem can be obtained as

$$\begin{aligned} &\left\langle \frac{D\hat{J}}{D\gamma}, \psi \right\rangle_{\mathscr{L}^\infty(\Omega),\mathscr{L}^1(\Omega)} \\ &= \int_0^T \int_\Omega \left(\beta_1 \frac{\partial A}{\partial \gamma} - \frac{\partial \mathbf{f}}{\partial \gamma} \cdot \mathbf{u}_a \right) \psi \, \mathrm{d}\Omega \mathrm{d}t \\ &= \int_\Omega \int_0^T \left(\beta_1 \frac{\partial A}{\partial \gamma} - \frac{\partial \mathbf{f}}{\partial \gamma} \cdot \mathbf{u}_a \right) \psi \, \mathrm{d}t \mathrm{d}\Omega \\ &= \int_\Omega \left[\int_0^T \left(\beta_1 \frac{\partial A}{\partial \gamma} - \frac{\partial \mathbf{f}}{\partial \gamma} \cdot \mathbf{u}_a \right) \mathrm{d}t \right] \psi \, \mathrm{d}\Omega \end{aligned} \tag{2.79}$$

and

$$\begin{aligned} &\left\langle \frac{D\hat{J}}{D\gamma}, \psi \right\rangle_{\mathscr{L}^\infty(\partial\Omega),\mathscr{L}^1(\partial\Omega)} \\ &= \int_0^T \int_{\partial\Omega} \beta_2 \frac{\partial B}{\partial \gamma} \psi \, \mathrm{d}\Gamma \mathrm{d}t \\ &= \int_{\partial\Omega} \left[\int_0^T \beta_2 \frac{\partial B}{\partial \gamma} \mathrm{d}t \right] \psi \, \mathrm{d}\Gamma \end{aligned} \tag{2.80}$$

where $\forall \psi \in \mathscr{C}^\infty(\bar{\Omega})$. Therefore, according to Eq. 2.12, the adjoint derivatives can be expressed as

$$\begin{aligned} \left.\frac{D\hat{J}}{D\gamma}\right|_\Omega &= \int_0^T \left(\beta_1 \frac{\partial A}{\partial \gamma} + \frac{\partial \alpha}{\partial \gamma} \mathbf{u} \cdot \mathbf{u}_a \right) \mathrm{d}t, \quad \text{in } \Omega \\ \left.\frac{D\hat{J}}{D\gamma}\right|_{\partial\Omega} &= \int_0^T \beta_2 \frac{\partial B}{\partial \gamma} \mathrm{d}t, \quad \text{on } \partial\Omega \end{aligned} \tag{2.81}$$

2.3.2.2 For Expression Defined on [T_1, T_2] ⊆ [0, *T*]

In some cases, the expression in Eq. 2.8 is defined on the subset $[T_1, T_2]$ instead of $[0, T]$:

$$\begin{aligned} J(\mathbf{u}, p; \gamma) = & \int_{T_1}^{T_2} \int_{\Omega} \beta_1 A(\mathbf{u}, \nabla \mathbf{u}, p; \gamma) \, \mathrm{d}\Omega \mathrm{d}t \\ & + \int_{T_1}^{T_2} \int_{\partial \Omega} \beta_2 B(\mathbf{u}, p; \gamma) \, \mathrm{d}\Gamma \mathrm{d}t \end{aligned} \tag{2.82}$$

where $[T_1, T_2]$ is a subset of $[0, T]$. Then Eq. 2.82 can be rewritten as

$$\begin{aligned} J(\mathbf{u}, p; \gamma) = & \int_0^{T_1} \int_{\Omega} 0 \, \mathrm{d}\Omega \mathrm{d}t + \int_0^{T_1} \int_{\partial \Omega} 0 \, \mathrm{d}\Gamma \mathrm{d}t \\ & + \int_{T_1}^{T_2} \int_{\Omega} \beta_1 A(\mathbf{u}, \nabla \mathbf{u}, p; \gamma) \, \mathrm{d}\Omega \mathrm{d}t + \int_{T_1}^{T_2} \int_{\partial \Omega} \beta_2 B(\mathbf{u}, p; \gamma) \, \mathrm{d}\Gamma \mathrm{d}t \\ & + \int_{T_2}^{T} \int_{\Omega} 0 \, \mathrm{d}\Omega \mathrm{d}t + \int_{T_2}^{T} \int_{\partial \Omega} 0 \, \mathrm{d}\Gamma \mathrm{d}t \end{aligned} \tag{2.83}$$

where A and B are set to 0 in $(0, T_1)$ and (T_1, T_2), respectively. Then Eqs. 2.76 and 2.77 can be rewritten as

$$\begin{aligned} & \langle J_{\mathbf{u}}(\mathbf{u}, p; \gamma), \mathbf{w} \rangle_{\mathscr{U}_\Omega^{d*}, \mathscr{U}_\Omega^d} \\ = & \int_0^{T_1} \int_{\Omega} 0 \, \mathrm{d}\Omega \mathrm{d}t + \int_0^{T_1} \int_{\Gamma_N} 0 \, \mathrm{d}\Gamma \mathrm{d}t + \int_0^{T_1} \int_{\Gamma_N} 0 \, \mathrm{d}\Gamma \mathrm{d}t \\ & + \int_{T_1}^{T_2} \int_{\Omega} \beta_1 \left[\frac{\partial A}{\partial \mathbf{u}} \cdot \mathbf{w} - \left(\nabla \cdot \frac{\partial A}{\partial \nabla \mathbf{u}} \right) \cdot \mathbf{w} \right] \mathrm{d}\Omega \mathrm{d}t + \int_{T_1}^{T_2} \int_{\Gamma_N} \beta_1 \frac{\partial A}{\partial \nabla \mathbf{u}} \mathbf{n} \cdot \mathbf{w} \, \mathrm{d}\Gamma \mathrm{d}t \\ & + \int_{T_1}^{T_2} \int_{\Gamma_N} \beta_2 \frac{\partial B}{\partial \mathbf{u}} \cdot \mathbf{w} \, \mathrm{d}\Gamma \mathrm{d}t + \int_{T_2}^{T} \int_{\Omega} 0 \, \mathrm{d}\Omega \mathrm{d}t + \int_{T_2}^{T} \int_{\Gamma_N} 0 \, \mathrm{d}\Gamma \mathrm{d}t + \int_{T_2}^{T} \int_{\Gamma_N} 0 \, \mathrm{d}\Gamma \mathrm{d}t \end{aligned} \tag{2.84}$$

and

$$\begin{aligned} & \langle J_p(\mathbf{u}, p; \gamma), r \rangle_{\mathscr{P}_\Omega^*, \mathscr{P}_\Omega} \\ = & \int_0^{T_1} \int_{\Omega} 0 \, \mathrm{d}\Omega \mathrm{d}t + \int_{T_1}^{T_2} \int_{\Omega} \beta_1 \frac{\partial A}{\partial p} r \, \mathrm{d}\Omega \mathrm{d}t + \int_{T_2}^{T} \int_{\Omega} 0 \, \mathrm{d}\Omega \mathrm{d}t \\ & + \int_0^{T_1} \int_{\Gamma_D} 0 \, \mathrm{d}\Gamma \mathrm{d}t + \int_{T_1}^{T_2} \int_{\Gamma_D} \beta_2 \frac{\partial B}{\partial p} r \, \mathrm{d}\Gamma \mathrm{d}t + \int_{T_2}^{T} \int_{\Gamma_D} 0 \, \mathrm{d}\Gamma \mathrm{d}t \end{aligned} \tag{2.85}$$

Therefore, the adjoint equations of the Navier-Stokes equations can be written as
When $t \in (T_2, T)$

$$
\begin{aligned}
-\rho\frac{\partial \mathbf{u}_a}{\partial t} - \eta\nabla\cdot\left(\nabla\mathbf{u}_a + \nabla\mathbf{u}_a^{\mathrm{T}}\right) - \rho\left(\mathbf{u}\cdot\nabla\right)\mathbf{u}_a + \rho\left(\nabla\mathbf{u}\right)\cdot\mathbf{u}_a& \\
+\nabla p_a &= \frac{\partial \mathbf{f}}{\partial \mathbf{u}}\mathbf{u}_a,\ \text{in}\ (T_2, T)\times\Omega \\
-\nabla\cdot\mathbf{u}_a &= \frac{\partial \mathbf{f}}{\partial p}\cdot\mathbf{u}_a,\ \text{in}\ (T_2, T)\times\Omega \\
\mathbf{u}_a\left(T, \mathbf{x}\right) &= \mathbf{0},\ \text{in}\ \Omega \\
\mathbf{u}_a &= \mathbf{0},\ \text{on}\ (T_2, T)\times\Gamma_D \\
\left[-p_a\mathbf{I} + \eta\left(\nabla\mathbf{u}_a + \nabla\mathbf{u}_a^{\mathrm{T}}\right)\right]\mathbf{n} &= -\rho\left(\mathbf{u}\cdot\mathbf{n}\right)\mathbf{u}_a,\ \text{on}\ (T_2, T)\times\Gamma_N
\end{aligned}
\tag{2.86}
$$

When $t \in (T_1, T_2)$

$$
\begin{aligned}
-\rho\frac{\partial \mathbf{u}_a}{\partial t} - \eta\nabla\cdot\left(\nabla\mathbf{u}_a + \nabla\mathbf{u}_a^{\mathrm{T}}\right) - \rho\left(\mathbf{u}\cdot\nabla\right)\mathbf{u}_a + \rho\left(\nabla\mathbf{u}\right)\cdot\mathbf{u}_a& \\
+\nabla p_a &= -\beta_1\left(\frac{\partial A}{\partial \mathbf{u}} - \nabla\cdot\frac{\partial A}{\partial \nabla\mathbf{u}}\right) + \frac{\partial \mathbf{f}}{\partial \mathbf{u}}\mathbf{u}_a,\ \text{in}\ (T_1, T_2)\times\Omega \\
-\nabla\cdot\mathbf{u}_a &= -\beta_1\frac{\partial A}{\partial p} + \frac{\partial \mathbf{f}}{\partial p}\cdot\mathbf{u}_a,\ \text{in}\ (T_1, T_2)\times\Omega \\
\mathbf{u}_a &= -\frac{\partial B}{\partial p}\mathbf{n},\ \text{on}\ (T_1, T_2)\times\Gamma_D \\
\left[-p_a\mathbf{I} + \eta\left(\nabla\mathbf{u}_a + \nabla\mathbf{u}_a^{\mathrm{T}}\right)\right]\mathbf{n} &= -\rho\left(\mathbf{u}\cdot\mathbf{n}\right)\mathbf{u}_a - \beta_1\frac{\partial A}{\partial \nabla\mathbf{u}}\mathbf{n} - \beta_2\frac{\partial B}{\partial \mathbf{u}},\ \text{on}\ (T_1, T_2)\times\Gamma_N
\end{aligned}
\tag{2.87}
$$

When $t \in (0, T_1)$

$$
\begin{aligned}
-\rho\frac{\partial \mathbf{u}_a}{\partial t} - \eta\nabla\cdot\left(\nabla\mathbf{u}_a + \nabla\mathbf{u}_a^{\mathrm{T}}\right) - \rho\left(\mathbf{u}\cdot\nabla\right)\mathbf{u}_a + \rho\left(\nabla\mathbf{u}\right)\cdot\mathbf{u}_a& \\
+\nabla p_a &= \frac{\partial \mathbf{f}}{\partial \mathbf{u}}\mathbf{u}_a,\ \text{in}\ (0, T_1)\times\Omega \\
-\nabla\cdot\mathbf{u}_a &= \frac{\partial \mathbf{f}}{\partial p}\cdot\mathbf{u}_a,\ \text{in}\ (0, T_1)\times\Omega \\
\mathbf{u}_a &= \mathbf{0},\ \text{on}\ (0, T_1)\times\Gamma_D \\
\left[-p_a\mathbf{I} + \eta\left(\nabla\mathbf{u}_a + \nabla\mathbf{u}_a^{\mathrm{T}}\right)\right]\mathbf{n} &= -\rho\left(\mathbf{u}\cdot\mathbf{n}\right)\mathbf{u}_a,\ \text{on}\ (0, T_1)\times\Gamma_N
\end{aligned}
\tag{2.88}
$$

By solving Eqs. 2.86, 2.87 and 2.88 sequentially, the adjoint variables of $\mathbf{u}$ and p can be obtained. Because the solution of Eq. 2.86 is $\mathbf{u}_a = \mathbf{0}$ and $p_a = 0$, one only needs to solved Eqs. 2.87 and 2.88 with the initial condition $\mathbf{u}_a\left(T_2, \mathbf{x}\right) = \mathbf{0}$. The adjoint derivatives corresponding to expression 2.82 can be expressed as

$$
\begin{aligned}
\left.\frac{D\hat{J}}{D\gamma}\right|_{\Omega} &= \int_{T_1}^{T_2}\left(\beta_1\frac{\partial A}{\partial \gamma} + \frac{\partial \alpha}{\partial \gamma}\mathbf{u}\cdot\mathbf{u}_a\right)\mathrm{d}t + \int_0^{T_1} + \frac{\partial \alpha}{\partial \gamma}\mathbf{u}\cdot\mathbf{u}_a\,\mathrm{d}t,\ \text{in}\ \Omega \\
\left.\frac{D\hat{J}}{D\gamma}\right|_{\partial\Omega} &= \int_{T_1}^{T_2}\beta_2\frac{\partial B}{\partial \gamma}\,\mathrm{d}t,\ \text{on}\ \partial\Omega
\end{aligned}
\tag{2.89}
$$

Therefore, the adjoint analysis of the objective or design constraints defined on the subinterval $[T_1, T_2] \subseteq [0, T]$ is implemented on $[0, T_2]$ instead of $[0, T]$, although the unsteady flow problem is defined on the time interval $[0, T]$. Subsequently, the corresponding adjoint equations and adjoint derivatives are transient equations and integrations on $[0, T_2]$ instead of $[0, T]$, respectively.

2.4 Appendix for Section 2.2

2.4.1 Adjoint Equations for Level Set Method-Based Topology Optimization

The optimization problem (Eq. 2.45) constrained by the unsteady Navier-Stokes equations can be written into the following abstract form:

$$\min \ J(\mathbf{u}, p; \phi) \quad \text{s.t. } e(\mathbf{u}, p; \phi) = 0, \ \phi \in \mathscr{K} \tag{2.90}$$

where $\mathscr{K}$ is the set of feasible level set function ϕ; $e(\cdot)$ is the operator corresponding to the weak form of the Navier-Stokes equations. The weak form of the Navier-Stokes equations in Eq. 2.39 is

$$\begin{aligned} & e(\mathbf{u}, p; \phi) \\ = & \rho \int_\Omega (\mathbf{u} \cdot \mathbf{u}_a)\big|_{t=T} \mathrm{d}\Omega - \rho \int_\Omega \mathbf{u}_0 \cdot \mathbf{u}_a\big|_{t=0} \mathrm{d}\Omega - \rho \int_0^T \int_\Omega \mathbf{u} \cdot \frac{\partial \mathbf{u}_a}{\partial t} \mathrm{d}\Omega \mathrm{d}t \\ & + \eta \int_0^T \int_\Omega \nabla \mathbf{u} : \nabla \mathbf{u}_a \, \mathrm{d}\Omega \mathrm{d}t + \int_0^T \int_\Omega \rho (\mathbf{u} \cdot \nabla) \mathbf{u} \cdot \mathbf{u}_a \, \mathrm{d}\Omega \mathrm{d}t \\ & - \int_0^T \int_\Omega p \nabla \cdot \mathbf{u}_a \, \mathrm{d}\Omega \mathrm{d}t - \int_0^T \int_\Omega \mathbf{f} \cdot \mathbf{u}_a \, \mathrm{d}\Omega \mathrm{d}t - \int_0^T \int_\Omega p_a \nabla \cdot \mathbf{u} \, \mathrm{d}\Omega \mathrm{d}t \\ & + \int_0^T \int_{\Gamma_D} [(p\mathbf{I} - \eta \nabla \mathbf{u}) \mathbf{n} \cdot \mathbf{u}_a]\big|_{\mathbf{u}=\mathbf{u}_D} \mathrm{d}\Gamma \mathrm{d}t + \int_0^T \int_\Omega H(\phi) \mathbf{u} \cdot \mathbf{u}_a \, \mathrm{d}\Omega \mathrm{d}t \\ = & 0, \ \forall \mathbf{u}_a \in \mathscr{U}_\Omega^d \ \forall p_a \in \mathscr{P}_\Omega \end{aligned} \tag{2.91}$$

where $\mathscr{U}_\Omega^d := \left(\mathscr{L}^2((0, T); \mathscr{H}(\Omega))\right)^d$, $\mathscr{P}_\Omega := \mathscr{L}^2\left((0, T); \mathscr{L}^2(\Omega)\right)$; $\mathbf{u} \in \mathscr{U}_\Omega^d$, $p \in \mathscr{P}_\Omega$, $\mathbf{f} \in \mathscr{U}_\Omega^{d*} := \left(\mathscr{L}^2((0, T); \mathscr{H}^*(\Omega))\right)^d$, $\mathbf{u}_0 \in \left(\mathscr{H}_0^1(\Omega)\right)^d$ and $\phi \in \mathscr{H}(\Omega)$; $d =$ is the spatial dimension; $\mathscr{L}^2(\Omega)$ is second order integrable Lebesgue space; $\mathscr{H}^*(\Omega)$ is the dual space of Hilbert space $\mathscr{H}(\Omega)$; $\mathscr{H}_0^1(\Omega) = \{\mathbf{u} \in \mathscr{H}(\Omega) \mid \nabla \cdot \mathbf{u} = 0\}$; $\mathscr{L}^1(\Omega)$ is the first order integrable Lebesgue space. On the further treatment of the integral on Γ_D in Eq. 2.91, one can be refer to [30].

According to the Lagrangian multiplier method [20], the augmented Lagrangian objective functional $\hat{J}(\mathbf{u}, p; \phi)$ for the optimization problem in Eq. 2.45 can be expressed as

$$\begin{aligned}\hat{J}(\mathbf{u}, p; \phi) =& J(\mathbf{u}, p; \phi) + e(\mathbf{u}, p; \phi) - \lambda \left(\int_{\Omega} H(-\phi)\, \mathrm{d}\Omega - V_r V_0 \right) \\ &+ \frac{\Lambda}{2} \left(\int_{\Omega} H(-\phi)\, \mathrm{d}\Omega - V_r V_0 \right)^2\end{aligned} \tag{2.92}$$

where the volume constraint is treated with the Lagrangian multiplier based quadratic penalty method [20]; λ and Λ are the Lagrangian multiplier and the quadratic penalty parameter, respectively. According to Karush-Kuhn-Tucker conditions [5], the variational of $\hat{J}(\mathbf{u}, p; \phi)$

$$\delta \hat{J} = \frac{\partial \hat{J}}{\partial \mathbf{u}} \cdot \delta \mathbf{u} + \frac{\partial \hat{J}}{\partial p} \delta p + \frac{\partial \hat{J}}{\partial \phi} \delta \phi \tag{2.93}$$

should be zero corresponding to the optimal distribution of the level set function. Therefore, there are

$$\frac{\partial \hat{J}}{\partial \mathbf{u}} \cdot \delta \mathbf{u} = \mathbf{0}; \ \frac{\partial \hat{J}}{\partial p} \delta p = 0; \ \frac{\partial \hat{J}}{\partial \phi} \delta \phi = 0 \tag{2.94}$$

By setting $\tilde{A} = \beta_1 H(-\phi) A + \beta_2 \tau(\phi) \|\nabla \phi\| B$ for Eq. 2.47, one can obtain the variational of J to $\mathbf{u}$:

$$\begin{aligned}\frac{\partial J}{\partial \mathbf{u}} \cdot \delta \mathbf{u} =& \int_0^T \int_{\Omega} \frac{\partial \tilde{A}}{\partial \mathbf{u}} \cdot \delta \mathbf{u}\, \mathrm{d}\Omega \mathrm{d}t + \int_0^T \int_{\partial\Omega} \beta_3 \frac{\partial C}{\partial \mathbf{u}} \cdot \delta \mathbf{u}\, \mathrm{d}\Omega \mathrm{d}t \\ &+ \int_0^T \int_{\Gamma_N} \frac{\partial \tilde{A}}{\partial \nabla \mathbf{u}} \mathbf{n} \cdot \delta \mathbf{u}\, \mathrm{d}\Gamma \mathrm{d}t - \int_0^T \int_{\Omega} \left(\nabla \cdot \frac{\partial \tilde{A}}{\partial \nabla \mathbf{u}} \right) \cdot \delta \mathbf{u}\, \mathrm{d}\Omega \mathrm{d}t\end{aligned} \tag{2.95}$$

and the variational of J to p:

$$\frac{\partial J}{\partial p} \delta p = \int_0^T \int_{\Omega} \frac{\partial \tilde{A}}{\partial p} \delta p\, \mathrm{d}\Omega \mathrm{d}t + \int_0^T \int_{\partial\Omega} \beta_3 \frac{\partial C}{\partial p} \delta p\, \mathrm{d}\Gamma \mathrm{d}t \tag{2.96}$$

The Lebesgue measure of $(0, T) \times \Gamma$ in Q is zero, then

$$\int_0^T \int_{\Omega} H(\phi)\, \delta \mathbf{u} \cdot \mathbf{u}_a\, \mathrm{d}\Omega \mathrm{d}t = 0 \tag{2.97}$$

Based on the partial integration approach, the Gauss theory [5] and $\delta \mathbf{u} = \mathbf{0}$ on Γ_D, the following transformations are obtained:

$$\begin{aligned}\int_0^T \int_\Omega \nabla \delta \mathbf{u} : \nabla \mathbf{u}_a \, \mathrm{d}\Omega \mathrm{d}t &= \int_0^T \int_\Omega \nabla \cdot (\nabla \mathbf{v} \cdot \delta \mathbf{u}) \, \mathrm{d}\Omega \mathrm{d}t - \int_0^T \int_\Omega \Delta \mathbf{u}_a \cdot \delta \mathbf{u} \, \mathrm{d}\Omega \mathrm{d}t \\ &= \int_0^T \int_{\Gamma_N} (\nabla \mathbf{u}_a) \, \mathbf{n} \cdot \delta \mathbf{u} \, \mathrm{d}\Gamma \mathrm{d}t - \int_0^T \int_\Omega \Delta \mathbf{u}_a \cdot \delta \mathbf{u} \, \mathrm{d}\Omega \mathrm{d}t\end{aligned} \tag{2.98}$$

$$\begin{aligned}-\int_0^T \int_\Omega p_a \nabla \cdot \delta \mathbf{u} \, \mathrm{d}\Omega \mathrm{d}t &= -\int_0^T \int_\Omega \nabla \cdot (p_a \delta \mathbf{u}) \, \mathrm{d}\Omega \mathrm{d}t + \int_0^T \int_\Omega \nabla p_a \cdot \delta \mathbf{u} \, \mathrm{d}\Omega \mathrm{d}t \\ &= -\int_0^T \int_{\partial\Omega} p_a \mathbf{n} \cdot \delta \mathbf{u} \, \mathrm{d}\Gamma \mathrm{d}t + \int_0^T \int_\Omega \nabla p_a \cdot \delta \mathbf{u} \, \mathrm{d}\Omega \mathrm{d}t \\ &= -\int_0^T \int_{\Gamma_N} p_a \mathbf{n} \cdot \delta \mathbf{u} \, \mathrm{d}\Gamma \mathrm{d}t + \int_0^T \int_\Omega \nabla p_a \cdot \delta \mathbf{u} \, \mathrm{d}\Omega \mathrm{d}t\end{aligned} \tag{2.99}$$

$$\begin{aligned}\int_0^T \int_\Omega (\mathbf{u} \cdot \nabla) \, \delta \mathbf{u} \cdot \mathbf{u}_a \, \mathrm{d}\Omega \mathrm{d}t &= \int_0^T \int_\Omega \nabla \cdot [\mathbf{u} \cdot (\mathbf{v} \cdot \delta \mathbf{u})] \, \mathrm{d}\Gamma \mathrm{d}t - \int_0^T \int_\Omega (\mathbf{u} \cdot \nabla) \, \mathbf{u}_a \cdot \delta \mathbf{u} \, \mathrm{d}\Omega \mathrm{d}t \\ &= \int_0^T \int_{\partial\Omega} (\mathbf{u} \cdot \mathbf{n}) \, \mathbf{v} \cdot \delta \mathbf{u} \, \mathrm{d}\Gamma \mathrm{d}t - \int_0^T \int_\Omega (\mathbf{u} \cdot \nabla) \, \mathbf{u}_a \cdot \delta \mathbf{u} \, \mathrm{d}\Omega \mathrm{d}t \\ &= \int_0^T \int_{\Gamma_N} (\mathbf{u} \cdot \mathbf{n}) \, \mathbf{u}_a \cdot \delta \mathbf{u} \, \mathrm{d}\Gamma \mathrm{d}t - \int_0^T \int_\Omega (\mathbf{u} \cdot \nabla) \, \mathbf{u}_a \cdot \delta \mathbf{u} \, \mathrm{d}\Omega \mathrm{d}t\end{aligned} \tag{2.100}$$

$$\int_0^T \int_\Omega (\delta \mathbf{u} \cdot \nabla) \, \mathbf{u} \cdot \mathbf{u}_a \, \mathrm{d}\Omega \mathrm{d}t = \int_0^T \int_\Omega (\nabla \mathbf{u}) \, \mathbf{u}_a \cdot \delta \mathbf{u} \, \mathrm{d}\Omega \mathrm{d}t \tag{2.101}$$

Because $\mathbf{u}$ is known on Σ_D, and p can be expressed by $\nabla \mathbf{u}$ on Σ_N according to Eq. 2.44, the functional C in Eq. 2.46 can be expressed as

$$C(\mathbf{u}, p; \gamma) = \begin{cases} C(p; \gamma), \text{ on } \Sigma_D \\ C(\mathbf{u}, \nabla \mathbf{u}; \gamma), \text{ on } \Sigma_N \end{cases} \tag{2.102}$$

Substituting Eqs. 2.95–2.102 into the first two formulas of Eq. 2.94, the following equations are obtained:

$$\int_\Omega \rho \, (\delta \mathbf{u} \cdot \mathbf{v}) \big|_{t=T} \, \mathrm{d}\Omega = 0 \tag{2.103}$$

$$\int_0^T \int_\Omega \left[-\rho \frac{\partial \mathbf{v}}{\partial t} - \eta \Delta \mathbf{v} - \rho \left(\mathbf{u} \cdot \nabla \right) \mathbf{v} + \rho \left(\nabla \mathbf{u} \right) \cdot \mathbf{v} + \nabla p_a - \frac{\partial \mathbf{f}}{\partial \mathbf{u}} \cdot \mathbf{v} + \frac{\partial \tilde{A}}{\partial \mathbf{u}} - \nabla \cdot \frac{\partial \tilde{A}}{\partial \nabla \mathbf{u}} \right] \cdot \delta \mathbf{u} \, \mathrm{d}\Omega \mathrm{d}t + \int_0^T \int_{\Gamma_N} \left[\left(-p_a \mathbf{I} + \eta \nabla \mathbf{u}_a \right) \mathbf{n} + \rho \left(\mathbf{u} \cdot \mathbf{n} \right) \mathbf{u}_a + \frac{\partial \tilde{A}}{\partial \nabla \mathbf{u}} \mathbf{n} + \beta_3 \frac{\partial C}{\partial \mathbf{u}} \right] \cdot \delta \mathbf{u} \, \mathrm{d}\Gamma \mathrm{d}t = 0 \tag{2.104}$$

$$\int_0^T \int_\Omega \left(-\nabla \cdot \mathbf{v} + \frac{\partial \tilde{A}}{\partial p} - \frac{\partial \mathbf{f}}{\partial p} \cdot \mathbf{v} \right) \delta p \, \mathrm{d}\Omega \mathrm{d}t + \int_0^T \int_{\Gamma_D} \left(\mathbf{v} + \beta_3 \frac{\partial C}{\partial p} \mathbf{n} \right) \delta p \, \mathrm{d}\Gamma \mathrm{d}t = 0 \tag{2.105}$$

Based on Eqs. 2.103–2.105 and the arbitrariness of $\delta \mathbf{u}$ and δp, the adjoint equations of the Navier-Stokes equations for the optimization problem in Eq. 2.45 can be obtained as in Eq. 2.48.

2.4.2 *Variational of Augmented Lagrangian Objective Functional to Level Set Function*

According to Eq. 2.92 and the third formula of Eq. 2.94, the variational of the augmented Lagrangian objective functional $\hat{J}$ to the level set function ϕ can be obtained as

$$\begin{aligned} \delta \hat{J} &= \frac{\partial \hat{J}}{\partial \phi} \delta \phi \\ &= \frac{\partial J}{\partial \phi} \delta \phi + \lambda \int_\Omega \tau(\phi) \, \delta \phi \, \mathrm{d}\Omega + \int_0^T \int_\Omega \tau(\phi) \mathbf{u} \cdot \mathbf{u}_a \delta \phi \, \mathrm{d}\Omega \mathrm{d}t \\ &\quad - \Lambda \int_\Omega \tau(\phi) \, \delta \phi \, \mathrm{d}\Omega \left(\int_\Omega H\left(-\phi\right) \, \mathrm{d}\Omega - V_r V_0 \right) \end{aligned} \tag{2.106}$$

In addition,

$$\begin{aligned} \delta \left\| \nabla \phi \right\| &= \delta \left(\frac{(\nabla \phi)^2}{\left\| \nabla \phi \right\|} \right) \\ &= \frac{2 \nabla \phi \cdot \nabla (\delta \phi)}{\left\| \nabla \phi \right\|} - \frac{(\nabla \phi)^2 \cdot \delta \left\| \nabla \phi \right\|}{\left\| \nabla \phi \right\|^2} \\ &= \frac{2 \nabla \phi \cdot \nabla (\delta \phi)}{\left\| \nabla \phi \right\|} - \delta \left\| \nabla \phi \right\| \end{aligned} \tag{2.107}$$

Therefore,

$$\delta\|\nabla\phi\| = \frac{\nabla\phi\cdot\nabla(\delta\phi)}{\|\nabla\phi\|} \tag{2.108}$$

Based on the similar transformation in [31] and Eq. 2.108, the following transformation can be obtained:

$$\begin{aligned}
&\frac{\partial J}{\partial\phi}\delta\phi\\
&= -\int_0^T\int_\Omega \tau(\phi)A\,\delta\phi\,\mathrm{d}\Omega\mathrm{d}t + \int_0^T\int_\Omega \delta\big(\tau(\phi)\|\nabla\phi\|\big)B\,\mathrm{d}\Omega\mathrm{d}t\\
&= -\int_0^T\int_\Omega \tau(\phi)A\,\delta\phi\,\mathrm{d}\Omega\mathrm{d}t + \int_0^T\int_\Omega \big[\delta\big(\tau(\phi)\big)\|\nabla\phi\| + \tau(\phi)\,\delta\|\nabla\phi\|\big]B\,\mathrm{d}\Omega\mathrm{d}t\\
&= -\int_0^T\int_\Omega \tau(\phi)A\,\delta\phi\,\mathrm{d}\Omega\mathrm{d}t + \int_0^T\int_\Omega \left[\delta\big(\tau(\phi)\big)\|\nabla\phi\| + \tau(\phi)\frac{\nabla\phi\cdot\nabla(\delta\phi)}{\|\nabla\phi\|}\right]B\,\mathrm{d}\Omega\mathrm{d}t\\
&= -\int_0^T\int_\Omega \tau(\phi)A\,\delta\phi\,\mathrm{d}\Omega\mathrm{d}t + \int_0^T\int_\Omega \Big[\delta\big(\tau(\phi)\big)\|\nabla\phi\|\\
&\quad + \nabla\cdot\left(\tau(\phi)\frac{\nabla\phi}{\|\nabla\phi\|}\delta\phi\right) - \nabla\cdot\left(\tau(\phi)\frac{\nabla\phi}{\|\nabla\phi\|}\right)\delta\phi\Big]B\,\mathrm{d}\Omega\mathrm{d}t\\
&= -\int_0^T\int_\Omega \tau(\phi)A\,\delta\phi\,\mathrm{d}\Omega\mathrm{d}t + \int_0^T\int_\Omega \big[\delta\big(\tau(\phi)\big)\|\nabla\phi\| + \nabla\cdot\big(\tau(\phi)\mathbf{n}_\Gamma\,\delta\phi\big)\\
&\quad - \delta\phi\tau(\phi)\nabla\cdot\mathbf{n}_\Gamma - \delta\big(\tau(\phi)\big)\|\nabla\phi\|\big]B\,\mathrm{d}\Omega\mathrm{d}t\\
&= -\int_0^T\int_\Omega \tau(\phi)A\,\delta\phi\,\mathrm{d}\Omega\mathrm{d}t + \int_0^T\int_\Omega \big[\nabla\cdot\big(\tau(\phi)\mathbf{n}_\Gamma\,\delta\phi\big) - \delta\phi\tau(\phi)\nabla\cdot\mathbf{n}_\Gamma\big]B\,\mathrm{d}\Omega\mathrm{d}t\\
&= -\int_0^T\int_\Omega \tau(\phi)A\,\delta\phi\,\mathrm{d}\Omega\mathrm{d}t - \int_0^T\int_\Omega \tau(\phi)B\nabla\cdot\mathbf{n}_\Gamma\,\delta\phi\,\mathrm{d}\Omega\mathrm{d}t\\
&\quad + \int_0^T\int_\Omega \big[\nabla\cdot\big(\tau(\phi)\mathbf{n}_\Gamma B\,\delta\phi\big) - \nabla B\cdot\mathbf{n}_\Gamma\tau(\phi)\,\delta\phi\big]\,\mathrm{d}\Omega\mathrm{d}t\\
&= -\int_0^T\int_\Omega \tau(\phi)A\,\delta\phi\,\mathrm{d}\Omega\mathrm{d}t + \int_0^T\int_{\partial\Omega} \tau(\phi)B\,\delta\phi\,\mathbf{n}_\Gamma\cdot\mathbf{n}\,\mathrm{d}\Gamma\mathrm{d}t\\
&\quad - \int_0^T\int_\Omega \nabla B\cdot\mathbf{n}_\Gamma\tau(\phi)\,\delta\phi\,\mathrm{d}\Omega\mathrm{d}t - \int_0^T\int_\Omega \tau(\phi)B\nabla\cdot\mathbf{n}_\Gamma\,\delta\phi\,\mathrm{d}\Omega\mathrm{d}t\\
&= -\int_0^T\int_\Omega \tau(\phi)A\,\delta\phi\,\mathrm{d}\Omega\mathrm{d}t - \int_0^T\int_\Omega \nabla B\cdot\mathbf{n}_\Gamma\tau(\phi)\,\delta\phi\,\mathrm{d}\Omega\mathrm{d}t\\
&\quad - \int_0^T\int_\Omega \tau(\phi)B\kappa\,\delta\phi\,\mathrm{d}\Omega\mathrm{d}t
\end{aligned} \tag{2.109}$$

where $\mathbf{n}_\Gamma = \nabla\phi/\|\nabla\phi\|$ and $\kappa = \nabla\cdot\mathbf{n}_\Gamma$. By substituting Eq. 2.109 into Eq. 2.106, the adjoint sensitivity of the optimization problem in Eq. 2.45 is obtained as:

$$
\begin{aligned}
\delta \hat{J} = &-\int_0^T \int_\Omega \tau(\phi) A\, \delta\phi \,\mathrm{d}\Omega \mathrm{d}t - \int_0^T \int_\Omega \nabla B \cdot \mathbf{n}_\Gamma \tau(\phi)\, \delta\phi \,\mathrm{d}\Omega \mathrm{d}t \\
&- \int_0^T \int_\Omega \tau(\phi) B\kappa\, \delta\phi \,\mathrm{d}\Omega \mathrm{d}t + \int_0^T \int_\Omega \tau(\phi) \mathbf{u} \cdot \mathbf{w} \delta\phi \,\mathrm{d}\Omega \mathrm{d}t \\
&+ \lambda \int_\Omega \tau(\phi)\, \delta\phi \,\mathrm{d}\Omega - \Lambda \left(\int_\Omega H(-\phi)\, \mathrm{d}\Omega - V_r V_\Omega \right) \int_\Omega \tau(\phi)\, \delta\phi \,\mathrm{d}\Omega \\
= &-\int_\Omega \Bigg[\int_0^T (\beta_1 A + \beta_2 \nabla B \cdot \mathbf{n}_\Gamma + \beta_2 B\kappa - \mathbf{u} \cdot \mathbf{u}_a)\, \mathrm{d}t - \lambda \\
&+ \Lambda \left(\int_\Omega H(-\phi)\, \mathrm{d}\Omega - V_r V_0 \right) \Bigg] \tau(\phi) \delta\phi \,\mathrm{d}\Omega
\end{aligned}
\tag{2.110}
$$

References

1. R.L. Panton, *Incompressible Flow* (Wiley, 1984)
2. T. Borrvall, J. Petersson, Topology optimization of fluid in Stokes flow. Int. J. Numer. Meth. Fluids **41**, 77–107 (2003)
3. A. Gersborg-Hansen, O. Sigmund, R.B. Haber, Topology optimization of channel flow problems. Struct. Multidiscip. Optim. **29**, 1–12 (2005)
4. L.H. Olesen, F. Okkels, H. Bruus, A high-level programming-language implementation of topology optimization applied to steady-state Navier-Stokes flow. Int. J. Numer. Meth. Eng. **65**, 975–1001 (2006)
5. M. Hinze, R. Pinnau, M. Ulbrich, S. Ulbrich, *Optimization with PDE Constraints* (Springer, 2009)
6. M.B. Giles, N.A. Pierce, An introduction to the adjoint approach to design. Flow Turbul. Combust. **65**, 393–415 (2000)
7. B. Mohammadi, O. Pironneau, *Applied Shape Optimization for Fluids* (OXFORD, 2010)
8. J. Nocedal, S.J. Wright, *Numerical Optimization* (Springer, 1998)
9. K. Svanberg, The method of moving asymptotes: a new method for structural optimization. Int. J. Numer. Meth. Eng. **24**, 359–373 (1987)
10. http://www.comsol.com
11. H.C. Elman, D.J. Silvester, A.J. Wathen, *Finite Elements and Fast Iterative Solvers: with Applications in Incompressible Fluid Dynamics* (OXFORD, 2006)
12. U.M. Ascher, L.R. Petzold, *Computer Methods for Ordinary Differential Equations and Differential-Algebraic Equations* (SIAM, 1998)
13. D.J. Laser, J.G. Santiago, A review of micropumps. J. Micromech. Microeng. **14**, 35–64 (2004)
14. P. Chen, X. Feng, J. Sun, Y. Wang, W. Du, B.F. Liu, Hydrodynamic gating for sample introduction on a microfluidic chip. Lab chip **10**, 1472–1475 (2010)
15. N. Aage, T.H. Poulsen, A. Gersborg-Hansen, O. Sigmund, Topology optimization of large scale stokes flow problems. Struct. Multidisc. Optim. **35**, 175–180 (2008)
16. A. Gersborg-Hansen, M. Berggren, B. Dammann, Topology optimizatiopn of mass distribution problems in Stokes flow. Solid Mech. Appl. **137**, 365–374 (2006)
17. C.H. Li Paul, *Microfluidic Lab-on-a-Chip for Chemical and Biological Analysis and Discovery* (CRC Press, 2005)
18. J. Ducrée, S. Haeberle, S. Lutz, S. Pausch, F. Stetten, R. Zengerle, Design and fabrication of a centrifugally driven microfluidic disk for fully integrated metabolic assays on whole blood. J. Micromech. Microeng. **17**, 103–115 (2007)

19. V.J. Challis, J.K. Guest, Level set topology optimization of fluids in Stokes flow. Int. J. Numer. Meth. Eng. **79**, 1284–1308 (2009)
20. J. Nocedal, S.J. Wright, *Numerical Optimization*, 2nd edn. (Springer, New York, 2000)
21. S. Osher, J.A. Sethian, Front propagating with curvature dependent speed: algorithms based on Hamilton-Jacobi formulations. J. Comput. Phys. **78**, 12–49 (1988)
22. G. Hauke, T.J.R. Hughes, A unified approach to compressible and incompressible flows. Comp. Meth. Appl. Mech. Eng. **113**, 389–395 (1994)
23. K. Ito, K. Kunisch, *Lagrange Multiplier Approach to Variational Problems and Applications* (SIAM, 2008)
24. S. Osher, R. Fedkiw, *Level Set Methods and Dynamic Implicit Surfaces* (Springer, New York, 2003)
25. V.J. Challis, A discrete level-set topology optimization code written in Matlab. Struct. Multidiscip. Optim. **41**(3), 453–464 (2009)
26. J.K. Guest, J.H. Proevost, Topology optimization of creeping fluid flows using a Darcy-Stokes finite element. Int. J. Numer. Meth. Eng. **66**, 461–484 (2006)
27. S. Kreissl, G. Pingen, K. Maute, Topology optimization for unsteady flow. Int. J. Numer. Meth. Eng. (2011). doi:10.1002/nme.3151
28. K. Yosida, *Functional Analysis* (Springer, Berlin, 1980)
29. E. Zeidler, *Nonlinear Functional Analysis and its Applications. I, Fixed-Point Theorems* (Springer, 1986)
30. C. Othmer, A continuous adjoint formulation for the computation of topological and surface sensitivities of ducted flows. Int. J. Numer. Meth. Fluids **58**, 861–877 (2008)
31. S. Zhou, Q. Li, A variational level set method for the topology optimization of steady-state Navier-Stokes flow. J. Comput. Phys. **227**, 10178–10195 (2008)

Chapter 3
Topology Optimization for Fluid Flows with Body Forces

This chapter presents the topology optimization method for the steady and unsteady incompressible Navier-Stokes flows driven by body forces that influence the optimal shape and topology of fluid flows. The considered body forces include the constant force (e.g. the gravity) and the centrifugal and Coriolis forces. In the density method-based topology optimization, the artificial friction force with design variable interpolated porosity is added into the Navier-Stokes equations as the conventional method, and the physical body forces in the Navier-Stokes equations are penalized using the power-law approach. In the level set method-based topology optimization, the topological sensitivity along with the shape sensitivity is computed based on the asymptotic analysis of the optimization problem, to nucleate new zero level set and weaken the dependence on the initial distribution of the level set function. Numerical examples on the topology optimization of the fluidic channel, mass distribution of the flow and local velocity control for the flows with body forces, are presented to demonstrate that the robustness of the topology optimization method for flows with body forces.

3.1 Density Method-Based Topology Optimization for Flows with Body Forces

3.1.1 Topology Optimization Problem for Flows with Body Forces

The incompressible flows of Newtonian fluid with body forces are described by the incompressible Navier-Stokes equations as [1]:

Y. Deng et al., *Topology Optimization Theory for Laminar Flow*,
https://doi.org/10.1007/978-981-10-4687-2_3

$$\rho \frac{\partial \mathbf{u}}{\partial t} - \eta \nabla \cdot \left(\nabla \mathbf{u} + \nabla \mathbf{u}^{\mathrm{T}} \right) + \rho \left(\mathbf{u} \cdot \nabla \right) \mathbf{u} + \nabla p = \mathbf{f}, \text{ in } (0, T) \times \Omega$$
$$-\nabla \cdot \mathbf{u} = 0, \text{ in } (0, T) \times \Omega \tag{3.1}$$

where $t \in (0, T)$ is the time and Ω is the space domain; $\mathbf{u}$ is the fluidic velocity; p is the fluidic pressure; ρ is the fluidic density; η is the dynamic viscosity; $\mathbf{f}$ is the body force loaded on the fluid. The considered body forces $\mathbf{f}$, typically including the constant body force (e.g. the gravity) and centrifugal and Coriolis forces, are not induced by the multiphysical effect but the inertia of the fluid. For the steady flow, Eq. 3.1 can be reduced by the local derivative of the velocity $\frac{\partial \mathbf{u}}{\partial t} = \mathbf{0}$. For the unsteady flow, the initial condition is needed

$$\mathbf{u}(0, \mathbf{x}) = \mathbf{u}_0(\mathbf{x}), \quad \text{in } \Omega \tag{3.2}$$

where $\mathbf{u}_0(\mathbf{x})$ satisfies the incompressible condition $\nabla \cdot \mathbf{u}_0 = 0$. The commonly used boundary conditions for the incompressible Navier-Stokes equations include the Dirichlet boundary condition:

$$\mathbf{u} = \mathbf{u}_D, \quad \text{on } (0, T) \times \Gamma_D \tag{3.3}$$

and the Neumann boundary condition:

$$\left[-p\mathbf{I} + \eta \left(\nabla \mathbf{u} + \nabla \mathbf{u}^{\mathrm{T}} \right) \right] \cdot \mathbf{n} = \mathbf{g}, \quad \text{on } (0, T) \times \Gamma_N \tag{3.4}$$

where $\mathbf{u}_D$ and $\mathbf{g}$ are the specified velocity and stress distribution on the boundaries Γ_D and Γ_N; $\mathbf{n}$ is the unitary outward normal on $\partial\Omega$. Specifically, the no-slip boundary is a particular Dirichlet boundary condition where $\mathbf{u}_D = \mathbf{0}$, and the open-boundary can be expressed by the Neumann boundary condition as $\mathbf{g} = \mathbf{0}$. As in the topology optimization of the flows without physical body forces, the artificial friction force is added into the body force term of the Navier-Stokes equations for the topology optimization of the flows driven by body forces as [2–4]:

$$\mathbf{f}_a = -\alpha \mathbf{u} \tag{3.5}$$

where α is the impermeability of a porous medium. α is the interpolation of the design variable γ representing the distribution of the solid and fluidic phases [2–4]:

$$\alpha(\gamma) = \alpha_{min} + (\alpha_{max} - \alpha_{min}) q \frac{1 - \gamma}{q + \gamma} \tag{3.6}$$

where α_{min} and α_{max} are the minimal and maximal values of α, respectively; q is a real and positive parameter used to adjust the convexity of the interpolation function in Eq. 3.6. The value of γ can vary between 0 and 1, which correspond to the artificial solid and fluidic phases respectively. For both the steady and unsteady incompressible Navier-Stokes flows, the design variable γ is time independent, because the layout of fluidic flows is the fixed distribution of the materials. Usually, α_{min} is chosen to be 0, and α_{max} is chosen to be a finite but large number to ensure the numerical stability of the optimization and approximate the solid phase with negligible permeability [3, 4]. The fluidic velocity should approach to zero in the approximated solid phase after evolving the design variable. If the physical body force is added to the artificial friction force directly to implement the topology optimization of the flow with body force, the physical body force will drive the fluid to flow and increase the fluidic velocity in the approximated solid with low permeability. This causes the gray regions in the result. In order to avoid the fluidic flow driven by the physical body force in the approximated solid phase, the physical body force is penalized using the power-law approach:

$$\mathbf{f}_{bp} = \mathbf{f}_b \gamma^n \tag{3.7}$$

where $\mathbf{f}_b$ is the physical body force; n is the penalty parameter, which is chosen based on numerical experiments. Based on the penalization of the body force, the body force is weaken in the approximated solid region ($\gamma = 0$) and maintained in the fluidic region ($\gamma = 1$) after evolving the design variable. Therefore, the value of the penalty parameter should satisfy $n \geq 1$. On the other hand, n can not be chosen too big, because the value of γ close to 1 but less than 1, representing the approximated liquid phase, can be over penalized to approach 0 by a big value of n. Therefore, the body force in Eq. 3.1 is the composition of the artificial friction force and the penalized physical body force:

$$\mathbf{f} = -\alpha \mathbf{u} + \mathbf{f}_b \gamma^n \tag{3.8}$$

Then the topology optimization problem for the incompressible Navier-Stokes flow with body forces can be formulated to be

$$\begin{aligned} & \text{Find } \gamma \in [0, 1] \text{ to minimize or maximize } J(\mathbf{u}, p; \gamma), \text{ subject to} \\ & \begin{cases} \rho \dfrac{\partial \mathbf{u}}{\partial t} - \eta \nabla \cdot \left(\nabla \mathbf{u} + \nabla \mathbf{u}^{\mathrm{T}} \right) + \rho (\mathbf{u} \cdot \nabla) \mathbf{u} + \nabla p = \mathbf{f} \\ - \nabla \cdot \mathbf{u} = 0 \\ \displaystyle\int_{\Omega} \gamma \, d\Omega \leq V_r V_0 \end{cases} \end{aligned} \tag{3.9}$$

where the optimization problem is regularized by a volume constraint as in [2–4]; $\frac{\partial \mathbf{u}}{\partial t} = \mathbf{0}$ is satisfied for the steady flows; $\mathbf{f}$ is the composed force $-\alpha \mathbf{u} + \mathbf{f}_b \gamma^n$; $V_0 = \int_\Omega 1 \, \mathrm{d}\Omega$ is the volume of the whole design domain and $V_r \in (0, 1]$ is the upper bound of the volume fraction. The general optimization objective functionals, which include both the domain and boundary integrations about the unknowns of the Navier-Stokes equations, are chosen to be

$$J(\mathbf{u}, p; \gamma) = \int_\Omega \beta_1 A(\mathbf{u}, \nabla \mathbf{u}, p; \gamma) \, \mathrm{d}\Omega + \int_{\partial\Omega} \beta_2 B(\mathbf{u}, p; \gamma) \, \mathrm{d}\Gamma \tag{3.10}$$

for the steady flow, and

$$J(\mathbf{u}, p; \gamma) = \int_0^T \int_\Omega \beta_1 A(\mathbf{u}, \nabla\mathbf{u}, p; \gamma)\, \mathrm{d}\Omega \mathrm{d}t + \int_0^T \int_{\partial\Omega} \beta_2 B(\mathbf{u}, p; \gamma)\, \mathrm{d}\Gamma \mathrm{d}t \tag{3.11}$$

for the unsteady flow, where β_1 and β_2 are space-independent parameters.

3.1.2 Sensitivity Analysis

According to the adjoint method for the Navier-Stokes equations in [5–8], the adjoint analysis of the topology optimization problem for the Navier-Stokes flows can be implemented. Based on the Karush-Kuhn-Tucker conditions for the optimization problem constrained by the partial differential equation [6, 9], the adjoint equations of the optimization problem in Eq. 3.9 can be obtained as

$$\begin{aligned} -\eta\nabla\cdot\left(\nabla\mathbf{u}_a + \nabla\mathbf{u}_a^{\mathrm{T}}\right) - \rho(\mathbf{u}\cdot\nabla)\mathbf{u}_a + \rho(\nabla\mathbf{u})\cdot\mathbf{u}_a & \\ +\nabla p_a &= -\beta_1\left(\frac{\partial A}{\partial\mathbf{u}} - \nabla\cdot\frac{\partial A}{\partial\nabla\mathbf{u}}\right) + \left(\frac{\partial\mathbf{f}_b}{\partial\mathbf{u}}\gamma^n - \alpha\right)\mathbf{u}_a, \text{ in } \Omega \\ -\nabla\cdot\mathbf{u}_a &= -\beta_1\frac{\partial A}{\partial p} + \frac{\partial\mathbf{f}_b}{\partial p}\cdot\mathbf{u}_a\gamma^n, \text{ in } \Omega \\ \mathbf{u}_a &= -\frac{\partial B}{\partial p}\mathbf{n}, \text{ on } \Gamma_D \\ \left[-p_a\mathbf{I} + \eta\left(\nabla\mathbf{u}_a + \nabla\mathbf{u}_a^{\mathrm{T}}\right)\right]\mathbf{n} &= -\rho(\mathbf{u}\cdot\mathbf{n})\mathbf{u}_a - \beta_1\frac{\partial A}{\partial\nabla\mathbf{u}}\mathbf{n} - \beta_2\frac{\partial B}{\partial\mathbf{u}}, \text{ on } \Gamma_N \end{aligned} \tag{3.12}$$

for the steady flow, and

$$\begin{aligned} -\rho\frac{\partial\mathbf{u}_a}{\partial t} - \eta\nabla\cdot\left(\nabla\mathbf{u}_a + \nabla\mathbf{u}_a^{\mathrm{T}}\right) - \rho(\mathbf{u}\cdot\nabla)\mathbf{u}_a + \rho(\nabla\mathbf{u})\cdot\mathbf{u}_a + \nabla p_a & \\ = -\beta_1\left(\frac{\partial A}{\partial\mathbf{u}} - \nabla\cdot\frac{\partial A}{\partial\nabla\mathbf{u}}\right) & \\ + \left(\frac{\partial\mathbf{f}_b}{\partial\mathbf{u}}\gamma^n - \alpha\right)\mathbf{u}_a, \text{ in } (0, T)\times\Omega & \\ -\nabla\cdot\mathbf{u}_a = -\beta_1\frac{\partial A}{\partial p} + \frac{\partial\mathbf{f}_b}{\partial p}\cdot\mathbf{u}_a\gamma^n, \text{ in } (0, T)\times\Omega & \\ \mathbf{u}_a(T, \mathbf{x}) = \mathbf{0}, \text{ in } \Omega & \\ \mathbf{u}_a = -\frac{\partial B}{\partial p}\mathbf{n}, \text{ on } (0, T)\times\Gamma_D & \\ \left[-p_a\mathbf{I} + \eta\left(\nabla\mathbf{u}_a + \nabla\mathbf{u}_a^{\mathrm{T}}\right)\right]\mathbf{n} = -\rho(\mathbf{u}\cdot\mathbf{n})\mathbf{u}_a - \beta_1\frac{\partial A}{\partial\nabla\mathbf{u}}\mathbf{n} - \beta_2\frac{\partial B}{\partial\mathbf{u}}, \text{ on } (0, T)\times\Gamma_N & \end{aligned} \tag{3.13}$$

for the unsteady flow, where $\mathbf{u}_a$ and p_a are the adjoint variables for the fluid velocity $\mathbf{u}$ and pressure p, respectively. The adjoint sensitivity of the optimization problem can be expressed as follows

$$\begin{aligned} \left.\frac{D\hat{J}}{D\gamma}\right|_{\Omega} &= \beta_1 \frac{\partial A}{\partial \gamma} - \left(n\mathbf{f}_b \gamma^{n-1} - \frac{\partial \alpha}{\partial \gamma}\mathbf{u} \right) \cdot \mathbf{u}_a, \text{ in } \Omega \\ \left.\frac{D\hat{J}}{D\gamma}\right|_{\partial\Omega} &= \beta_2 \frac{\partial B}{\partial \gamma}, \text{ on } \partial\Omega \end{aligned} \tag{3.14}$$

for the steady flow, and

$$\begin{aligned} \left.\frac{D\hat{J}}{D\gamma}\right|_{\Omega} &= \int_0^T \left[\beta_1 \frac{\partial A}{\partial \gamma} - \left(n\mathbf{f}_b \gamma^{n-1} - \frac{\partial \alpha}{\partial \gamma}\mathbf{u} \right) \cdot \mathbf{u}_a \right] \mathrm{d}t, \text{ in } \Omega \\ \left.\frac{D\hat{J}}{D\gamma}\right|_{\partial\Omega} &= \int_0^T \beta_2 \frac{\partial B}{\partial \gamma} \mathrm{d}t, \text{ on } \partial\Omega \end{aligned} \tag{3.15}$$

for the unsteady flow, respectively. In Eqs. 3.14 and 3.15, $\hat{J}$ represents the objective functional with the implicit relations substituted, where the implicit relations are defined by the Navier-Stokes equations with the body force term interpolated by the design variable. The above adjoint sensitivities define the most rapid descendant or ascendant directions for the minima or maxima optimization problems. The implementation of the adjoint sensitivity analysis for the integral constraint in the topology optimization problem is similar with that for the objective functional.

3.1.3 Numerical Implementation

The topology optimization for the Navier-Stokes flows driven by body forces is implemented using the gradient based iterative approach. The procedure for the iterative optimization includes the following steps (Table 3.1):

- the Navier-Stokes equations are solved with the given value of the design variable;
- the adjoint equations are solved based on the numerical solution of the Navier-Stokes equations;
- the adjoint sensitivity of the objective function and design constraint are computed;
- the design variable is updated by the method of moving asymptotes (MMA) [10].

The above steps are implemented iteratively until the stopping criteria are satisfied. In the above procedure, the Navier-Stokes equations and the corresponding adjoint equations are solved by the finite element method using the finite element package *Comsol Multiphysics* [11]. For the details on the implementation setting in *Comsol*, one can refer to [4]. During the optimization procedure, the Navier-Stokes equations and the adjoint equations are solved using Taylor-Hood elements [12], which

Table 3.1 Procedure of the iterative approach for solving the topology optimization problem

1.	Give the initial value of the design variable γ;
2.	Solve the Navier-Stokes equations by the finite element method;
3.	Solve the adjoint equations; compute the adjoint sensitivity and the corresponding objective value;
4.	Update the design variable by MMA;
5.	Check for convergence; if the stopping conditions are not satisfied, go to 2;
6.	Post processing

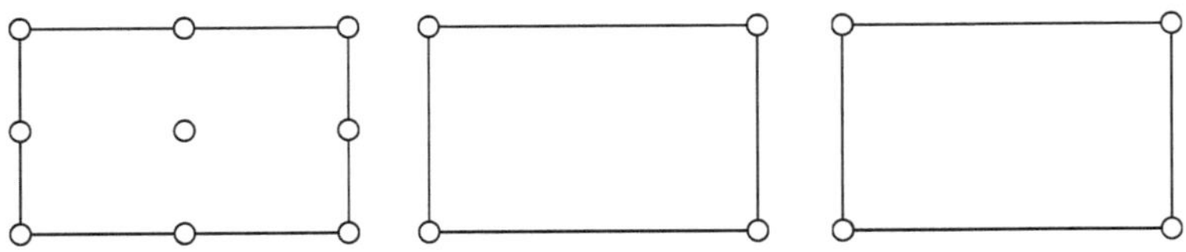
(a) **u** (b) p (c) γ

Fig. 3.1 The finite element nodes used to express the velocity, pressure and design variable on a rectangular element

interpolate the fluidic velocity quadratically and the pressure linearly. The design variable is interpolated linearly based on the corner nodes of the elements (Fig. 3.1). The stopping criteria are specified to be the change of the objective values in 5 consecutive iterations and the residual of the volume constraint satisfying

$$\begin{aligned} &\sum_{i=0}^{4} |J_{k-i} - J_{k-i-1}| / |J_k| < 1 \times 10^{-3} \\ &\left| \int_\Omega \gamma_k \, d\Omega / V_0 - \theta \right| / \theta < 1 \times 10^{-3} \end{aligned} \tag{3.16}$$

in the k-th iteration, where J_k and γ_k are respectively the objective value and the distribution of the design variable in the k-th iteration.

3.1.4 Numerical Examples

The topology optimization of several steady and unsteady Navier-Stokes flows, respectively driven by the constant body force and centrifugal and Coriolis forces, are performed to demonstrate the capability and utility of the presented topology optimization method for the flows with body forces. The density and viscosity of the fluid are set to be unitary, if there is no specification. The Reynolds number is calculated to be

$$\mathrm{Re} = \rho U_{max} L/\eta \tag{3.17}$$

where L is the width of the inlet, and U_{max} is the maximal scalar value of the velocity. The initial condition of the transient Navier-Stokes equations for the unsteady flow is set to be $\mathbf{u}_0 = \mathbf{0}$ in Eq. 3.2. The open boundary conditions are posed on the inlets and outlets; the no slip boundary condition is imposed on the walls; and the continuous condition is imposed on the interior boundaries. In the topology optimization problem of the flows with body forces (Eq. 3.9), four parameters, which are α_{max}, q, n and V_r, need to be fixed. The values of α_{max} and q are chosen based on numerical experiments, and more details are available in [4]. Based on the theory in Sects. 3.1.1 and 3.1.2, the topology optimization of the fluidic channels for steady and unsteady flows is implemented in Sect. 3.1.4.1. Numerical tests on the choice of the value of the penalty parameter n in Eq. 3.7 are carried out, and the difference of the optimal topologies corresponding to the boundary condition driven flow and body force driven flow is demonstrated. The mass distribution of the flows driven by the body forces is performed in Sect. 3.1.4.3. And the local velocity control of the flows driven by the body forces is discussed in Sect. 3.1.4.4. All the computation are performed on a DELL workstation (DELL PRECISION T5500, two Intel Xeon Quad X5550 CPUs).

3.1.4.1 Topology Optimization of Fluidic Channel

For the topology optimization of fluidic channels, the optimization can be implemented by maximizing the work done by body forces and the flux at the inlet of the flow. Therefore, the objective functionals in Eqs. 3.10 and 3.11 are fixed by setting:

$$A = \mathbf{f} \cdot \mathbf{u}, \text{ in } \Omega; \quad B = \begin{cases} -\mathbf{u} \cdot \mathbf{n}, \text{ on } \Gamma_i \\ \quad 0 \text{ , on } \partial\Omega \backslash \Gamma_i \end{cases} \tag{3.18}$$

where Γ_i is the inlet of the flow driven by body forces. In the objective (Eqs. 3.10 and 3.11), the parameters β_1 and β_2 depend on specified examples and can be adjusted by the designer based on the design necessity or numerical experiments, and the effect of the choice of these two parameters will be demonstrated by the unsteady flows driven by the centrifugal and Coriolis forces.

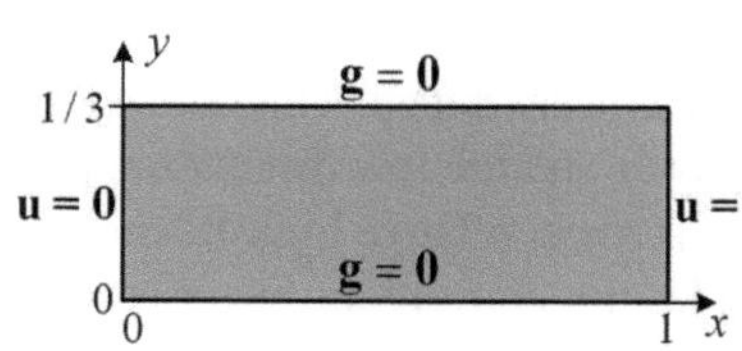

Fig. 3.2 Design domain for the short channel with steady flows driven by body forces. The *top* and *bottom* boundaries of the domain are set to be open boundary, and the *left* boundaries are set to be no slip type

Table 3.2 Parameter settings in the topology optimization of the short channel

β_1	β_2	V_r	α_{min}	α_{max}	q
1	0.1	0.50	0	1×10^4	1

Steady flows driven by body forces

For the steady flows driven by body forces, a short channel is optimized by the topology optimization method. The design domain is shown in Fig. 3.2 and discretized by 60×20 rectangular elements. The parameter choice are listed in Table 3.2. Firstly, the numerical tests on the choice of the penalty parameter n in Eq. 3.7 are performed for the flows driven by the downward constant body forces. For different scalar values of the constant body force, the topology optimization problem is solved. The obtained topologies of the optimized short channel, the corresponding objective values, the objective values calculated without interpolation in the same configuration and the tolerances are listed in Table 3.3. Table 3.3 shows that there are many unreasonable areas with small length scale in the optimal topologies corresponding to $n = 6$. Then $n = 6$ is regarded as a unreasonable choice, and $[1, 5]$ is chosen to be the reasonable interval for the penalty value. The penalty is helpful to decrease the body force in the area with low permeability and make the optimization objective value more close to its corresponding value calculated without interpolation in the same configuration. This can be confirmed by the values of the tolerance in Table 3.3. On the other hand, the tolerance also demonstrate that reasonable higher value of penalty means stronger penalty of the body force and smaller tolerance between the optimization objective value and the corresponding value calculated without interpolation in the same configuration. Therefore, we chose nonlinear penalty instead of the linear penalty. When the penalty value is chosen to be relative big, the iterative optimization procedure will become less robust, and the design variable with value close to 1 but less than 1, representing the approximated liquid phase, can be over penalized to approach 0 by the big penalty value. Then the penalty of the body force will be excessive and this will result in unreasonable topology. In sum, the eclectic penalty value 3 is chosen. Based on the choice of the penalty value, the optimal topologies of the short channel for flows driven by constant body forces with different scalar values are the ones corresponding to $n = 3$ in Table 3.3. And the distribution of the velocity vectors in the optimal topologies are shown in Fig. 3.3, where the Reynolds numbers are calculated according to Eq. 3.17. In Fig. 3.3, the optimal topology of the short channel is a straight channel when the flow is driven by the constant body force with low scalar value; as the scalar value of the body force increasing, the optimal topology of the short channel becomes diffuser-shaped. This is because that the diffuser-shaped flow has low dissipation power and the low dissipation power is helpful to increase the work done by the body force at relative high Reynolds number. For the optimal topology shown in Fig. 3.3a, the convergent histories of the objective values and volume constraint are plotted in Fig. 3.4, and snapshots for the distribution of the

Table 3.3 List of the optimal topologies of the short channel for different values of the penalty parameter n, the corresponding objective values J, the objective values J_r calculated without interpolation in the same configuration and the tolerance $|J - J_r|/J_r$ between the objective value J and J_r. $|\mathbf{f}|$ is the scalar value of the constant body force

(a) $|\mathbf{f}| = 10^1$

	$n=1$	$n=2$	$n=3$	$n=4$	$n=5$	$n=6$
Optimal	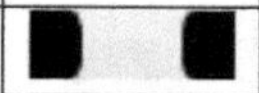					
J	4.5178×10^{-1}	4.5144×10^{-1}	4.5117×10^{-1}	4.5085×10^{-1}	4.5055×10^{-1}	7.2154×10^{-1}
J_r	4.3483×10^{-1}	4.3483×10^{-1}	4.4208×10^{-1}	4.4345×10^{-1}	4.5668×10^{-1}	3.4820×10^{-1}
$\|J-J_r\|/J_r$	0.0390	0.0382	0.0206	0.0167	0.0134	1.0722

(b) $|\mathbf{f}| = 10^2$

	$n=1$	$n=2$	$n=3$	$n=4$	$n=5$	$n=6$
Optimal					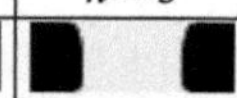	
J	4.2658×10^{1}	4.2678×10^{1}	4.2673×10^{1}	4.2658×10^{1}	4.2643×10^{1}	6.3262×10^{1}
J_r	4.1975×10^{1}	4.1997×10^{1}	4.2021×10^{1}	4.2134×10^{1}	4.3047×10^{1}	3.0333×10^{1}
$\|J-J_r\|/J_r$	0.0163	0.0162	0.0155	0.0124	0.0094	1.0856

(c) $|\mathbf{f}| = 10^3$

	$n=1$	$n=2$	$n=3$	$n=4$	$n=5$	$n=6$
Optimal	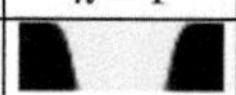	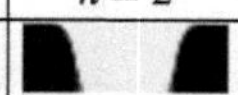	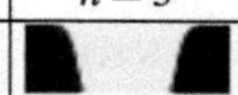	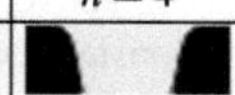	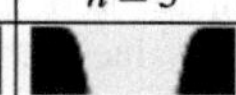	
J	2.5061×10^{3}	2.4935×10^{3}	2.4952×10^{3}	2.4983×10^{3}	2.5000×10^{3}	3.0916×10^{3}
J_r	2.6573×10^{3}	2.6161×10^{3}	2.6128×10^{3}	2.6012×10^{3}	2.5538×10^{3}	1.7700×10^{3}
$\|J-J_r\|/J_r$	0.0569	0.0469	0.0450	0.0396	0.0211	0.7467

(d) $|\mathbf{f}| = 10^4$

	$n=1$	$n=2$	$n=3$	$n=4$	$n=5$	$n=6$
Optimal			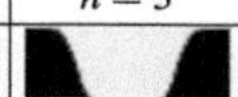		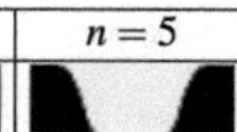	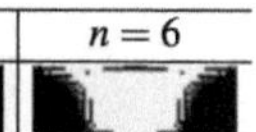
J	8.7051×10^{4}	8.7014×10^{4}	8.7090×10^{4}	8.7172×10^{4}	8.7256×10^{4}	8.7575×10^{4}
J_r	8.1918×10^{4}	8.2162×10^{4}	8.2271×10^{4}	8.2485×10^{4}	8.2721×10^{4}	1.4543×10^{4}
$\|J-J_r\|/J_r$	0.0627	0.0591	0.0586	0.0568	0.0548	5.0218

design variable in the evolving progress are shown in Fig. 3.5. For the validation of the optimality of the optimal topologies obtained in Fig. 3.3, the objective values are computed and cross-compared for the flows driven by constant body forces with different scalar values (Table 3.4).

Secondly, the short channel is optimized for the flow in the rotating frame. In the rotating system, the body forces is expressed as the composition of centrifugal and Coriolis forces, i.e. $\mathbf{f} = \rho\,(\boldsymbol{\omega} \times \mathbf{r}) \times \boldsymbol{\omega} + 2\rho\mathbf{u} \times \boldsymbol{\omega}$, where $\boldsymbol{\omega}$ and $\mathbf{r}$ are the angular velocity vector and the radial vector of the point inside the flow relative to the rotating center respectively [13]. The design domain is shown in Fig. 3.2; the rotating center is set to be the point with coordinates $(1/2, 2/3)$; and the scalar value of the angular velocity is set to be 10. The obtained optimal topologies are shown in Figs. 3.6 and 3.7. The cross-comparison of the objective values of the optimal topologies in Figs. 3.6 and 3.7 is shown in Table 3.5. When only the centrifugal force is considered,

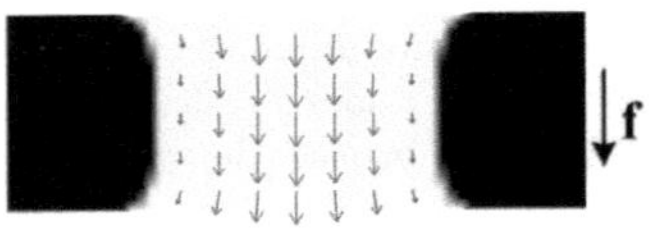

(a) $|\mathbf{f}| = 10^1$, Re $= 4.4 \times 10^{-1}$, $n = 3$

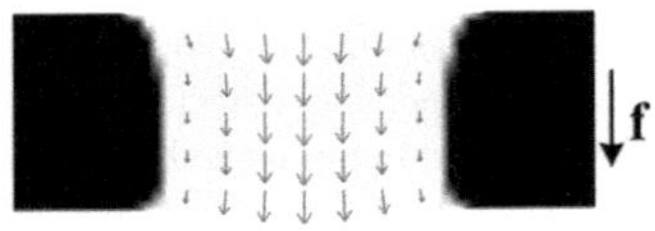

(b) $|\mathbf{f}| = 10^2$, Re $= 4.3 \times 10^0$, $n = 3$

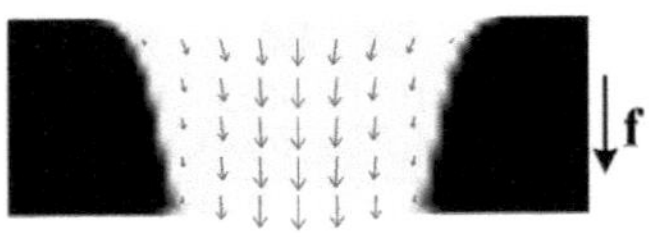

(c) $|\mathbf{f}| = 10^3$, Re $= 2.7 \times 10^1$, $n = 3$

(d) $|\mathbf{f}| = 10^4$, Re $= 1.0 \times 10^2$, $n = 3$

Fig. 3.3 Optimal short channels for the flows driven by the constant body forces with different scalar values. $|\mathbf{f}|$ is the scalar value of the constant body force. The *arrows* represent the distribution of the velocity vectors, where the length of the *arrows* are scaled to show the velocity field clearly in different cases

Table 3.4 Cross-comparison of the objective values corresponding to the optimal topologies in Fig. 3.3. $|\mathbf{f}|$ is the scalar value of the constant body force

	Figure 3.3a	Figure 3.3b	Figure 3.3c	Figure 3.3d
$\lvert\mathbf{f}\rvert = 10^1$	4.5117×10^{-1}	4.4625×10^{-1}	3.7143×10^{-1}	2.8765×10^{-1}
$\lvert\mathbf{f}\rvert = 10^2$	4.2658×10^1	4.2673×10^1	3.4621×10^1	2.6837×10^1
$\lvert\mathbf{f}\rvert = 10^3$	1.9708×10^3	2.2132×10^3	2.4952×10^3	2.4913×10^3
$\lvert\mathbf{f}\rvert = 10^4$	7.9906×10^4	8.1150×10^4	8.5974×10^4	8.7090×10^4

the optimal topology is symmetrical (Fig. 3.6); when the Coriolis force along with the centrifugal force is considered, the symmetry is broken up by the asymmetrical distribution of the composed force of the centrifugal and Coriolis forces (Fig. 3.7). The symmetry and asymmetry characteristics of the optimal topologies in Figs. 3.6 and 3.7 are both caused by the maximization of the work done by the body forces, which can be increased by achieving the consistency between the directions of the body force and fluidic velocity. For the flows driven by the centrifugal and Coriolis forces corresponding to different scalar values of the angular velocity, the optimal topologies of the short channel are shown in Fig. 3.8.

The topology optimization of the short channel demonstrates that the optimal topology of the steady flow is influenced by the type of the body force.

Unsteady flows driven by body forces

For the topology optimization of the unsteady flows driven by body forces, a four-terminal flow is used to investigate the feasibility of the optimization method. The design domain is shown in Fig. 3.9a, which is discretized by 120×120 rectangular elements. The optimization parameter values are listed in Table 3.6. When the gravity $\mathbf{f} = (0, -10)$ is loaded on the unsteady flow in the design domain for the time interval

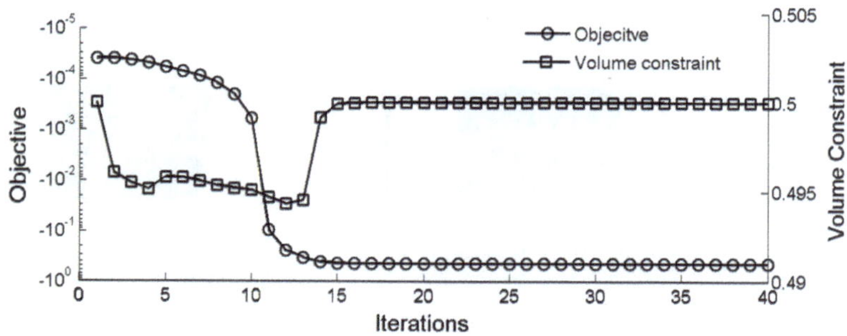

Fig. 3.4 Convergent histories of the reverse objective values and the volume constraint for the optimal topology in Fig. 3.3a. The CPU time cost is 0.054 h

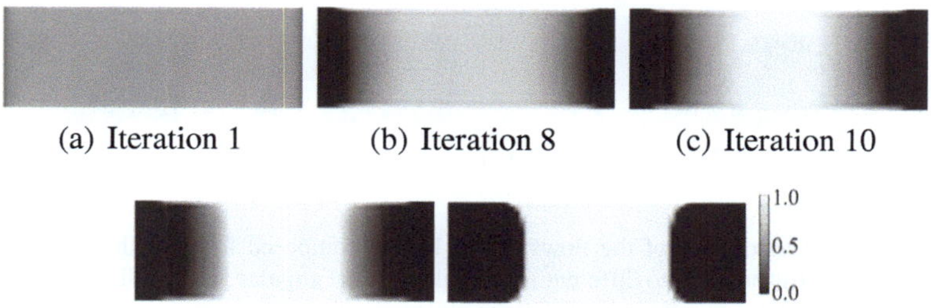

(a) Iteration 1 (b) Iteration 8 (c) Iteration 10

(d) Iteration 15 (e) Iteration 40

Fig. 3.5 Snapshots of the evolving progress for the optimal topology of the short channel in Fig. 3.3a

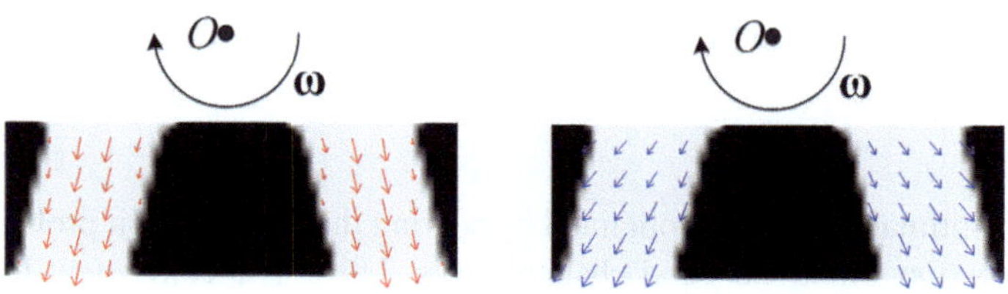

(a) Velocity vectors (b) Centrifugal force

Fig. 3.6 Optimal topology of the flow driven by only centrifugal force. **a** The velocity vector distribution in the optimal topology; **b** the centrifugal force distribution in the optimal topology. O is the rotating center at $(1/2, 2/3)$; $\boldsymbol{\omega}$ is the angular velocity in clockwise direction, and the corresponding scalar value is 10

Table 3.5 Cross-comparison of the objective values corresponding to the optimal topologies in Figs. 3.6 and 3.7

	Figure 3.6	Figure 3.7
Centrifugal force	2.8910×10^0	1.0423×10^0
Composed force	2.8841×10^0	1.1036×10^1

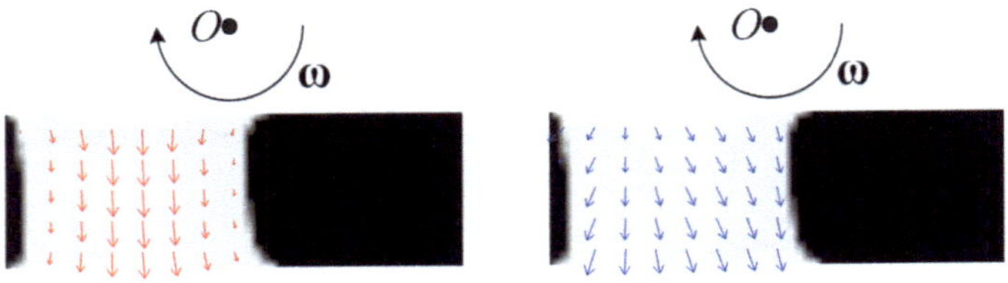

(a) Velocity vectors (b) Centrifugal-Coriolisforce

Fig. 3.7 Optimal topology of the flow driven by the composed force of the centrifugal and Coriolis forces. **a** The velocity vector distribution in the optimal topology; **b** the composed force distribution in the optimal topology. O is the rotating center at $(1/2, 2/3)$; $\boldsymbol{\omega}$ is the angular velocity in clockwise direction, and the corresponding scalar value is 10

(a) $|\omega| = 1$ (b) $|\omega| = 10$ (c) $|\omega| = 100$

Fig. 3.8 Optimal topologies of the flows driven by the composed force of the centrifugal and Coriolis forces corresponding to different scalar values of the angular velocity. The rotating center is at $(1/2, 2/3)$, and the angular velocity $\boldsymbol{\omega}$ is in clockwise direction

Table 3.6 Parameter settings in the topology optimization of the four-terminal flow

β_1	β_2	V_r	α_{min}	α_{max}	q	n
1	0	0.25	0	1×10^4	1	3

[0, 1], the optimal topology is obtained by solving the optimization problem, where there are two symmetrical bend channels (Fig. 3.9b). The convergent histories of the objective values and volume constraint are plotted in Fig. 3.10. The fluidic field stabilization progress is demonstrated by the velocity vector distribution at different time in the optimal four-terminal flow (Fig. 3.11). Figure 3.11 also shows that the flow bends to the direction of the gravity to achieve the consistency between the directions of the gravity and the velocity vectors and increase the work done by the gravity.

Furthermore, the four-terminal flow is located in a rotating system. The rotating center is at the point $(-0.5, 0.5)$. The scalar value of the angular velocity $\boldsymbol{\omega}$ is $30 \sin(t)$ in the clockwise direction, where the time interval is $[0, \pi]$. Because the angular velocity is time dependent, the term $\mathbf{r} \times \frac{\mathrm{d}\boldsymbol{\omega}}{\mathrm{d}t}$ should be included in the body force, i.e. the body force is $\mathbf{f} = \rho (\boldsymbol{\omega} \times \mathbf{r}) \times \boldsymbol{\omega} + 2\rho \mathbf{u} \times \boldsymbol{\omega} + \mathbf{r} \times \frac{\mathrm{d}\boldsymbol{\omega}}{\mathrm{d}t}$ in the rotating frame with time dependent angular velocity [13]. As the Coriolis force is omitted, the optimal topology is shown in Fig. 3.12a; as the centrifugal force and the Coriolis force are considered simultaneously, the optimal topology is shown in Fig. 3.12b.

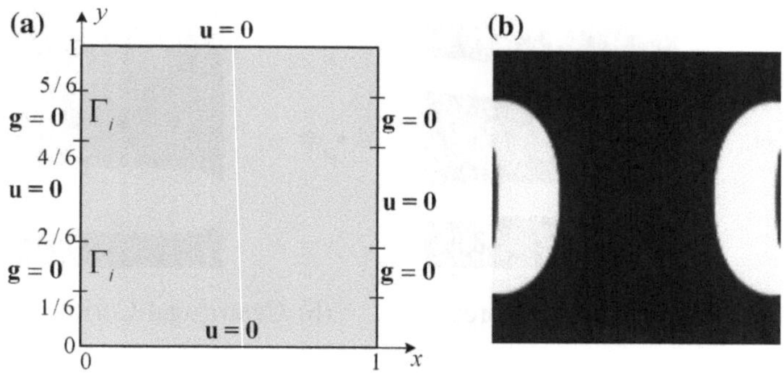

Fig. 3.9 **a** Design domain of the four-terminal flow, where the inlets and the outlets are set to be open boundary, and the other boundary are set to be no slip type; **b** optimal topology of the unsteady four-terminal flow driven by the gravity $\mathbf{f} = (0, -10)$. The objective value corresponding to the optimal topology is 4.1976×10^{-2}. The CPU time cost is 1.42 h

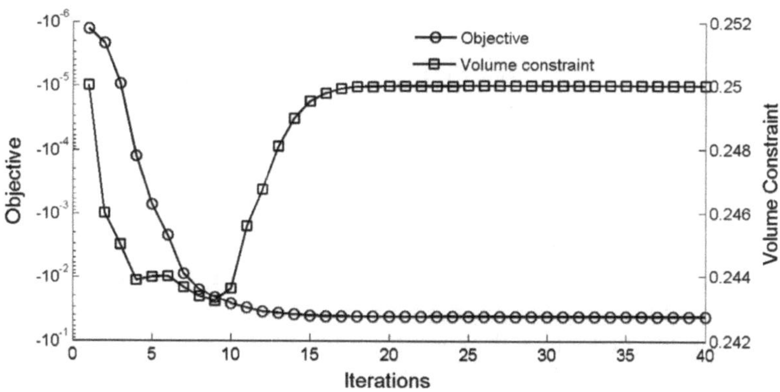

Fig. 3.10 Convergent histories of the reverse objective values and the volume constraint for the optimal topology in Fig. 3.9b

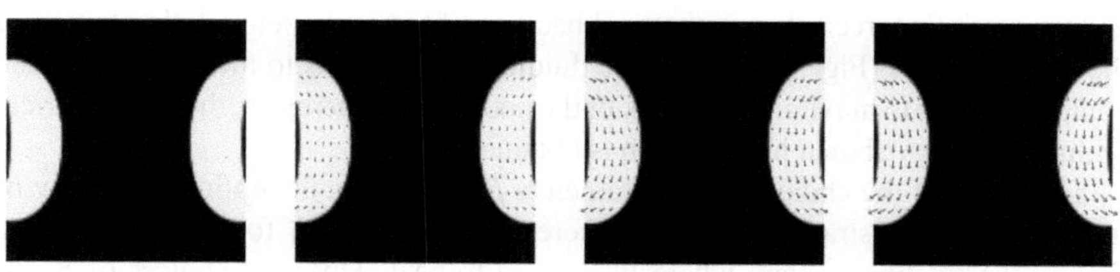

(a) $t = 0$ (b) $t = 0.02$ (c) $t = 0.04$ (d) $t \in (0.06, 1)$

Fig. 3.11 Velocity distribution in the optimal topology of the unsteady four-terminal flow driven by the gravity, where the flow reaches the steady state after $t = 0.06$

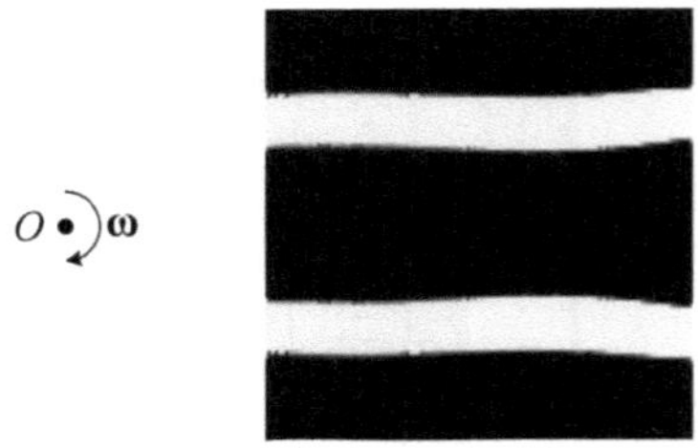

(a) Only centrifugal force

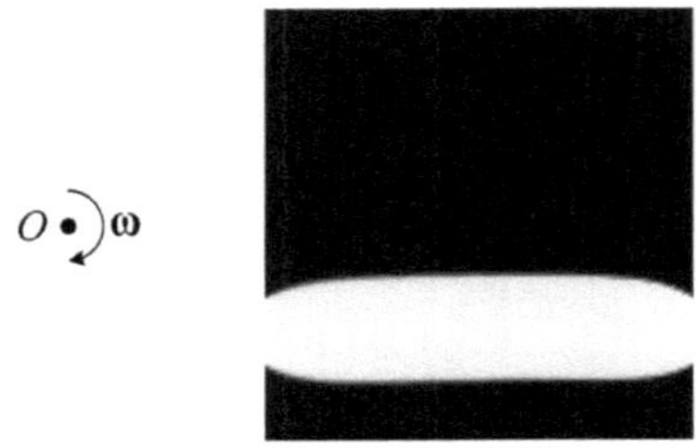

(b) Centrifugal-Coriolis force

Fig. 3.12 **a** Optimal topology of the four-terminal flow in the rotating system omitting the Coriolis force; **b** optimal topology of the four-terminal flow in the rotating system considering both the centrifugal force and the Coriolis force. O is the rotating center at $(-1/2, 1/2)$; $\boldsymbol{\omega}$ is the angular velocity with scalar value equal to $30\sin(t)$ $(t \in [0, \pi])$. The CPU time cost for these two results are 3.57 h and 3.76 h, respectively

Table 3.7 Cross-comparison of the objective values corresponding to the optimal topologies in Fig. 3.12a and b

	Figure 3.12a	Figure 3.12b
Centrifugal force	6.2635×10^2	4.3723×10^2
Composed force	6.2562×10^2	7.1377×10^2

The cross-comparison of the objective values corresponding to the optimal topologies in Fig. 3.12 is shown in Table 3.7. Because the rotating center is on the symmetrical axis of the design domain, the centrifugal force has a symmetrical distribution. Therefore, the symmetrical centrifugal force induces the symmetrical distribution of the fluidic field (Fig. 3.13) and the symmetrical optimal topology (Fig. 3.12a) is helpful to maximize the work done by the centrifugal force, when the Coriolis force is omitted. As the Coriolis force is considered, the composition of the centrifugal force and the Coriolis force are asymmetrical because of the asymmetrical distribution of the Coriolis force (Fig. 3.14). Then the fluidic velocity leans to the direction of the composed force to maximize the work of the body force. Therefore, the asymmetrical optimal topology shown in Fig. 3.12b is obtained.

The effect of the choice of the parameters β_1 and β_2 on the optimal topology of the flows is demonstrated by setting different values of β_2/β_1 for the four terminal flow in the rotating system, where β_1 is fixed to be 1. Different choices of β_2/β_1 mean different importance between the work done by the body forces and the flux at the inlet, and this influences the optimal topology of the four-terminal flow as shown in Fig. 3.15.

The topology optimization of the four-terminal flow demonstrates that the optimal topology of the unsteady flow is influenced by the type of the body force.

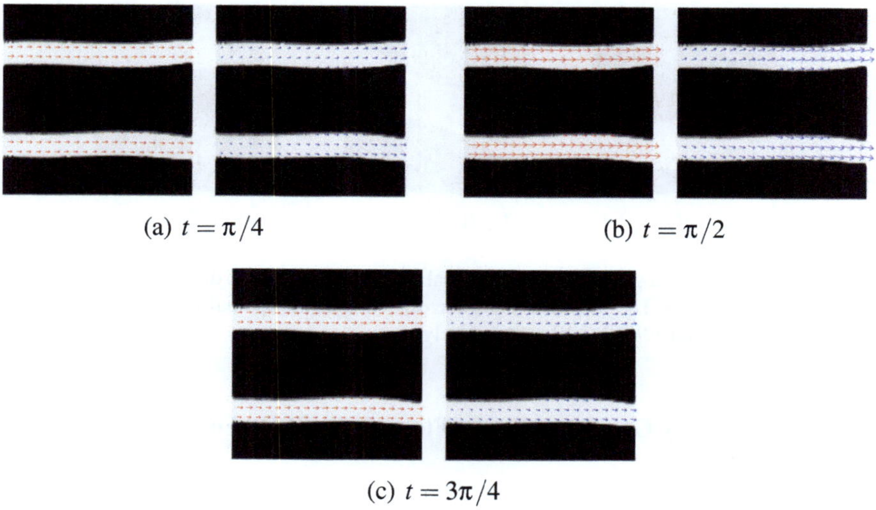

Fig. 3.13 Velocity and body force distribution in the optimal topology of the unsteady four-terminal flow driven by the centrifugal force, where the *red* and *blue arrows* represent the velocity and body force vectors respectively

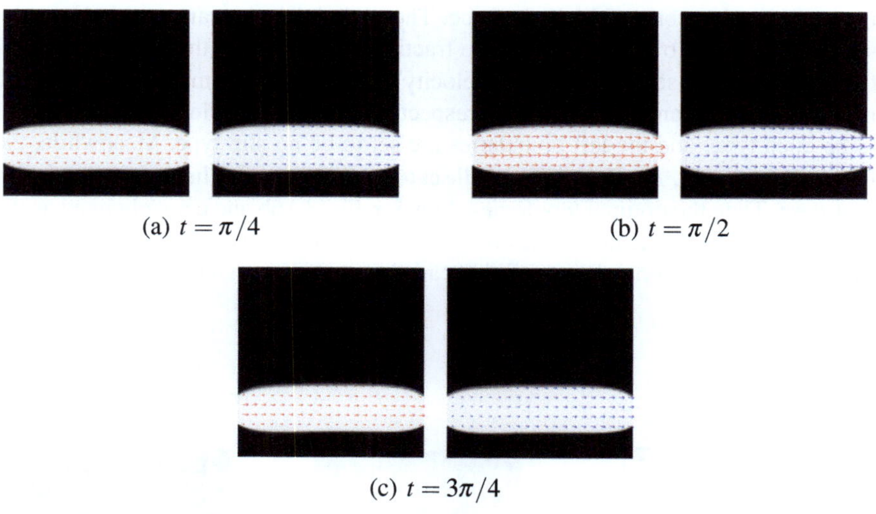

Fig. 3.14 Velocity and body force distribution in the optimal topology of the unsteady four-terminal flow driven by the composed force of the centrifugal and Coriolis forces, where the *red* and *blue arrows* represent the velocity and body force vectors respectively

(a) $\beta_2/\beta_1 = 0.1$ (b) $\beta_2/\beta_1 = 50$ (c) $\beta_2/\beta_1 = 90$

Fig. 3.15 Optimal topologies of the four-terminal flow corresponding to different values of β_2/β_1, where β_1 is fixed to be 1. The objective values corresponding to the optimal topologies are 7.1506×10^2, 5.2902×10^2 and 7.9518×10^2, respectively

3.1.4.2 Boundary Condition Driven Flow Versus Body Force Driven Flow

To demonstrate the difference of the optimal topologies corresponding to the boundary condition driven flow and body force driven flow, one steady flow in the design domain shown in Fig. 3.16a is investigated in this section. The design domain is discretized by 100×100 rectangular elements. The optimization objective is chosen to maximize the work of the body force. The optimization parameter choice is the same as in Table 3.6 except the volume fraction $V_r = 0.40$. In the boundary condition driven flow case, two parabolic velocity distribution with maximal value 1 are imposed on the boundaries Γ_1 and Γ_2 respectively; the boundaries Γ_3 and Γ_4 are set to be open type; and the left boundaries are set to be no slip type. Meanwhile, the four boundaries Γ_1, Γ_2, Γ_3 and Γ_4 are all set to be open type for the body force driven flow case. Then the optimal topologies shown in Fig. 3.16b and c are obtained as the gravity is loaded on the fluid. The comparison between the two results in Fig. 3.16b and c shows that the boundary condition can change the optimal topology of the flow with body force.

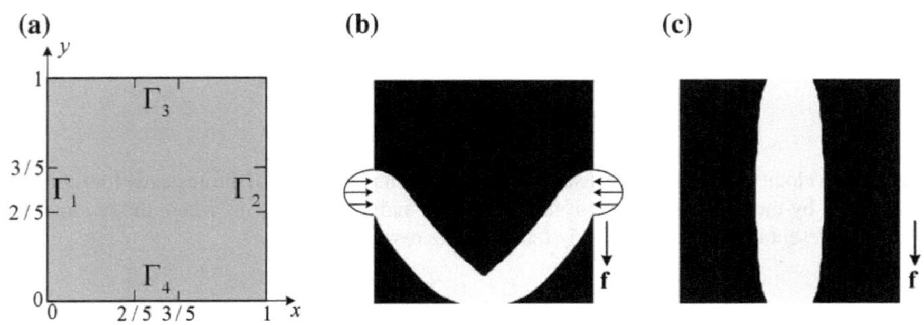

Fig. 3.16 **a** Design domain of the flow used to demonstrate the difference of the optimal topologies corresponding to the boundary condition driven flow and body force driven flow; **b** optimal topology of the flow with parabolic velocity distribution on the boundaries Γ_1 and Γ_2; **b** optimal topology of the flow driven by body force. The body force is the gravity $\mathbf{f} = (0, -10)$

3.1.4.3 Topology Optimization for Mass Distribution Problems in Flows Driven by Body Forces

The topology optimization for the mass distribution of flows without body forces has been discussed in [14, 15]. The mass distribution of flows can be performed by adding the flux constraint into the topology optimization problem. The flux constraint is implemented as:

$$\left(\frac{\int_{\Gamma_{ok}} \mathbf{u} \cdot \mathbf{n} \, \mathrm{d}\Gamma}{\int_{\Gamma_i} -\mathbf{u} \cdot \mathbf{n} \, \mathrm{d}\Gamma} - R_k \right)^2 \leq \varepsilon \tag{3.19}$$

for the steady flow, and

$$\left(\frac{\int_{T_1}^{T_2} \int_{\Gamma_{ok}} \mathbf{u} \cdot \mathbf{n} \, \mathrm{d}\Gamma \mathrm{d}t}{\int_{T_1}^{T_2} \int_{\Gamma_i} -\mathbf{u} \cdot \mathbf{n} \, \mathrm{d}\Gamma \mathrm{d}t} - R_{k[T_1,T_2]} \right)^2 \leq \varepsilon \tag{3.20}$$

for the unsteady flow, where Γ_i and Γ_{ok} are the inlet and the k-th outlet of the flow, respectively; R_k and $R_{k[T_1,T_2]}$ are the ratios of the flux between the k-th outlet and the inlet for the steady and unsteady flows, respectively; ε is the tolerance specified by the designer; $[T_1, T_2] \subseteq [0, T]$ is the time interval. The objective is chosen to maximize the work of the body forces by setting:

$$A = \mathbf{f} \cdot \mathbf{u}, \text{ in } \Omega; \; B = 0, \text{ on } \partial\Omega \tag{3.21}$$

Based on the above flux constraint, a splitter is optimized for the flow driven by body forces. Splitter is a device used to achieve the mass distribution at the outlets of the flow. For the splitter with steady flow driven by the body forces, the design domain is in Fig. 3.17a, where there are three outlets Γ_{o1}, Γ_{o2} and Γ_{o3}. The design domain is discretized by 70×70 rectangular elements. The ratios of the flux are set to be $R_1 = 0.6$, $R_2 = 0.2$ and $R_3 = 0.2$; the tolerance ε is set to be 1×10^{-4}. Then according to Eq. 3.19, the flux constraints are formulated as:

$$\begin{aligned} &\left(\frac{\int_{\Gamma_{o1}} \mathbf{u} \cdot \mathbf{n} \, \mathrm{d}\Gamma}{\int_{\Gamma_i} -\mathbf{u} \cdot \mathbf{n} \, \mathrm{d}\Gamma} - 0.6 \right)^2 \leq 1 \times 10^{-4}, \\ &\left(\frac{\int_{\Gamma_{o2}} \mathbf{u} \cdot \mathbf{n} \, \mathrm{d}\Gamma}{\int_{\Gamma_i} -\mathbf{u} \cdot \mathbf{n} \, \mathrm{d}\Gamma} - 0.2 \right)^2 \leq 1 \times 10^{-4}, \\ &\left(\frac{\int_{\Gamma_{o3}} \mathbf{u} \cdot \mathbf{n} \, \mathrm{d}\Gamma}{\int_{\Gamma_i} -\mathbf{u} \cdot \mathbf{n} \, \mathrm{d}\Gamma} - 0.2 \right)^2 \leq 1 \times 10^{-4} \end{aligned} \tag{3.22}$$

With the parameter setting as listed in Table 3.9, the topology optimization problem with flux constraints is solved. The optimal topologies for the flow driven by the gravity or the centrifugal and Coriolis forces are obtained as shown in Fig. 3.17b

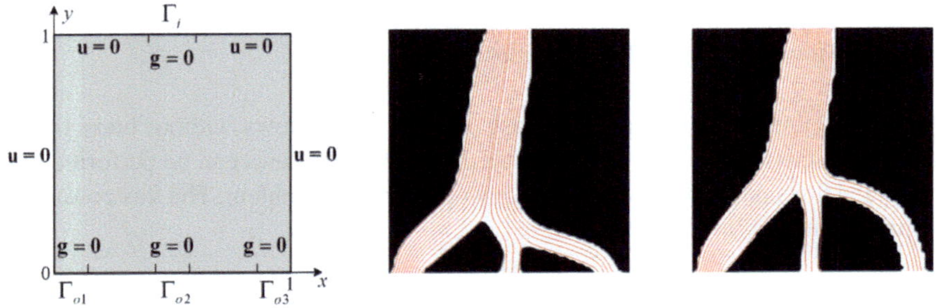

(a) Design domain (b) Gravity (c) Centrifugal-Coriolis forces

Fig. 3.17 **a** Design domain of the splitter for the steady flow driven by body forces; **b** optimal topology of the splitter for the flow driven by the gravity $\mathbf{f} = (0, -10)$; **c** optimal topology of the splitter for the flow driven by the centrifugal and Coriolis forces, where the rotating center is set to be the point $(1/2, 2)$ and the scalar value of the angular velocity in the clockwise direction is 30

and c, respectively. For the splitter with centrifugal and Coriolis forces, the rotating center is set to be the point $(1/2, 2)$ and the scalar value of the angular velocity in the clockwise direction is 30. The streamlines of the flows in Fig. 3.17b and c show the mass distribution at the three outlets of the splitter.

For the unsteady flow driven by body forces, the splitter is optimized to achieve the angular-velocity independent ratios of the flux between the outlets and the inlet. The unsteady flow is driven by the centrifugal and Coriolis forces with time dependent angular velocity. The design domain for the splitter is shown in Fig. 3.18a, which has four outlets. The design domain is discretized by 70×70 rectangular elements. The time dependent scalar value of the clockwise angular velocity $\boldsymbol{\omega}$ is shown in Fig. 3.18b, where the considered time interval is $[0, 1]$ and the angular velocity is increased in the time interval $[0.4, 0.6]$. The rotating center is set to be the point $(1/2, 6)$. In order to achieve the angular-velocity independent ratios of the flux, the ratios of the flux should be the same value no matter for the low or high angular velocity. As shown in Fig. 3.18b, the time intervals corresponding to the low and high angular velocity are $[0, 0.4]$ and $[0.6, 1]$, respectively. So the flux ratios should satisfy $R_{k[0,0.4]} = R_{k[0.6,1]}$ $(k = 1, 2, 3, 4)$. By setting $\varepsilon = 1 \times 10^{-4}$, the flux constraints added into the topology optimization problem of the splitter are:

$$
\begin{aligned}
&\left(\frac{\int_0^{0.4} \int_{\Gamma_{ok}} \mathbf{u} \cdot \mathbf{n} \, \mathrm{d}\Gamma \mathrm{d}t}{\int_0^{0.4} \int_{\Gamma_i} -\mathbf{u} \cdot \mathbf{n} \, \mathrm{d}\Gamma \mathrm{d}t} - R_{k[0,0.4]} \right)^2 \leq 1 \times 10^{-4}, \\
&\left(\frac{\int_{0.6}^{1} \int_{\Gamma_{ok}} \mathbf{u} \cdot \mathbf{n} \, \mathrm{d}\Gamma \mathrm{d}t}{\int_{0.6}^{1} \int_{\Gamma_i} -\mathbf{u} \cdot \mathbf{n} \, \mathrm{d}\Gamma \mathrm{d}t} - R_{k[0.6,1]} \right)^2 \leq 1 \times 10^{-4}; \; k = 1, 2, 3, 4
\end{aligned}
\tag{3.23}
$$

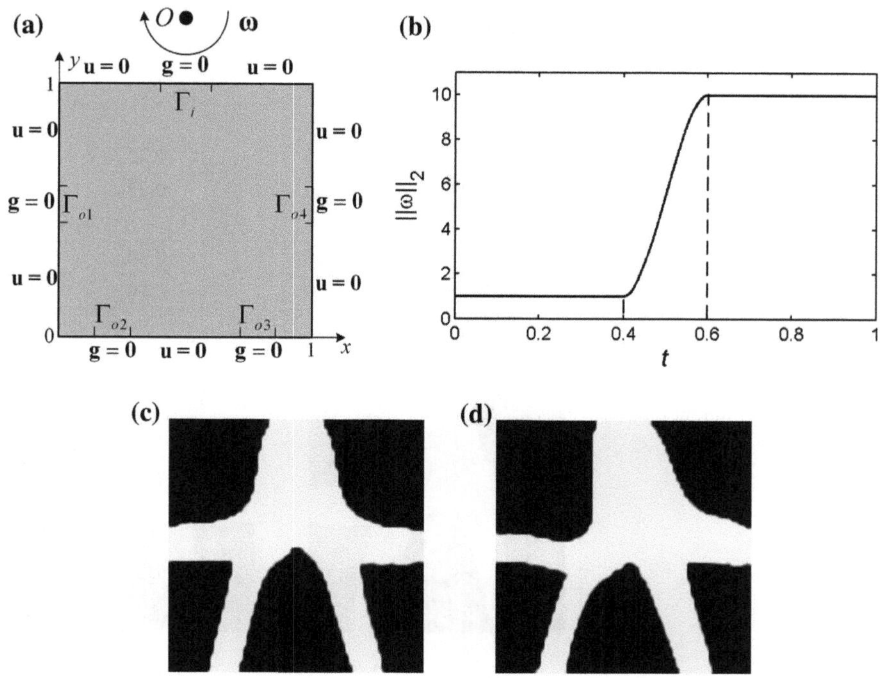

Fig. 3.18 **a** Design domain of the splitter for the unsteady flow driven by centrifugal and Coriolis forces, where the rotating center O is set to be the point $(1/2, 6)$ and the angular velocity is in the clockwise direction; **b** the time dependent scalar value of the angular velocity; **c** optimal topology corresponding to the ratios listed in Table 3.8a; **d** optimal topology corresponding to the ratios listed in Table 3.8b. The CPU time cost for the results in **c** and **d** are 4.15 h and 4.09 h, respectively

The other parameters are set as listed in Table 3.10. When the flux ratios are set as listed in Table 3.8a, the optimal topology of the splitter is obtained as shown in Fig. 3.18c; when the flux ratios are set as listed in Table 3.8b, the optimal topology of the splitter is obtained as shown in Fig. 3.18d. From the distribution of the streamlines in the optimal topologies at different time (Figs. 3.19 and 3.20), the angular-velocity independent ratios of the flux between the outlets and the inlet of the splitter can be confirmed.

3.1.4.4 Topology Optimization for Local Velocity Control in Flows Driven by Body Forces

For the topology optimization of flows with local velocity control, the objective is set to maximize the functional in Eqs. 3.10 or 3.11 with A and B as [3]:

$$A = \delta\left(dist\left(\mathbf{x}, l\right)\right) \mathbf{u} \cdot \mathbf{s}_l, \text{ in } \Omega; \; B = 0, \text{ on } \partial\Omega \tag{3.24}$$

Table 3.8 List of the flux ratios between the k-th outlet and the inlet for the unsteady flows in the design domain shown in Fig. 3.18a

	Γ_{o1}	Γ_{o2}	Γ_{o3}	Γ_{o4}
(a)				
$R_{k[0,0.4]}$	0.25	0.25	0.25	0.25
$R_{k[0.6,1]}$	0.25	0.25	0.25	0.25
(b)				
$R_{k[0,0.4]}$	0.10	0.20	0.30	0.40
$R_{k[0.6,1]}$	0.10	0.20	0.30	0.40

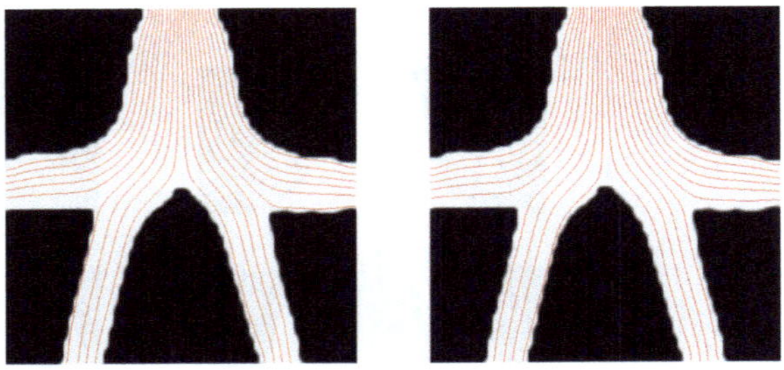

(a) $t = 0.2$ (b) $t = 0.8$

Fig. 3.19 Distribution of the streamlines at different time in the optimal topology shown in Fig. 3.18c

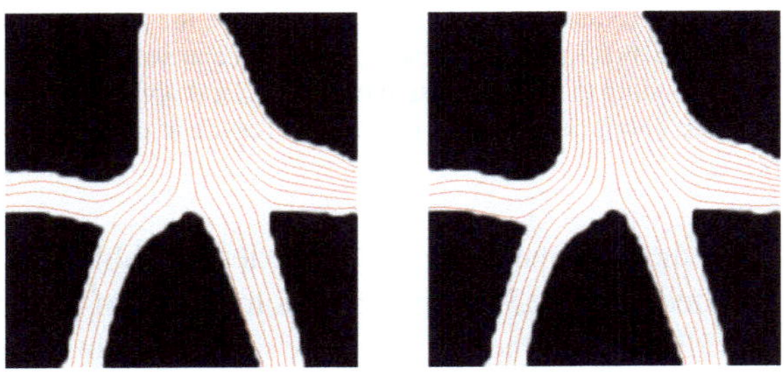

(a) $t = 0.2$ (b) $t = 0.8$

Fig. 3.20 Distribution of the streamlines at different time in the optimal topology shown in Fig. 3.18d

where $\delta(\cdot)$ is the Dirac function; $dist(\mathbf{x}, l)$ is the L_2 distance between the point $\forall \mathbf{x} \in \Omega$ and the line section l; the unit vector $\mathbf{s}_l$ represents the desired fluidic velocity direction on the line section l. Then the minimization of the objective functional means the consistency between the direction of the velocity on l and the direction of $\mathbf{s}_l$. The parameters β_1 is set to be 1. In order to ensure the smoothness of the fluidic channel, the constraint on the work of body forces is added into the topology optimization problem as:

$$\int_\Omega \mathbf{f} \cdot \mathbf{u} \, \mathrm{d}\Omega \geq W_0 \tag{3.25}$$

for the steady flow, and

Table 3.9 Parameter settings in the topology optimization of the splitter for the steady flows driven by the gravity or the centrifugal and Coriolis forces

β_1	β_2	V_r	α_{min}	α_{max}	q	n
1	0	0.30	0	1×10^4	1	3

Table 3.10 Parameter settings in the topology optimization of the splitter independent of the angular velocity in the centrifugal-Coriolis force field

β_1	β_2	V_r	α_{min}	α_{max}	q	n
1	0	0.40	0	1×10^4	1	3

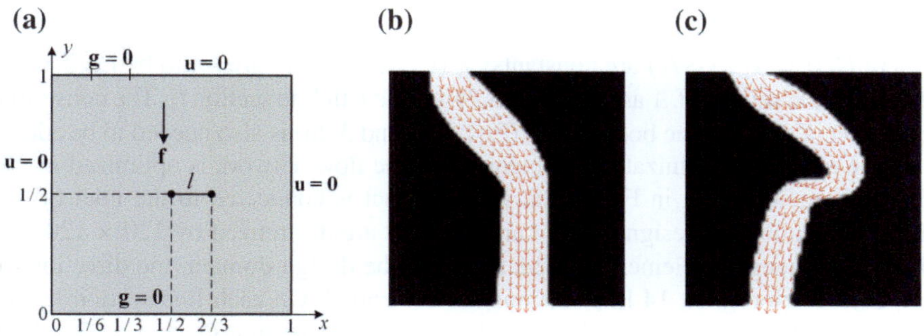

Fig. 3.21 **a** Design domain of the flow with local velocity control, where the desired direction $\mathbf{s}_l$ of the fluidic velocity on the line section l is the downward unit normal vector or leftward unit tangential vector of l; **b** optimal topology of the flow driven by the gravity with velocity control in the direction of the downward $\mathbf{s}_l$; **c** optimal topology of the flow driven by the gravity with velocity control in the direction of the leftward $\mathbf{s}_l$. The body force is the gravity $\mathbf{f} = (0, -10)$

$$\int_0^T \int_\Omega \mathbf{f} \cdot \mathbf{u} \, \mathrm{d}\Omega \mathrm{d}t \geq W_0 \tag{3.26}$$

for the unsteady flow, where W_0 is the low bound for the work done by body forces.

To demonstrate the method of local velocity control for the flow driven by body forces, the topology optimization of a steady flow driven by the gravity is implemented. The design domain is shown in Fig. 3.21a, which is discretized by 60×60 elements. The parameters are set as listed in Table 3.11. When the desired fluidic velocity direction on the line section l is the downward unit normal vector of l, the optimal topology is obtained as shown in Fig. 3.21b; when the desired direction is changed to be the leftward unit tangential vector of l, the corresponding optimal topology is shown in Fig. 3.21c. The distribution of the velocity vectors in Fig. 3.21b and c show that the local velocity control on the line section l is achieved by the reasonable bending of the fluidic channel.

Based on the method of local velocity control for the flow driven by body forces, the topology optimization of the flow networks can be performed. The flow networks

Table 3.11 Parameter settings in the topology optimization of the flow driven by the gravity with local velocity control on the line section l as shown in Fig. 3.21a

β_1	β_2	V_r	α_{min}	α_{max}	q	n	W_0
1	0	0.40	0	1×10^4	1	3	6×10^{-3}

have been designed by the topology optimization method for the steady flow without body forces in [15]. The topology optimization of networks for the flow driven by body forces can be implemented by minimizing the objective functional:

$$\max\left(C_1 J_1, C_2 J_2 \cdots C_N J_N\right) \tag{3.27}$$

where C_i $(i = 1, 2 \cdots N)$ are constants; J_i $(i = 1, 2 \cdots N)$ are set to be Eq. 3.10 or 3.11 with the setting of A and B in Eq. 3.24 for the i-th line section l_i. The constraint on the work done by the body forces (Eqs. 3.25 and 3.26) is also needed to be added into the topology optimization problem. Then the flow network is optimized in the design domain shown in Fig. 3.22a, where a duct is connected to the inlet of the design domain. The design domain and the duct are discretized by 120×120 and 40×40 rectangular elements, respectively. In the design domain, the direction of the fluidic velocity on 14 line sections are controlled. On each line section l_i, the desired fluidic velocity is the downward unit normal direction $\mathbf{n}_{l_i}$ (Fig. 3.22a). The constants C_i $(i = 1, 2 \cdots 14)$ are set to be $C_1 = C_2 = 1$, $C_3 = C_4 = C_5 = 1.5$, $C_6 = C_7 = C_8 = C_9 = 2$ and $C_{10} = C_{11} = C_{12} = C_{13} = C_{14} = 2.5$, respectively.

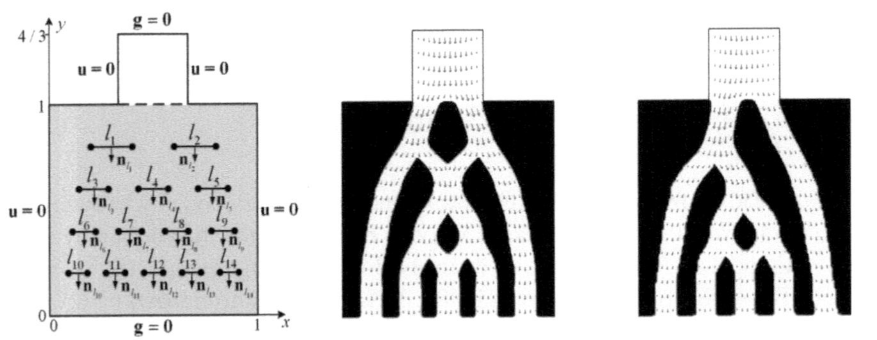

(a) Design domain (b) Gravity (c) Centrifugal and Coriolis forces

Fig. 3.22 **a** Design domain of the flow network; l_i $(i = 1, 2 \cdots 14)$ are the line sections; $\mathbf{n}_{l_i}$, the downward unit normal vector of l_i, is the desired direction of the velocity on l_i; **b** optimal topology of the network for the flow driven by the gravity $\mathbf{f} = (0, -10)$); **c** optimal topology of the network for the flow driven by the centrifugal and Coriolis forces, where the rotating center is at (1/2, 2) and the angular velocity with scalar value equal to 30 is in the clockwise direction. In **b** and **c**, the *red arrows* represents the distribution of the velocity vectors

When the flow is driven by the gravity, the symmetrical optimal topology of the network is shown in Fig. 3.22b, where the parameters are set to be the same as in Table 3.11. However, the optimal topology of the network is changed to be asymmetrical, when the flow is driven by the centrifugal and Coriolis forces (Fig. 3.22c). And the asymmetry is induced by the asymmetrical distribution of the composed force of the centrifugal and Coriolis forces as has been discussed in Sect. 3.1.4.1. For the optimal topology in Fig. 3.22c, the rotating center is set to be the point with coordinates $(1/2, 2)$; the scalar value of the clockwise angular velocity is 30; and the other parameters are set as listed in Table 3.11, except $W_0 = 5 \times 10^2$.

3.2 Level Set Method-Based Topology Optimization for Flow with Body Force

3.2.1 *Topology Optimization Problem for Flows with Body Forces*

In order to express the solid-liquid boundary implicitly, the incompressible Navier-Stokes equations can be coupled with the level set function that is expressed by a signed distance function ϕ defined on an optimization domain $\Omega \subset \mathbb{R}^d \to \mathbb{R}$ ($d = 2$ or 3 is the spatial dimension, and Ω is open). The solid and fluid regions can be distinguished as

$$\begin{cases} \phi(\mathbf{x}) > 0, & \forall \mathbf{x} \in \Omega_s \\ \phi(\mathbf{x}) < 0, & \forall \mathbf{x} \in \Omega_l \\ \phi(\mathbf{x}) = 0, & \forall \mathbf{x} \in \Gamma \end{cases} \tag{3.28}$$

where Ω_s, Ω_l and Γ are the solid region, fluid region and implicit boundary respectively; Ω_s and Ω_l are open, and they satisfy $\overline{\Omega}_s \cup \overline{\Omega}_l = \overline{\Omega}$ (Fig. 3.23). Then the incompressible Navier-Stokes equations can be modified by constraining the fluid velocity to be zero in the solid region Ω_s:

$$\begin{aligned} -\eta \Delta \mathbf{u} + \rho (\mathbf{u} \cdot \nabla) \mathbf{u} + \nabla p = \mathbf{f} \quad & \text{in } \Omega \\ -\nabla \cdot \mathbf{u} = 0 \quad & \text{in } \Omega \\ H(\phi) \mathbf{u} = \mathbf{0} \quad & \text{in } \Omega \end{aligned} \tag{3.29}$$

where η is the dynamic viscosity; ρ is the density of the fluid; $\mathbf{u}$ is the fluid velocity; p is the fluid pressure; and $\mathbf{f}$ is the body force. Particularly, if a domain $\Omega_c \subset \Omega$ (Ω_c is open) needs to be a solid region throughout an optimization procedure, one can constrain the fluid velocity to be zero in Ω_c (Fig. 3.23)

$$\mathbf{u} = \mathbf{0} \quad \text{in } \Omega_c \tag{3.30}$$

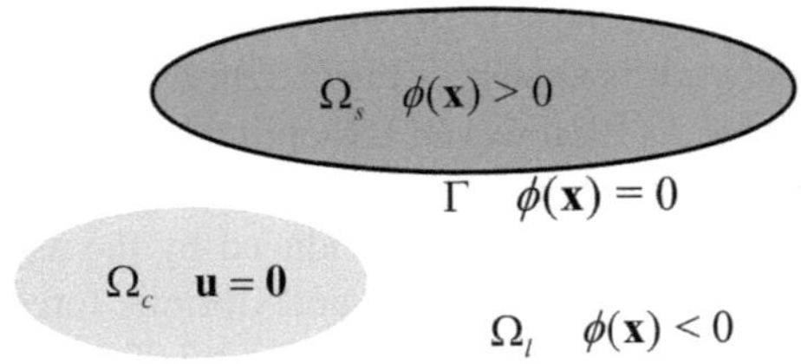

Fig. 3.23 Schematic for the regions distinguished based on the level set function

In the above, $H(\phi)$ is the Heaviside function [16]:

$$H(\phi) = \begin{cases} 1, & \phi \geq 0 \\ 0, & \phi < 0 \end{cases} \tag{3.31}$$

and the derivative of $H(\phi)$ to ϕ is the Dirac function $\tau(\phi)$. In most cases, the Heaviside function and the Dirac function need to be regularized from the numerical implementation point of view. $H(\phi)$ and $\tau(\phi)$ are approximated by the smoothed Heaviside function and Dirac function [11]:

$$H(\phi) = \begin{cases} 0, & \phi \leq -h \\ \dfrac{1}{2} + \dfrac{15\phi}{16h} - \dfrac{5\phi^3}{8h^3} + \dfrac{3\phi^5}{16h^5}, & -h < \phi \leq h \\ 1, & \phi > h \end{cases} \tag{3.32}$$

$$\tau(\phi) = \begin{cases} \dfrac{15}{16h}\left(1 - \dfrac{\phi^2}{h^2}\right)^2, & |\phi| \leq h \\ 0, & |\phi| > h \end{cases} \tag{3.33}$$

where h is the support size (Fig. 3.24). For more details on the regularization of the Heaviside function, one can refer to [17–19]. The support size h of the smoothed Heaviside function and the Dirac function is kept to be constant during the optimization procedure [20–23].

Usually, fluid flows have a known velocity or pressure distribution at the inlet boundary. For the velocity boundary case, the given velocity distribution $\mathbf{u}_{in}$ imposed on the inlet Γ_{in} of the computational domain Ω is

$$\mathbf{u} = \mathbf{u}_{in} \quad \text{on } \Gamma_{in} \tag{3.34}$$

where Γ_{in} belongs to the Dirichlet boundary Γ_D. For the pressure boundary case, the given pressure distribution p_0 with the zero viscous stress of flow is imposed on the inlet Γ_{in} of the computational domain Ω

$$\begin{aligned} p &= p_0 \quad \text{on } \Gamma_{in} \\ \eta\left(\nabla \mathbf{u} + \nabla \mathbf{u}^{\mathrm{T}}\right) \cdot \mathbf{n} &= \mathbf{0} \quad \text{on } \Gamma_{in} \end{aligned} \tag{3.35}$$

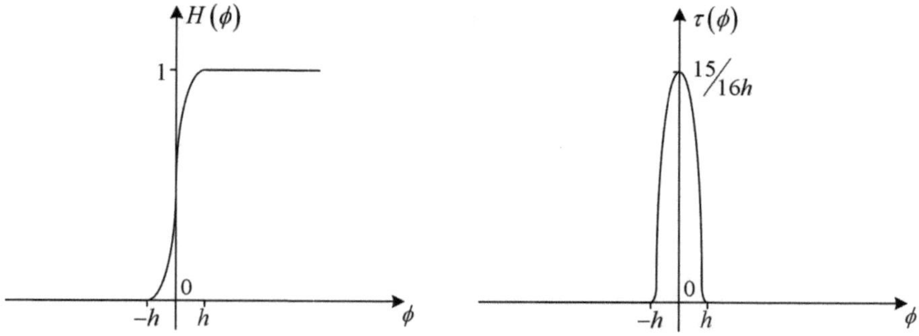

(a) Heaviside function (b) Dirac function

Fig. 3.24 Smoothed Heaviside function and Dirac function

In this case, the pressure condition on Γ_{in} belongs to the Dirichlet boundary condition, and the zero viscous stress condition on Γ_{in} belongs to the Neumann boundary condition. For both the velocity and pressure inlet boundary cases, the outlet boundary condition is usually set to be:

$$\left[-p\mathbf{I} + \eta\left(\nabla\mathbf{u} + \nabla\mathbf{u}^{\mathrm{T}}\right)\right] \cdot \mathbf{n} = \mathbf{g} \quad \text{on } \Gamma_{out} \tag{3.36}$$

where $\mathbf{g} = \mathbf{0}$ corresponds to the open boundary condition and Γ_{out} belongs to the Neumann boundary Γ_N. For the case that the flow is driven by the body force, the open boundary condition expressed by Eq. 3.36 can be imposed on both the inlet and outlet. The other boundaries of the computational domain Ω are no-slip type where $\mathbf{u} = \mathbf{0}$.

The optimization objective is chosen according to the description of the desired result and characteristics of the forward physical problems. For the flow with a fixed velocity distribution at the inlet, minimizing the dissipated power is equivalent to minimizing the pressure drop across the fluid system. Therefore, the characteristic of the flow can be measured by the dissipation power [2, 22, 24]

$$J_d\left(\nabla\mathbf{u}; \phi\right) = \int_\Omega \frac{1}{2}\eta H\left(-\phi\right)\left(\nabla\mathbf{u} + \nabla\mathbf{u}^{\mathrm{T}}\right) : \left(\nabla\mathbf{u} + \nabla\mathbf{u}^{\mathrm{T}}\right) \mathrm{d}\Omega \tag{3.37}$$

For the flow with a fixed pressure distribution at the inlet, the optimization of the flow can be preformed by maximizing the kinetic energy stored in the fluid system. Then the reverse of the kinetic energy can be chosen to be the objective

$$J_k(\mathbf{u}; \phi) = -\int_\Omega \frac{1}{2}\rho H\left(-\phi\right)\mathbf{u}^2\,\mathrm{d}\Omega \tag{3.38}$$

For the body force driven flow, both the expression in Eqs. 3.37 and 3.38 can be used as the objective of the optimization problem.

Then, the topology optimization problem for flows with body forces is expressed as

$$\text{Find } \phi \text{ to minimize } J(\mathbf{u}, \nabla\mathbf{u}, p; \phi) = \int_\Omega H(-\phi)\, A(\mathbf{u}, \nabla\mathbf{u}, p)\, \mathrm{d}\Omega + \int_\Gamma B(\mathbf{u}, p)\, \mathrm{d}\Gamma,$$
$$\text{subject to } \begin{cases} \int_\Omega H(-\phi)\, \mathrm{d}\Omega = V_r V_\Omega \\ -\eta\Delta\mathbf{u} + \rho(\mathbf{u}\cdot\nabla)\mathbf{u} + \nabla p = \mathbf{f}, \text{ in } \Omega \\ -\nabla\cdot\mathbf{u} = 0, \text{ in } \Omega \\ H(\phi)\,\mathbf{u} = \mathbf{0}, \text{ in } \Omega \\ \mathbf{u} = \mathbf{0}, \text{ in } \Omega_C \end{cases} \tag{3.39}$$

where $A(\mathbf{u}, \nabla\mathbf{u}, p)$ and $B(\mathbf{u}, p)$ are the functionals defined on the domain Ω and implicit boundary Γ, respectively; $V_r \in (0, 1)$ is the volume fraction of the fluid region; and V_Ω is the volume of the optimization domain Ω. The objective of the optimization problem in Eq. 3.39 can be transformed into the following formulation

$$J(\mathbf{u}, \nabla\mathbf{u}, p; \phi) = \int_\Omega [H(-\phi)A(\mathbf{u}, \nabla\mathbf{u}, p) + \tau(\phi)\|\nabla\phi\|B(\mathbf{u}, p)]\, \mathrm{d}\Omega \tag{3.40}$$

For the optimization problem in Eq. 3.39, the evolution of the level set function is performed by solving the Hamilton-Jacobin equation [23, 25]

$$\frac{\partial\phi}{\partial\theta} + v_n\|\nabla\phi\| + wG = 0 \tag{3.41}$$

where θ is the evolving time of the level set function; v_n is the normal evolving velocity of the level set function ϕ; G is the topological sensitivity; and w is the weight of the topological sensitivity. The reasonable weight of topological sensitivity is chosen based on numerical experiments. Based on the empirical insight, the new zero level set can not be nucleated if the weight of topological sensitivity is too low; numerical instability is produced if the weight of topological sensitivity is too high; and relative large weight should be chosen as the Reynolds number of the flow is increased. Based on the work of Osher and Santosa [16], the level set function is evolved along the steepest direction using the transformed formulation of Eq. 3.41

$$\delta\phi + v_n\|\nabla\phi\|\delta\theta + wG\delta\theta = 0 \tag{3.42}$$

where $\delta\theta$ is the positive evolution time. As pointed out by Jung et al. in [26], Eq. 3.42 is equivalent to Hamilton-Jacobi Eq. 3.41 if the change of the distance function is viewed as occurring continuously in time.

3.2.2 Sensitivity Analysis

The algorithm used to update the level set function belongs to the gradient descent method, where the gradient information includes the shape sensitivity and topological sensitivity. To derive these two types of sensitivities for the level set method-based topology optimization problem in Eq. 3.39, the shape sensitivity analysis and topological sensitivity analysis are implemented as follows.

3.2.2.1 Shape Sensitivity Analysis

The normal velocity is a measure of the continuum gradient of objective to the level set function. The Lagrangian multiplier based adjoint method [27–30] are used to perform the shape sensitivity analysis of the topology optimization problem in Eq. 3.39. According to the Lagrangian multiplier method, the Lagrangian corresponding to the objective in Eq. 3.39 is introduced as

$$\begin{aligned}\hat{J} =& J\left(\mathbf{u}, \nabla\mathbf{u}, p; \phi\right) + a\left(\mathbf{u}, \mathbf{u}_a\right)_\Omega + b\left(\mathbf{u}; \mathbf{u}, \mathbf{u}_a\right)_\Omega - \left(p, \nabla\cdot\mathbf{u}_a\right)_\Omega - \left(\mathbf{f}, \mathbf{u}_a\right)_\Omega \\ &- \left(\nabla\cdot\mathbf{u}, p_a\right)_\Omega + \left(H\left(\phi\right)\mathbf{u}, \mathbf{u}_a\right)_\Omega + \left(\mathbf{u}, \mathbf{u}_a\right)_{\Omega_c} - \lambda\left(\int_\Omega H\left(-\phi\right)\,\mathrm{d}\Omega - V_r V_\Omega\right) \\ &+ \frac{\Lambda}{2}\left(\int_\Omega H\left(-\phi\right)\,\mathrm{d}\Omega - V_r V_\Omega\right)^2 - \int_{\Gamma_N}\mathbf{g}\cdot\mathbf{u}_a\,\mathrm{d}\Gamma + \int_{\Gamma_D}\mathbf{u}\cdot\mathbf{u}_a\,\mathrm{d}\Gamma\end{aligned} \tag{3.43}$$

where $a(\mathbf{u}, \mathbf{u}_a)_\Omega = \eta\int_\Omega \nabla\mathbf{u} : \nabla\mathbf{u}_a\,\mathrm{d}\Omega$; $b(\mathbf{u}; \mathbf{u}, \mathbf{u}_a)_\Omega = \rho\int_\Omega(\mathbf{u}\cdot\nabla)\mathbf{u}\cdot\mathbf{u}_a\,\mathrm{d}\Omega$; $(\cdot,\cdot)_\Omega$ is the Hilbert inner product; $\mathbf{u}_a$ and p_a are the adjoint variables of the fluid velocity $\mathbf{u}$ and the pressure p, respectively. The velocity $\mathbf{u} \in \mathscr{H}_E(\Omega) := \left\{\mathbf{u} \in \mathscr{H}(\Omega)\,\middle|\,\mathbf{u} = \mathbf{u}_D \text{ on } \Gamma_D\right\}$, where $\mathbf{u}_D$ is the known velocity distribution on Γ_D and $\mathscr{H}(\Omega)$ is the first order Hilbert space; the adjoint variable $\mathbf{u}_a \in \mathscr{H}_{E_0}(\Omega) := \left\{\mathbf{w} \in \mathscr{H}(\Omega)\,\middle|\,\mathbf{w} = \mathbf{0} \text{ on } \Gamma_D\right\}$, and $p,\ p_a \in \mathscr{L}^2(\Omega)$, where $\mathscr{L}^2(\Omega)$ is the second order Lebesque space. Therefore, $(\mathbf{u}, \mathbf{u}_a)_{\Omega_c}$ and $\int_{\Gamma_D}\mathbf{u}\cdot\mathbf{u}_a\,\mathrm{d}\Gamma$ are equal to zero. The volume constraint is treated by the augmented Lagrangian method, where $\lambda \in \mathbb{R}$ is the Lagrangian multiplier and $\Lambda \in \mathbb{R}$ is the penalty parameter. According to the Lagrangian multiplier based adjoint method, the adjoint equations of the Navier-Stokes equations with body force can be expressed as (see Appendix 3.4.1 for more details)

$$\begin{aligned}-\eta\Delta\mathbf{u}_a + \rho\mathbf{u}_a\cdot(\nabla\mathbf{u}) - \rho\left(\mathbf{u}\cdot\nabla\right)\mathbf{u}_a + \nabla p_a =& -\frac{\partial\tilde{A}}{\partial\mathbf{u}} + \nabla\cdot\frac{\partial\tilde{A}}{\partial\nabla\mathbf{u}} + \frac{\partial\mathbf{f}}{\partial\mathbf{u}}\cdot\mathbf{u}_a, \text{ in } \Omega \\ -\nabla\cdot\mathbf{u}_a =& -\frac{\partial\tilde{A}}{\partial p} + \frac{\partial\mathbf{f}}{\partial p}\cdot\mathbf{u}_a, \text{ in } \Omega \\ \mathbf{u}_a =& \mathbf{0}, \text{ on } \Gamma_D \\ \left[-p_a\mathbf{I} + \eta\left(\nabla\mathbf{u}_a + \nabla\mathbf{u}_a^{\mathrm{T}}\right)\right]\cdot\mathbf{n} =& -\rho\left(\mathbf{u}\cdot\mathbf{n}\right)\mathbf{u}_a - \frac{\partial\tilde{A}}{\partial\nabla\mathbf{u}}\cdot\mathbf{n}, \text{ on } \Gamma_N\end{aligned} \tag{3.44}$$

where $\tilde{A} = H(-\phi) A + \tau(\phi) \|\nabla\phi\| B$. Compared with the adjoint analysis of the Navier-Stokes equations without body force [6, 31], Eq. 3.44 contains the derivatives of the body force to the unknown variables $\mathbf{u}$ and p. The shape sensitivity for the optimization problem in Eq. 3.39 is obtained as (see Appendix 3.4.2 for more details)

$$\begin{aligned} \delta\hat{J} = -\int_\Omega \Big[& A + \nabla B \cdot \mathbf{n}_\Gamma + B\kappa - \mathbf{u}\cdot\mathbf{u}_a - \lambda \\ & + \Lambda \left(\int_\Omega H(-\phi)\, \mathrm{d}\Omega - V_r V_\Omega \right) \Big] \tau(\phi)\delta\phi \,\mathrm{d}\Omega \end{aligned} \tag{3.45}$$

where $\mathbf{n}_\Gamma = \nabla\phi / \|\nabla\phi\|$ and $\kappa = \nabla \cdot \mathbf{n}_\Gamma$. To ensure the descendant of the objective, the shape sensitivity should satisfy $\delta\hat{J} < 0$. According to Eq. 3.42, the variational of the level set function is $\delta\phi = -v_n \|\nabla\phi\| \delta\theta$ in the shape sensitivity analysis. Because $\tau(\phi)$, $\|\nabla\phi\|$ and $\delta\theta$ are all nonnegative, the normal velocity for evolving the level set function can be set to be

$$v_n = -A - \nabla B \cdot \mathbf{n}_\Gamma - B\kappa + \mathbf{u}\cdot\mathbf{u}_a + \lambda - \Lambda \left(\int_\Omega H(-\phi)\, \mathrm{d}\Omega - V_r V_\Omega \right) \tag{3.46}$$

3.2.2.2 Topological Sensitivity Analysis

Topological sensitivity measures the change of a cost function with respect to a topological modification of a domain. The most simple way of modifying the topology consists in creating a small hole in the domain. In structural optimization, creating a hole means removing some material. However, in fluid optimization, where the domain represents the fluid, creating a hole means inserting an obstacle [32]. In the level set method, topological sensitivity measures the change of the objective value by incorporating small holes corresponding to the new level set to change the topology of the fluid domain. Such perturbation is continuous in the $\mathscr{L}^1$-distance of sets [23]. Topological sensitivity is defined as

$$d_{\mathscr{T}} \hat{J}(\Omega)(\mathbf{x}) = \lim_{r\to 0} \frac{\hat{J}(\Omega_{r,\mathbf{x}}) - \hat{J}(\Omega)}{\|B_{r,\mathbf{x}} \cap \Omega\|_{\mathscr{L}}}, \quad \forall \mathbf{x} \in \overline{\Omega} \tag{3.47}$$

where $B_{r,\mathbf{x}} = \left\{ \mathbf{y} \middle| \|\mathbf{y} - \mathbf{x}\|_{\mathscr{L}} < r \right\}$, $\Omega_{r,\mathbf{x}} = \Omega \setminus \overline{B_{r,\mathbf{x}}}$ and $\|\cdot\|_{\mathscr{L}}$ denotes the Lebesgue measure of a set. Based on the asymptotic analysis of Eq. 3.47 for the Navier-Stokes equations [33], the topological sensitivity can be expressed as

$$d_{\mathscr{T}} \hat{J}(\Omega)(\mathbf{x}) = \begin{cases} 4\pi\rho\eta \mathbf{u}\cdot\mathbf{u}_a + d_{\mathscr{T}} J(\Omega)(\mathbf{x}) - (\lambda - \Lambda V) d_{\mathscr{T}} V(\Omega)(\mathbf{x}), & \text{in 2D} \\ 6\pi\rho\eta \mathbf{u}\cdot\mathbf{u}_a + d_{\mathscr{T}} J(\Omega)(\mathbf{x}) - (\lambda - \Lambda V) d_{\mathscr{T}} V(\Omega)(\mathbf{x}), & \text{in 3D} \end{cases} \tag{3.48}$$

where $\mathbf{u}_a$ is the adjoint variable of $\mathbf{u}$, $V = \int_\Omega H(-\phi)\,\mathrm{d}\Omega - V_r V_\Omega$ is the residual of the volume constraint. In Eq. 3.48, $d_{\mathscr{T}} J(\Omega)(\mathbf{x})$ depends on the embody of the objective J. For the topology optimization problem in Eq. 3.39, $d_{\mathscr{T}} J(\Omega)(\mathbf{x})$ can be determined by the asymptotic expansion of J

$$J(\Omega_{r,\mathbf{x}}) - J(\Omega) = f(r) d_{\mathscr{T}} J(\Omega)(\mathbf{x}) + o(f(r)) \tag{3.49}$$

where $f(r)$ is a positive function satisfying $\lim_{r\to 0} f(r) = 0$. According to [33], $f(r)$ should be chosen as $-1/\ln r$ in two dimensional case or r in three dimensional case. Therefore, the topological sensitivity for the objectives in Eqs. 3.37 and 3.38 can be expressed as:

$$d_{\mathscr{T}} J_d(\Omega)(\mathbf{x}) = \begin{cases} 4\pi\rho\eta\mathbf{u}\cdot\mathbf{u}, \text{ in 2D} \\ 6\pi\rho\eta\mathbf{u}\cdot\mathbf{u}, \text{ in 3D} \end{cases} \tag{3.50}$$

and

$$d_{\mathscr{T}} J_k(\Omega)(\mathbf{x}) = 0 \tag{3.51}$$

respectively. The topological sensitivity related with the residual of volume constraint can be expressed as:

$$d_{\mathscr{T}} V(\Omega)(\mathbf{x}) = \begin{cases} -\pi, \text{ in 2D} \\ -\frac{4}{3}\pi, \text{ in 3D} \end{cases} \tag{3.52}$$

Therefore, the topological sensitivity for the optimization problem in Eq. 3.43 are

$$d_{\mathscr{T}} \hat{J}_d(\Omega)(\mathbf{x}) = \begin{cases} 4\pi\rho\eta\mathbf{u}\cdot(\mathbf{u}_a + \mathbf{u}) \\ \quad + \pi\left[\lambda - \Lambda\left(\int_\Omega H(-\phi)\,\mathrm{d}\Omega - V_r V_\Omega\right)\right], \text{ in 2D} \\ 6\pi\rho\eta\mathbf{u}\cdot(\mathbf{u}_a + \mathbf{u}) \\ \quad + \frac{4\pi}{3}\left[\lambda - \Lambda\left(\int_\Omega H(-\phi)\,\mathrm{d}\Omega - V_r V_\Omega\right)\right], \text{ in 3D} \end{cases} \tag{3.53}$$

when the objective is the dissipation power in Eq. 3.37; and

$$d_{\mathscr{T}} \hat{J}_k(\Omega)(\mathbf{x}) = \begin{cases} 4\pi\rho\eta\mathbf{u}\cdot\mathbf{u}_a + \pi\left[\lambda - \Lambda\left(\int_\Omega H(-\phi)\,\mathrm{d}\Omega - V_r V_\Omega\right)\right], \text{ in 2D} \\ 6\pi\rho\eta\mathbf{u}\cdot\mathbf{u}_a + \frac{4\pi}{3}\left[\lambda - \Lambda\left(\int_\Omega H(-\phi)\,\mathrm{d}\Omega - V_r V_\Omega\right)\right], \text{ in 3D} \end{cases} \tag{3.54}$$

when the objective is the reverse of the kinetic energy in Eq. 3.38. The topology of the fluid domain can be modified by nucleating the level set surface. Numerically, the above procedure can be implemented by solving Eq. 3.41 and setting G to be $d_{\mathcal{T}}\hat{J}(\Omega)(\mathbf{x})$. For more details on the asymptotic analysis of the Navier-Stokes equations for the topological sensitivity, one can refer to Amstutz's work in [33, 34].

3.2.3 Numerical Implementation

The flowchart of the optimization procedure includes the following steps (Fig. 3.25):

- the initial distribution of the level set function ϕ, the initial values of the Lagrangian multiplier λ_0 and the penalty parameter Λ_0 are given;
- the velocity $\mathbf{u}$ and the pressure p are computed by solving the Navier-Stokes equations 3.29, and the corresponding adjoint variables u_a and p_a are computed by solving the adjoint equations 3.44;
- the normal velocity v_n and the topological sensitivity G are computed;
- the level set function ϕ is evolved by solving the Hamilton-Jacobin Eq. 3.41;
- the level set function ϕ is reinitialized after several iterations;
- the iterative optimization is stopped when the change of the objective values and the volume constraint are less than the user-specified tolerance 1×10^{-3} in five consecutive iterations.

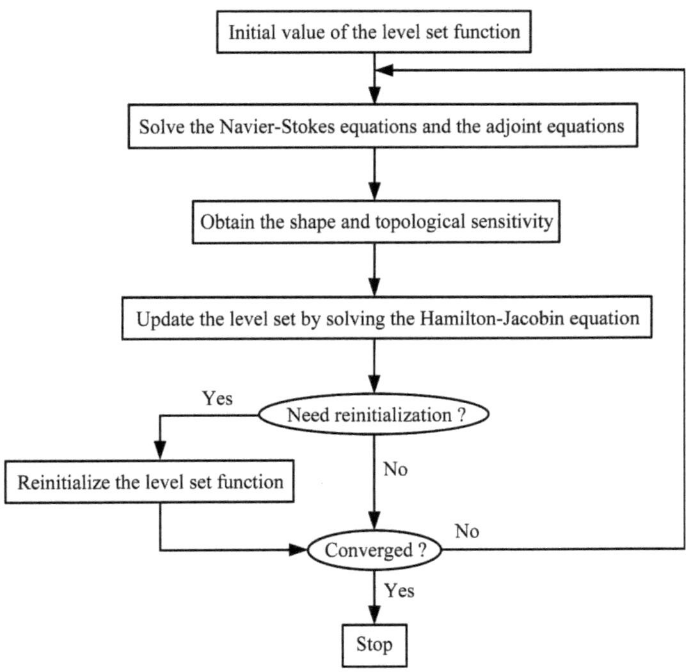

Fig. 3.25 Flowchart of the optimization procedure for solving the optimization problem in Eq. 3.39

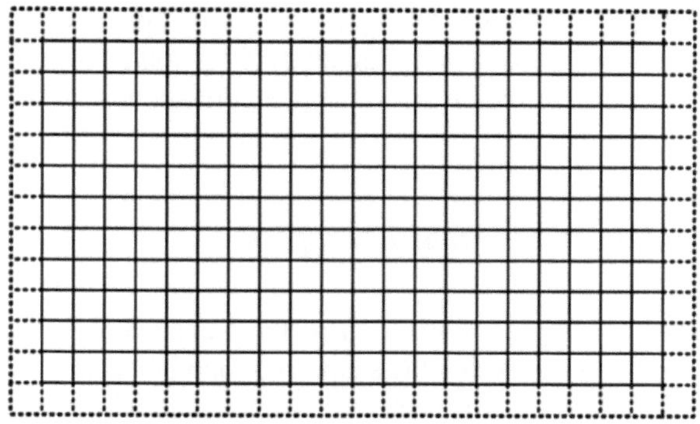

Fig. 3.26 Meshes used for solving the Navier-Stokes equations (*solid mesh*) and evolving the level set function (composition of the *solid* and *dash meshes*)

The level set function is evolved on a grid mesh with ghost elements (the composition of the dash and solid mesh in Fig. 3.26). The mesh for solving the Navier-Stokes equations is a set of elements embedded in the grid mesh for evolving the level set function (solid mesh in Fig. 3.26). The Navier-Stokes equations and the adjoint equations are solved by the finite element package *COMSOL Multiphysics* [11]. During the optimization procedure, the Navier-Stokes equations and the adjoint equations are solved by the Taylor-Hood elements [35], which interpolate the fluid velocity quadratically and the fluid pressure linearly. The level set function is interpolated by the linear elements. The Lagrangian multiplier λ and the penalty parameter Λ are updated as [28, 29]

$$\begin{aligned} \lambda_k &= \lambda_{k-1} - \Lambda_{k-1} \left(\int_\Omega H\left(-\phi_{k-1}\right) \mathrm{d}\Omega - V_r V_\Omega \right) \\ \Lambda_k &= \frac{1}{\alpha_\Lambda} \Lambda_{k-1}, \quad \alpha_\Lambda \in (0, 1) \end{aligned} \tag{3.55}$$

To evolve the level set function, the Hamilton-Jacobin equation is solved by the upwind finite difference method. The time step for the finite difference scheme is chosen based on the CFL stability condition [36]

$$\Delta\theta \leq \beta \frac{h_E}{\max\left\{|v_n|\right\}}, \quad \beta \in (0, 1) \tag{3.56}$$

where h_E is the size of the elements and β is chosen as 0.1. The evolved level set function is reinitialized after several iterations. The reinitialization of the level set function can be performed by solving the Eikonal equation [37]

$$\begin{cases} \|\nabla\phi\| = 1 \\ sgn(\phi) = sgn(\phi_0) \end{cases} \tag{3.57}$$

where sgn denotes the sign value, which is either 1, -1 or 0; ϕ_0 is the unreinitialized level set function. Equation 3.57 is solved by computing the Euclidean distance transform of the binary value that corresponds to the level set function [25].

3.2.4 Numerical Examples

The topology optimization of the Navier-Stokes flow by the variational level set method considering the shape and topological sensitivities is validated firstly. Then the optimization is respectively implemented for the Navier-Stokes flows with constant, nonuniform and solution-dependent body forces, including the unsteady flows with body forces. The density ρ and the viscosity η are set to be unit in the following numerical examples. The signed distance function is reinitialized after every five iterations. The Heaviside function and the Dirac function are approximated with the support size $h = 1.5h_E$, where h_E is the size of the elements [36]. Based on numerical experiments, the initial values of α_Λ, the Lagrangian multiplier λ and the penalty parameter Λ in Eq. 3.55 are chosen as 9×10^{-1}, -1×10^{-2} and 1×10^{-3}, respectively.

3.2.4.1 Validation of Topology Optimization by Considering Shape and Topological Sensitivities for Navier-Stokes Flow

To validate the topology optimization of the Navier-Stokes flow using the variational level set method considering the shape and topological sensitivities, a four-terminal device is optimized. The obtained results are compared with those obtained by Olesen et al. using the density method [24]. For the convenience of comparison, all of the data are chosen the same as those in [24], including the optimization domain (Fig. 3.27a), the boundary conditions (parabolic velocity distribution at the inlets, zero pressure and normal flow at the outlets), the objective function (Eq. 3.37), and the volume fraction ($V_r = 0.4$) for the volume constraint. The optimization domain is discretized by 70×100 rectangular elements. The Reynolds number is defined as $\text{Re} = \rho U_{max} L / \eta$, where U_{max} is the maximal value of the parabolic velocity distribution imposed on the inlet and L is the width of the inlet. The initial level set function and the corresponding zero contours are shown in Fig. 3.27b and c. By solving the optimization problem in Eq. 3.39 for the flows with Reynolds numbers 20 and 200 respectively, the optimized results are obtained as shown in Fig. 3.28. By observing the evolving procedure of the level set, it is clear that the topological sensitivity can nucleate new level set effectively (Figs. 3.29 and 3.32). Snapshots for the unitary shape sensitivity and topological sensitivity distributions are shown in Figs. 3.30, 3.31, 3.33 and 3.34, respectively. As the Reynolds number is increased, the dissipation of the fluid stream caused by the bending channel grows. When the inertia effect dominates the flow with a large Reynolds number, larger velocity gradients appear in the bend channels. This increases the dissipation compared to the

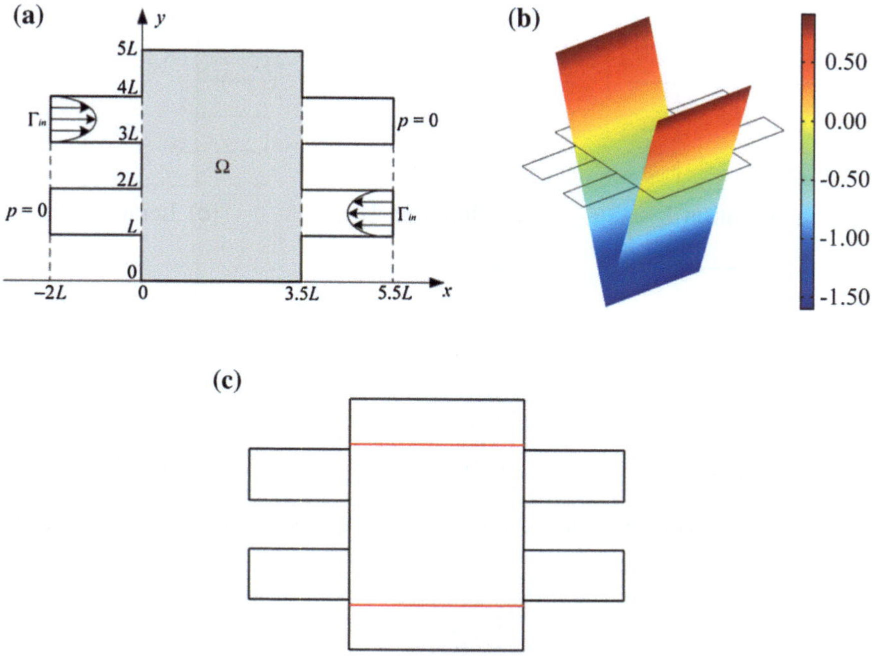

Fig. 3.27 **a** Optimization domain (*gray region*) and computational domain (*gray region* and four ducts) of the four-terminal device; **b** initial distribution of the level set function; **c** zero level set of the initial level set function

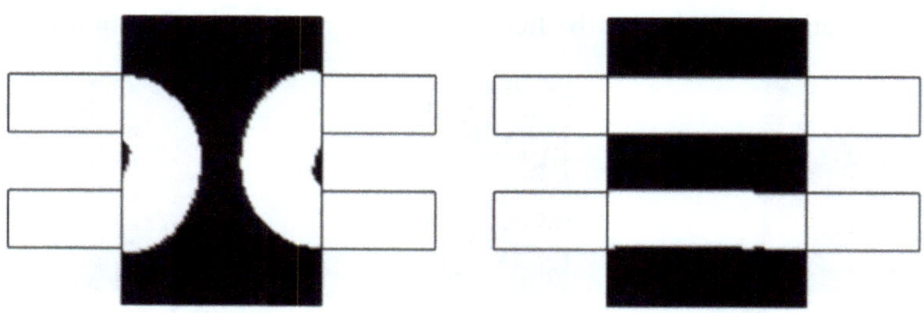

(a) Re = 20 (b) Re = 200

Fig. 3.28 Optimized topology of the four-terminal device with Reynolds numbers equal to 20 and 200, respectively

low Reynolds number case. Therefore, the optimized four-terminal device has two bending channels for the flow with low Reynolds number, and has two paralleling straight channels for the larger Reynolds number case. This is consistent with the conclusion obtained by Gersborg-Hansen et al. in [3].

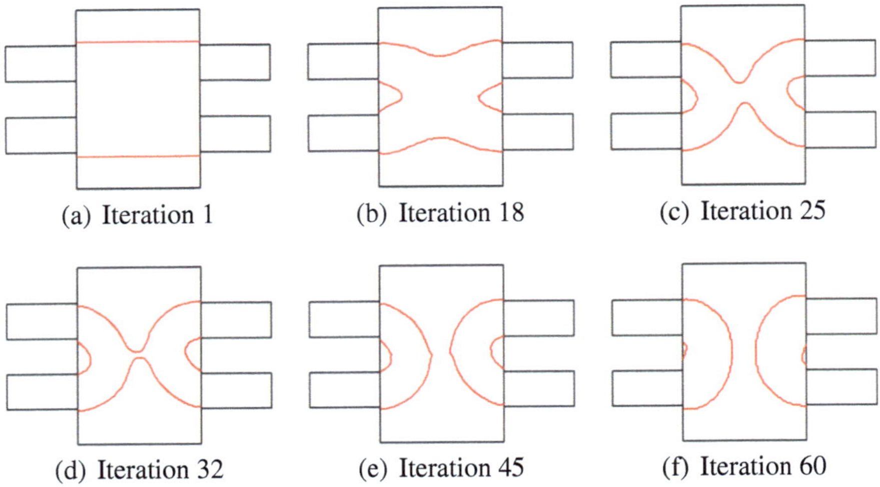

Fig. 3.29 Snapshots for the evolution and nucleation of the level set in the optimization procedure of the result in Fig. 3.28a

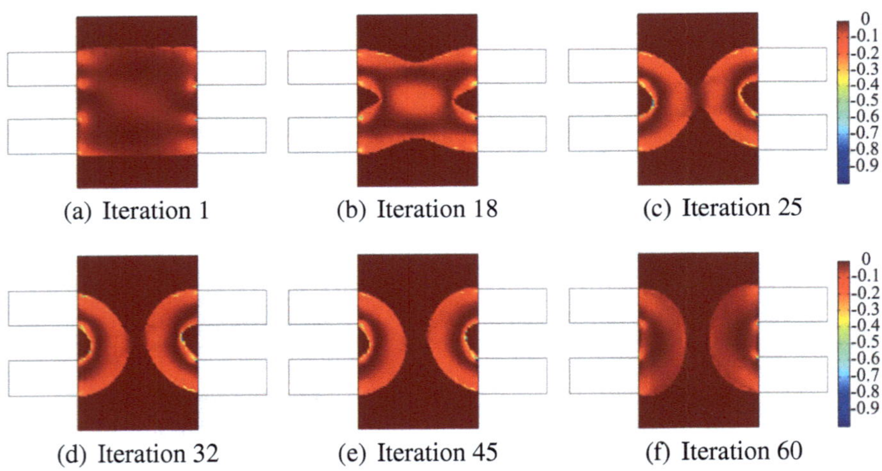

Fig. 3.30 Snapshots for the shape sensitivity in the optimization procedure of the result in Fig. 3.28a

3.2.4.2 Constant Body Force

Horizontal channel in gravity

A constant body force, e.g. the gravity, is the simplest body force in the Navier-Stokes equations. For the constant body force case, the horizontal flows with velocity and pressure boundary conditions imposed on the inlet are optimized respectively. The optimization domain are shown in Fig. 3.35a, where the domain is discretized by 360×120 rectangular elements. The initial level set function is set to be the solution

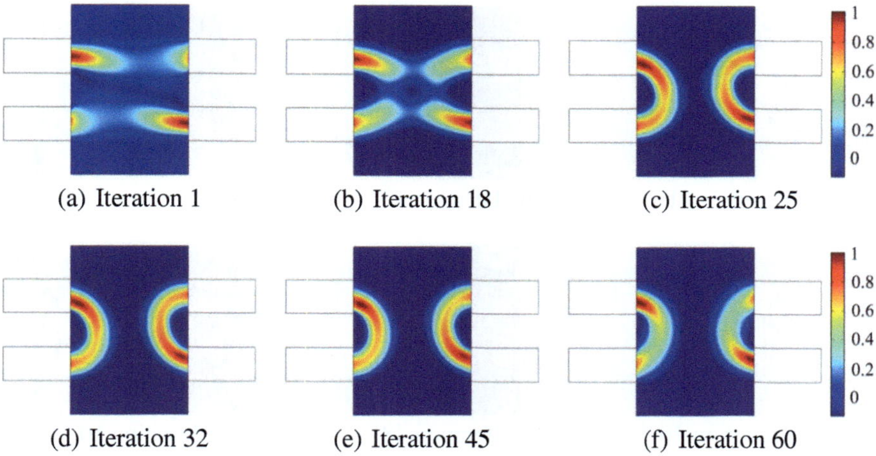

Fig. 3.31 Snapshots for the topological sensitivity in the optimization procedure of the result in Fig. 3.28a

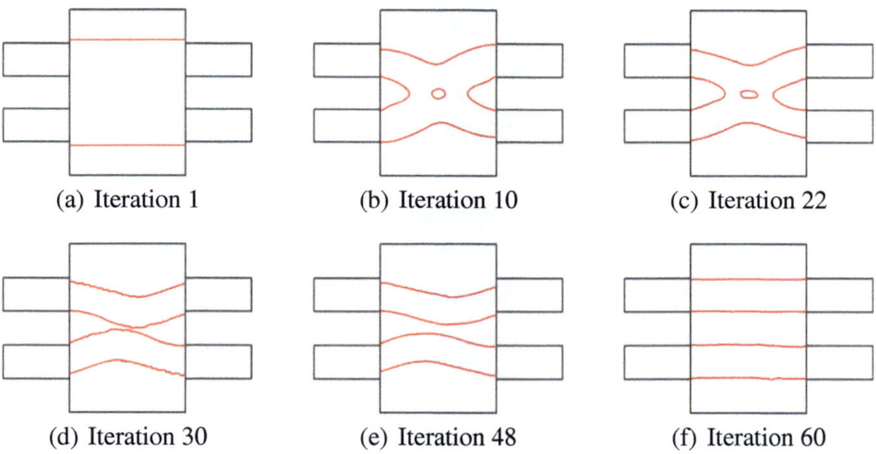

Fig. 3.32 Snapshots for the evolution and nucleation of the level set in the optimization procedure of the result in Fig. 3.28b

of Eq. 3.57, where the initial value ϕ_0 is specified as -1 (Fig. 3.35b). Because the value of ϕ_0 is less than zero, the initial level set function has no zero level set. The zero level set will be nucleated with the help of the topological sensitivity. The volume fraction of the fluid region V_r is set to be 0.4. The constant body force $\mathbf{f}$ is set to be the gravity $\mathbf{g} = (0, -10)$. The optimized channel is shown in Fig. 3.36a for the flow with a specified parabolic velocity distribution at the inlet, where the maximal value of the specified velocity is 1 and the objective is chosen to be the dissipation power (Eq. 3.37). The convergent history of the objective value and the volume constraint is shown in Fig. 3.37. By comparing with the optimized channel for the case without the

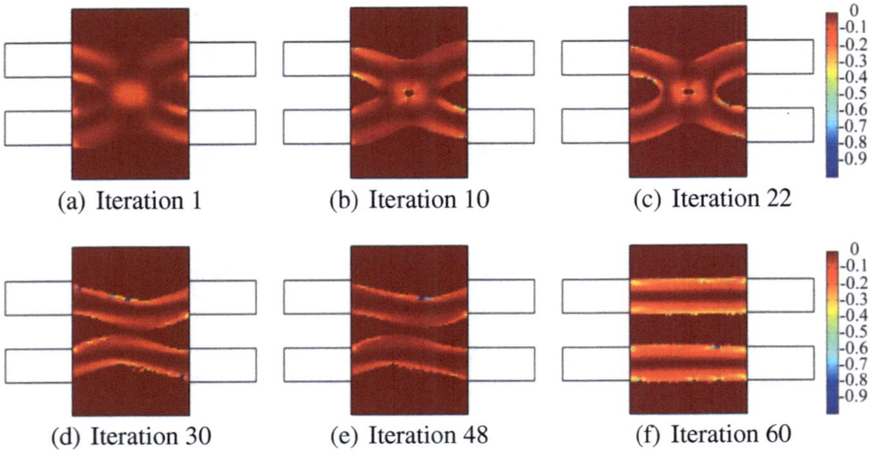

Fig. 3.33 Snapshots for the shape sensitivity in the optimization procedure of the result in Fig. 3.28b

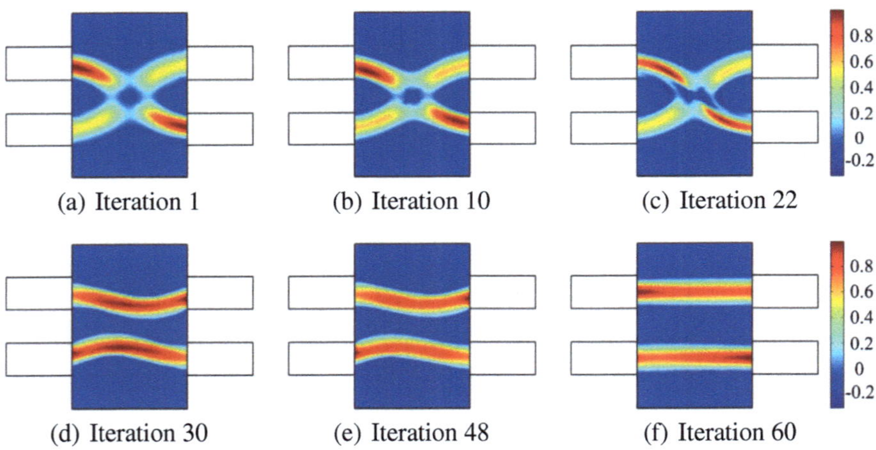

Fig. 3.34 Snapshots for the topological sensitivity in the optimization procedure of the result in Fig. 3.28b

gravity (Fig. 3.36b), the gravity causes the bending of the horizontal channel. For the case of inlet boundary with a given pressure distribution (Eq. 3.35), where the value of pressure is chosen to be 30, so that the flux at the inlet is the same as the velocity boundary case. The optimized channel is shown in Fig. 3.38 by using the reverse of the kinetic energy (Eq. 3.38) as objective. In the above numerical examples, the weight of the topological sensitivity w is set to be 2. The results in Figs. 3.36a and 3.38 show that the optimized channels bend to the direction of the gravity, where extra work is imposed on the fluid. For the flow with velocity boundary condition, the work decreases the pressure drop between the inlet and outlet. This is equivalent to decrease the value of the dissipated power inside the optimized channel. For the flow

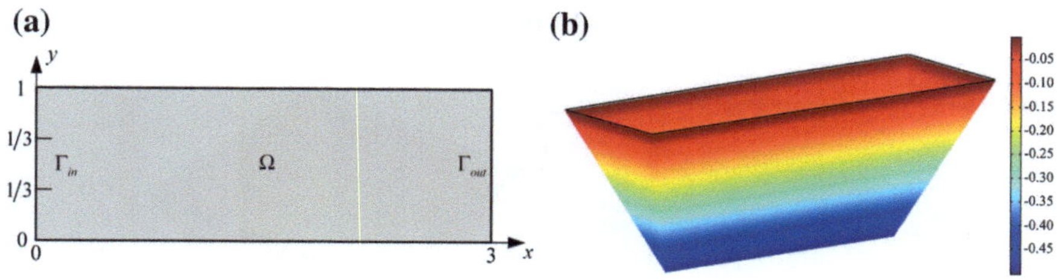

Fig. 3.35 **a** Optimization domain of the flow in the gravity; **b** initial distribution of the level set function

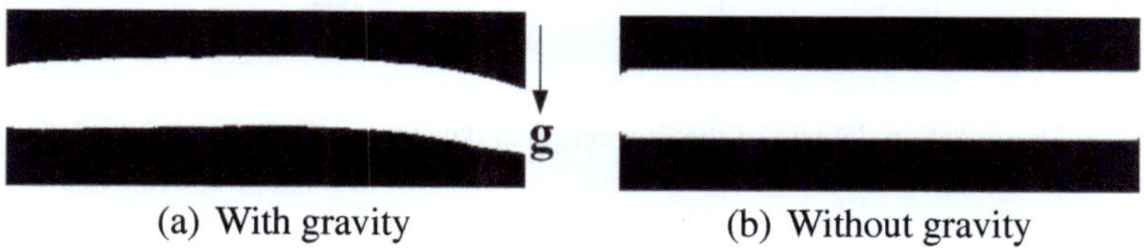

Fig. 3.36 Optimized topologies of the horizontal channel with or without the gravity

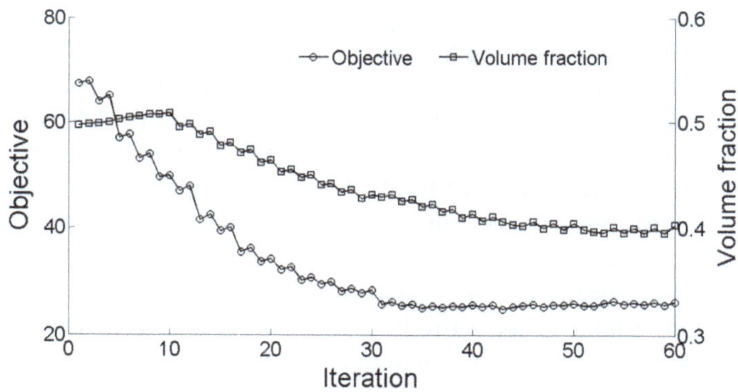

Fig. 3.37 Convergent histories of the objective value and volume constraint for the optimized flow in the gravity as shown in Fig. 3.36a

Fig. 3.38 Optimized topology of the horizontal channel for the flow with a given pressure distribution imposed on the inlet of the optimization domain in the gravity

with pressure boundary condition, the work done by the gravity is translated into the kinetic energy of the fluid partly, which helps to increase the kinetic energy of the flow. Therefore, similar topologies are obtained for both the cases of inlet boundary with known velocity and pressure distribution.

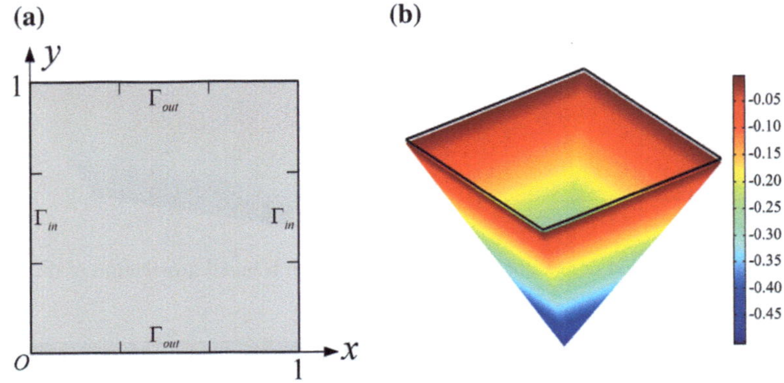

Fig. 3.39 **a** Optimization domain of the four-terminal device in the gravity; **b** initial distribution of the level set function

Four-terminal device in gravity

This numerical example performs the optimization of a four-terminal device in the gravity. The four-terminal device has two inlets with given parabolic velocity distributions, and the other two terminals are outlets with open boundary condition (Fig. 3.39a). The maximal value of the velocity imposed on the inlets is 1. The optimization domain is discretized by 60×60 rectangular elements. The initial level set function is set to be the solution of Eq. 3.57, where the initial value ϕ_0 is specified as -1 (Fig. 3.39b). Because the value of ϕ_0 is less than zero, the initial level set function has no zero level set. The zero level set will be nucleated with the help of the topological sensitivity. The weight of the topological sensitivity w and the volume fraction of fluid V_r are set to be 2 and 0.4, respectively. After solving the optimization problem in Eq. 3.39 with dissipation power (Eq. 3.37) as objective, the optimized topology of the four-terminal device and the corresponding distribution

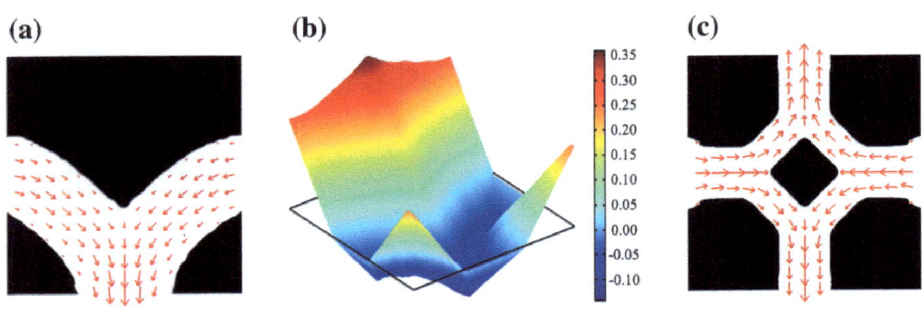

Fig. 3.40 **a** Optimized topology of the four-terminal device considering the gravity and the corresponding distribution of the fluid velocity vectors; **b** level set function corresponding to the optimized topology; **c** optimized topology of the four-terminal device without the gravity and the corresponding distribution of the fluid velocity vectors

of the level set function are shown in Fig. 3.40a and b. By solving the optimization problem with the same parameter settings for the four-terminal device without body force, the optimal topology is shown in Fig. 3.40c. The comparison between the optimized topologies in Fig. 3.40a and c demonstrates that the existence of the gravity results in the change of the topology of the optimized channel. The fluid prone to flow along the direction of the gravity. Therefore, it is really necessary to consider the body force when the design domain has multiple inlets and outlets.

Siphon in gravity

In the following, a siphon is designed using the topology optimization method. A siphon is a device in which the flow is driven by the gravity. The optimization domain, used for evolving the level set function, includes two parts Ω_s and Ω_c. Ω_c is the region that is specified as the solid domain ($\mathbf{u} = \mathbf{0}$), and Ω_s is the free optimization region during the optimization procedure. The computational domain for the distribution of the fluid velocity and pressure includes four parts, Ω_d, Ω_s, Ω_c and Ω_r (Fig. 3.41a), where Ω_r is the reservoir and Ω_d is the outlet duct. Based on the above setup, the Hamilton-Jacobin Eq. 3.41 can be solved by the finite difference method on the regular rectangular domain. The optimization domain, the reservoir and the outlet duct are discretized by 100×100, 25×50, and 50×10 rectangular elements, respectively. The initial level set function is set to be the solution of Eq. 3.57, where the initial value ϕ_0 is specified as -1 (Fig. 3.41b). By setting the weight of the topological sensitivity w and the volume fraction of the siphon V_r to be 5 and 0.2 respectively, the optimized siphon (Fig. 3.42a) is obtained by solving the optimization problem in Eq. 3.39, where the objective is the reverse of the kinetic energy (Eq. 3.38). The velocity and level set function distribution in the optimized results are shown in Fig. 3.42b and c, respectively. Because the inlet of the reservoir is specified to be higher than the outlet duct, a pressure drop is produced by the height difference in the gravity. The pressure drop results in the fluid flowing from the inlet of the reservoir

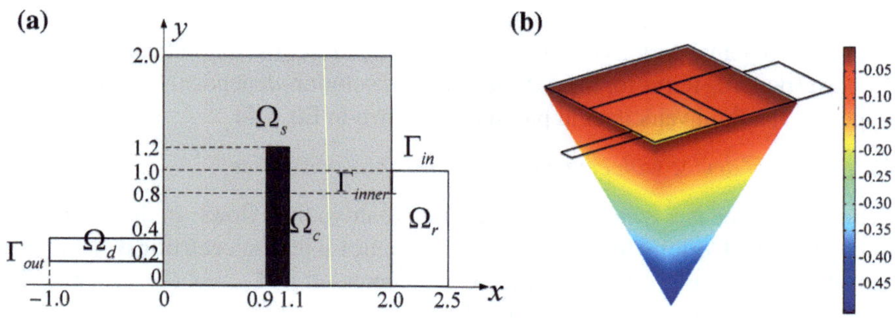

Fig. 3.41 **a** Optimization domain of the siphon in the gravity. Ω_r is a reservoir to supply the liquid flowing in the siphon; Ω_d is a duct connected to the outlet of the siphon; $\Omega_s \cup \Omega_c$ is the optimization region; the gap Ω_c is a *solid region*, which is used to ensure the tiptop of the siphon higher than the inlet of the reservoir Ω_r; Γ_{in} and Γ_{out} are the inlet and outlet respectively; Γ_{inner} and the left exterior boundaries are set to be no slip type; **b** initial distribution of the level set function defined in $\Omega_s \cup \Omega_c$

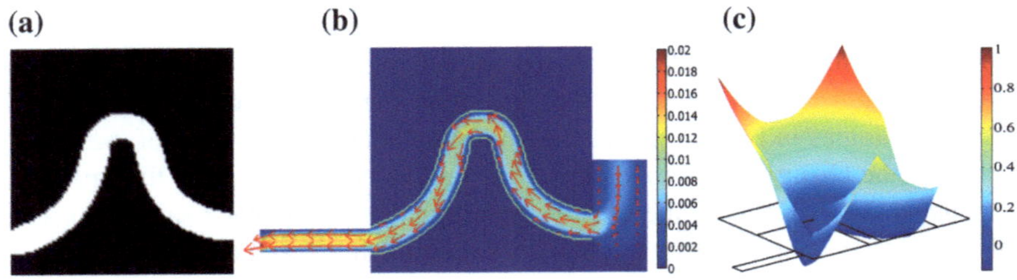

Fig. 3.42 **a** Optimized topology of the siphon in the gravity; **b** distribution of the fluid velocity; **c** distribution of the level set function

to the outlet duct. This is the siphonal flow occurring in the gravity. Because the solid domain Ω_c is higher than the inlet of the reservoir, the optimized channel balances the effect between the uphill flow and the maximal kinetic energy by tuning the shape and length of the fluid channel. In the siphonal flow, the gravitational potential energy of the fluid is translated into the kinetic energy of the fluid and dissipation of the flow. Therefore, more kinetic energy of the fluid means lower dissipation of the flow. Because the optimization objective is to maximize the kinetic energy, the dissipation power of the flow in the optimized siphon is minimized automatically.

3.2.4.3 Nonuniform and Solution-Dependent Body Forces

Optimization of the Navier-Stokes flows with nonuniform and solution-dependent body forces is considered in this section. The nonuniform and solution-dependent body forces are usually used as the driven force in fluid devices, such as the centrifugal chips in microfluidics [38, 39]. The nonuniform body force is a function of the spacial coordinate. Because the derivatives of the nonuniform body force with respect to the velocity and pressure are zeros, the nonuniform body force has no appearance in the adjoint equations. The solution-dependent body force depends on the unknowns of physical problems, such as the fluid velocity. Therefore, the adjoint equations have the corresponding terms for the derivatives of the solution-dependent body force with respect to the fluid velocity and pressure as shown in Eq. 3.44.

Four-terminal device with centrifugal and Coriolis body force

Centrifugal force is a typical body force used to drive fluid flows. In the centrifugal force field, the Coriolis force exists and companies with the centrifugal force. The centrifugal force and the Coriolis force are expressed as $\mathbf{f}_{cen} = (\boldsymbol{\omega} \times \mathbf{r}) \times \boldsymbol{\omega}$ and $\mathbf{f}_{cor} = 2\mathbf{u} \times \boldsymbol{\omega}$ respectively, where $\boldsymbol{\omega}$ is the angular velocity vector and $\mathbf{r}$ is the radial vector of the point inside the flow relative to the rotating center [40]. The centrifugal force and the Coriolis force are nonuniform body force and solution-dependent body force respectively. A four-terminal device in the centrifugal and Coriolis force field is optimized in this numerical example. The optimization domain and the boundary conditions imposed on the inlets and outlets of the four-terminal device are shown

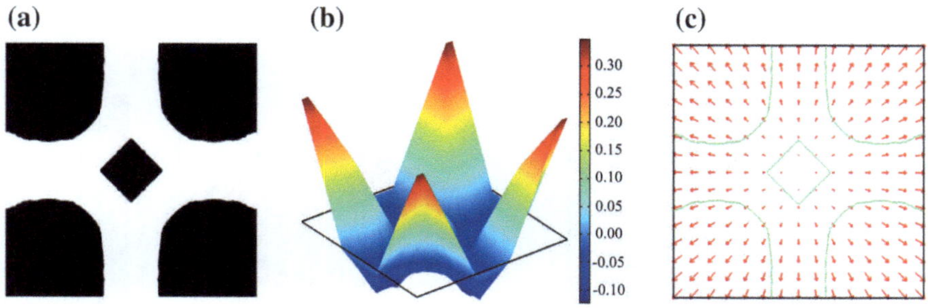

Fig. 3.43 **a** Optimized topology of the four-terminal device in centrifugal force field; **b** level set function corresponding to the optimized topology; **c** distribution of the centrifugal force

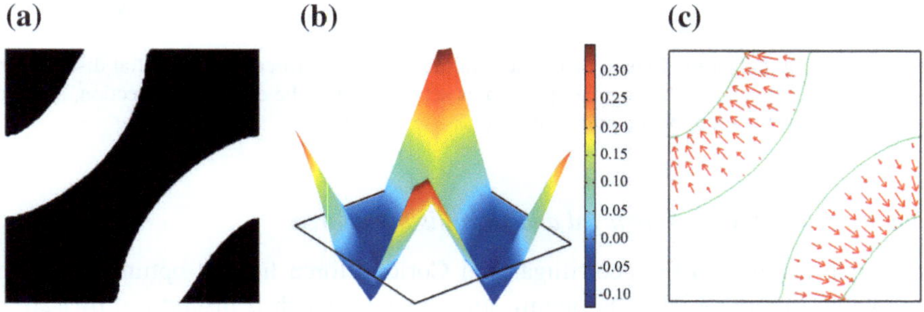

Fig. 3.44 **a** Optimized topology of the four-terminal device in the centrifugal-Coriolis force field; **b** level set function corresponding to the optimized topology; **c** distribution of the Coriolis force (To manifest the direction of the Coriolis force, the *arrows* has a different scaling from the centrifugal force)

in Fig. 3.39a. The rotating center is set to be the point with the coordinate (0.5, 0.5) which is the center of the optimization domain. The scalar value of the angular speed $\boldsymbol{\omega}$ is set to be 20. The other optimization settings are the same as that in Sect. 3.2.4.2. After solving the optimization problem in Eq. 3.39, the optimized topologies of the four-terminal device are obtained corresponding to the centrifugal force and the centrifugal-Coriolis force respectively (Fig. 3.43a and Fig. 3.44a). The corresponding level set functions are shown in Fig. 3.43b and Fig. 3.44b. In this numerical example, the rotating center is set to be the center of the optimization domain. Therefore, the centrifugal force distributes symmetrically in the outward direction of the radius vector (Fig. 3.43c). However, because the Coriolis force vector is vertical to the direction of the fluid velocity and along the rotating direction, the distribution of the Coriolis force is not symmetrical (Fig. 3.44c). When only the centrifugal force is considered, the symmetrical optimization domain and symmetrical distribution of the body force determine the symmetrical result in Fig. 3.43a commonly. As the centrifugal force and the Coriolis force are considered simultaneously, the Coriolis force results in the unsymmetrical topology of the four-terminal device in Fig. 3.44a.

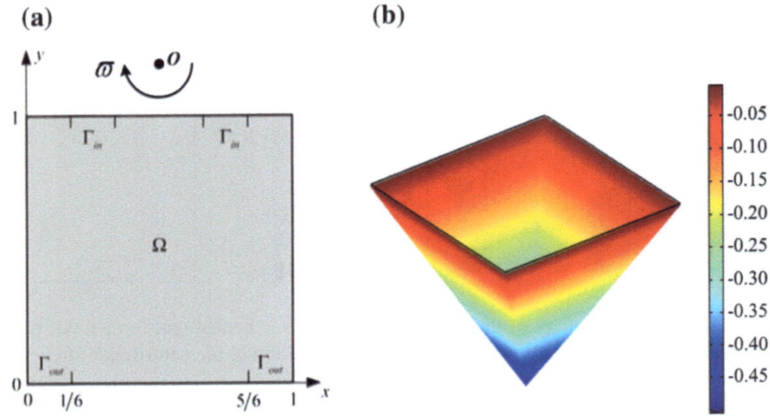

Fig. 3.45 **a** Optimization domain of the flow in the centrifugal force field; **b** initial distribution of the level set function. The angular speed ϖ is set to be 30 in the clockwise direction, and the rotating center O is set to be the point with coordinate (0.5, 2)

Double channel with centrifugal and Coriolis body force

A double channel in the centrifugal and Coriolis force field is optimized in the following. Figure 3.45a is the optimization domain which is discretized by 120 × 120 rectangular elements. The inlets and outlets of the flow are all set to be open boundaries (Eq. 3.36). The initial level set function is set to be the solution of Eq. 3.57, where the initial value ϕ_0 is specified as -1 (Fig. 3.45b). The rotating center is set to be the point with coordinate of (0.5, 2). The scalar value of the angular speed $\boldsymbol{\omega}$ in the clockwise direction is set to be 30. The volume fraction of the fluid is set to be $V_r = 0.3$, and the weight of the topological sensitivity w is set to be 2. The optimized results for the cases with only centrifugal force and with both the

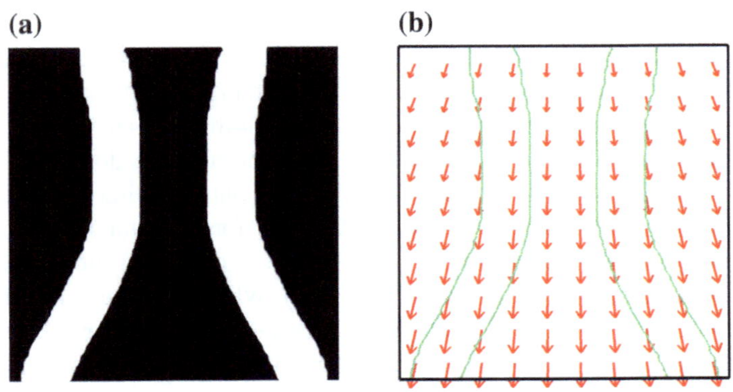

Fig. 3.46 **a** Optimized topology of the channel in the centrifugal force field; **b** distribution of the centrifugal force vector

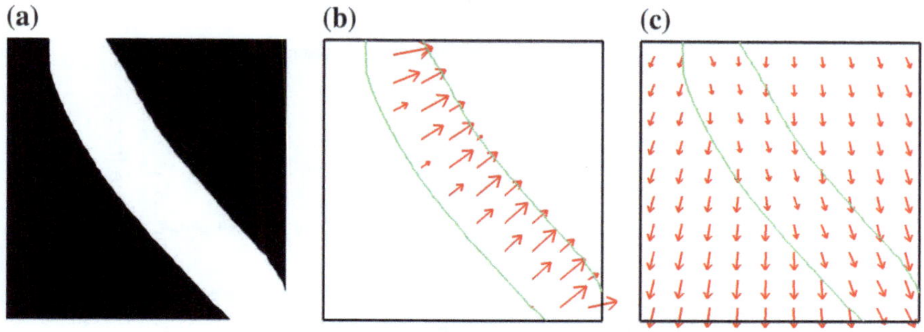

Fig. 3.47 **a** Optimized topology of the channel in the centrifugal and Coriolis force field; **b** distribution of the Coriolis force vector (To manifest the direction of the Coriolis force, the *arrows* has a different scaling from the centrifugal force); **c** composition of the centrifugal and Coriolis forces

centrifugal force and the Coriolis force are shown in Figs. 3.46a and 3.47a where the optimization objective is the dissipation power in Eq. 3.37. The convergent histories of the objective value and volume constraint are shown in Fig. 3.48. Snapshots for the evolving procedure of the level set are shown in Fig. 3.49. Similar to the numerical example of the four-terminal device in Sect. 3.2.4.2, the Coriolis force results in the change of the topology of the optimized fluid channel. Therefore, the Coriolis force deserves to be considered in the design of rotational fluid system.

3.2.4.4 Three-Dimensional Numerical Examples

3D channel with constant body force

For the 3D flow driven by constant body force, the flow in a hexahedron with constant body force is optimized. The optimization domain (Fig. 3.50a) is a hexahedron with length, width and height equal to 1, 1 and 0.5, respectively. And it is discretized by $30 \times 30 \times 10$ hexahedral elements. The inlet and outlet of the design domain are set to be open boundaries. The initial level set function is set to be the solution of Eq. 3.57, where the initial value ϕ_0 is specified as -1. The constant body force is given as $\mathbf{f} = (10, 10, 0)$. The weight of the topological sensitivity is set to be 3 and the volume fraction of fluid is set to be 0.4. The optimized channel is shown in Fig. 3.50b for the objective of the kinetic energy in Eq. 3.38. Snapshots for the evolving procedure of the level set are shown in Fig. 3.51. Figure 3.50b shows that the optimized channel has a definite lean to the direction of the constant body force, and this results in the work done by the body force. As has been discussed in Sect. 3.2.4.2 for the optimization of the flows in the gravity, the work of the constant body force helps to increase the kinetic energy of the fluid in this example too.

3D siphon in centrifugal-Coriolis force field

For the 3D flow that is driven by the nonuniform and solution-dependent body forces, a siphon driven by the centrifugal and Coriolis forces is optimized [38, 39]. The opti-

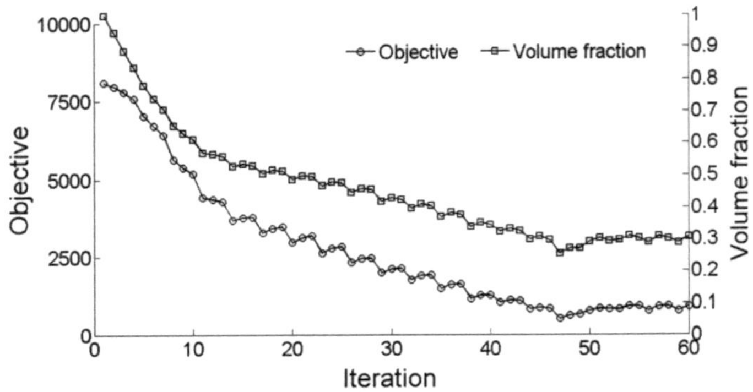

(a) Centrifugal force case

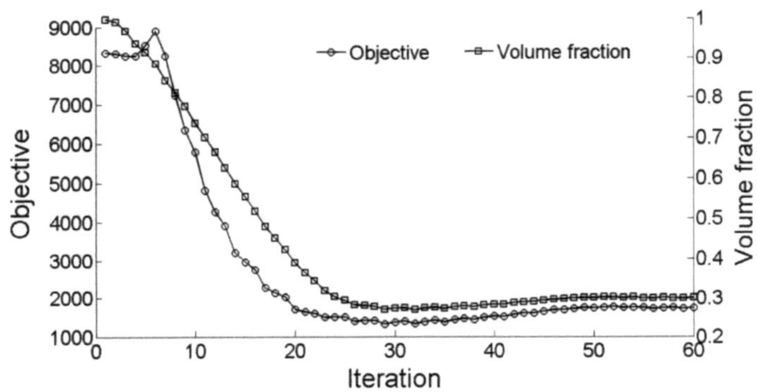

(b) Centrifugal-Coriolis force case

Fig. 3.48 Convergent histories of the objective value and volume constraint for the topology optimization of the flow with the centrifugal or centrifugal-Coriolis force

mization domain is $\Omega_s \cup \Omega_c$, where Ω_c is a gap filled with solid and enveloped by Ω_s (Fig. 3.52a and b). Similar to the 2D siphon example in Sect. 3.2.4.2, Ω_c is specified as a solid region by constraining the fluid velocity to be zero. The optimization domain, the reservoir and the duct are discretized by $100 \times 100 \times 10$, $25 \times 50 \times 6$ and $50 \times 10 \times 6$ hexahedral elements, respectively. The inlet and outlet of the flow are set to be open boundaries. The rotating axis is set to be the axis vertical to the xOy plane at the point (2.25, 2, 0) and the angular velocity is set to be (0, 0, -100). The initial level set function is set to be the solution of Eq. 3.57, where the initial value ϕ_0 is specified as -1. By solving the optimization problem with the weight of the topological sensitivity ω and the volume fraction of fluid equal to 20 and 0.2 respectively, the optimized siphon is obtained as shown in Fig. 3.52c and d from the different viewpoints, where the objective is the kinetic energy of the flow (Eq. 3.38). The siphonal flow in the centrifugal force field is similar to that in the gravity. Figure 3.52a and b

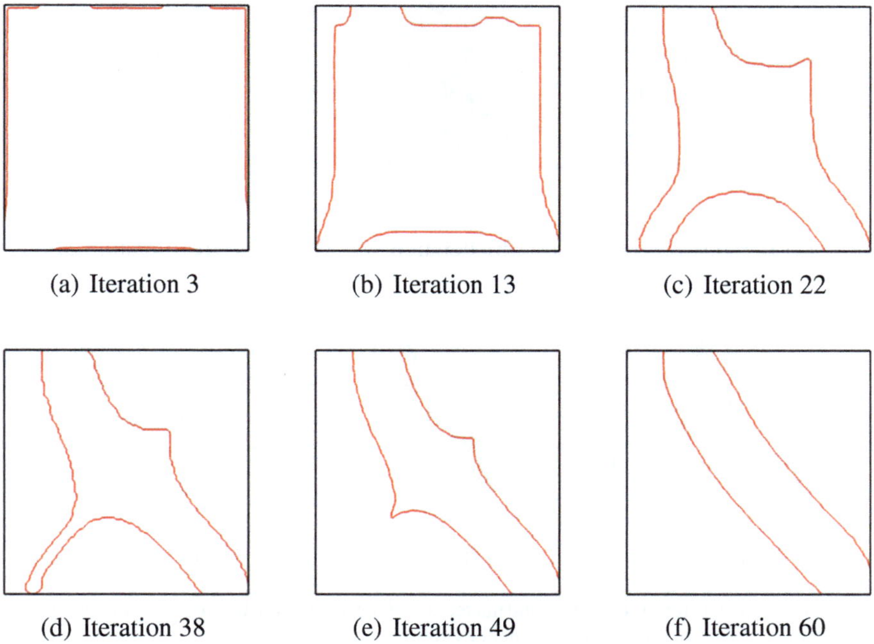

Fig. 3.49 Snapshots for the evolution and nucleation of the level set in the procedure of the topology optimization of the Navier-Stokes flow with centrifugal-Coriolis force

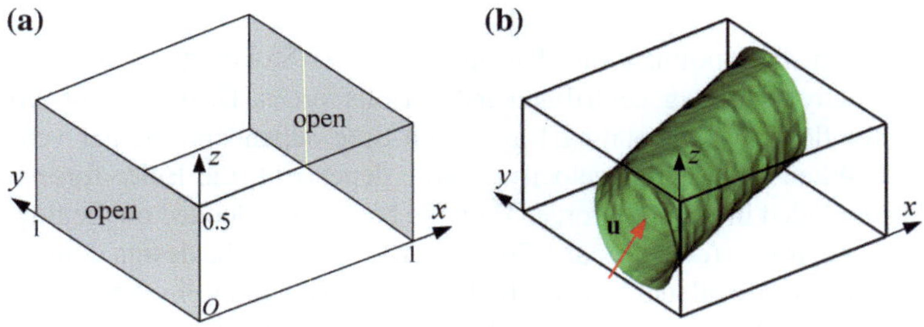

Fig. 3.50 **a** Optimization domain of the 3D channel in the constant body force field; **b** optimized topology of the channel. The artificial constant body force is set to be $\mathbf{f} = (10, 10, 0)$

show that there is a radius difference between the inlet and outlet of the siphon. A pressure drop is produced by the centrifugal force loaded on the fluid. This pressure drop drives the fluid flowing in the siphon. In the siphonal flow, the potential energy of the fluid is translated into the kinetic energy and viscous dissipation. Therefore, the optimized siphon tends to increase the kinetic energy and decrease the viscous dissipation in the optimization procedure.

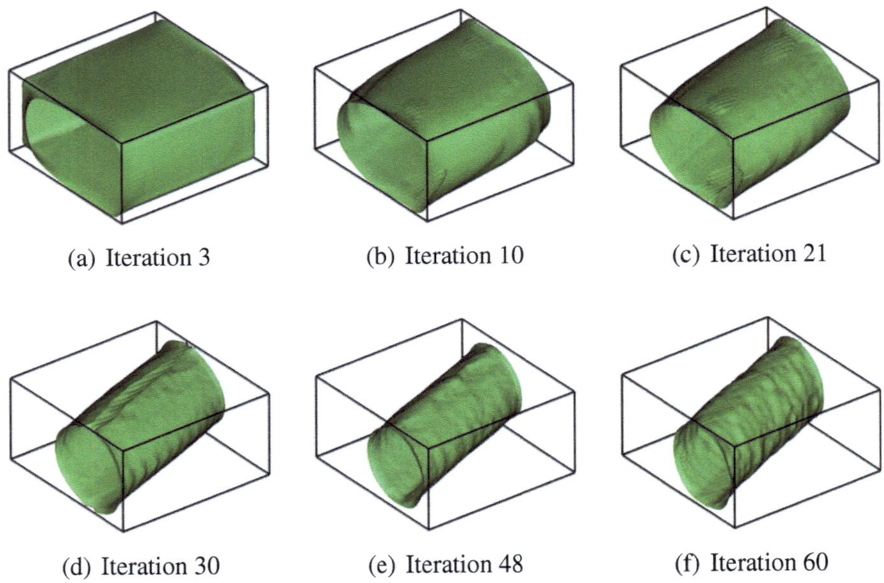

Fig. 3.51 Snapshots for the evolution and nucleation of the level set in the procedure of the topology optimization of the 3D flow with constant body force

3.2.4.5 Unsteady Flow with Body Forces

In this section, the optimization of unsteady Navier-Stokes flows is extended to flows with body forces, e.g. centrifugal and Coriolis forces. The following numerical example for flows in the rotating frame with time dependent angular velocity is presented. When the angular velocity is time dependent, the Euler force $\mathbf{r} \times \frac{d\boldsymbol{\omega}}{dt}$ should be included in the body force. Then the body force, loaded on the flow in the rotating frame, is $\mathbf{f} = (\boldsymbol{\omega} \times \mathbf{r}) \times \boldsymbol{\omega} + 2\mathbf{u} \times \boldsymbol{\omega} + \mathbf{r} \times \frac{d\boldsymbol{\omega}}{dt}$ [13]. The design domain of the channel with centrifugal and Coriolis body forces is shown in Fig. 3.53a. The initial distribution of the level set function and the corresponding zero level set are shown in Fig. 3.53b and c, respectively. The design domain is discretized by 120 × 120 rectangular elements. The open boundary condition is loaded on the inlet Γ_i and the outlet Γ_o. The other boundaries are set as no slip type. The volume fraction V_r is set as 0.3. The optimization objective is set to be the kinetic energy of the fluid. The angular velocity is set as $\boldsymbol{\omega} = (0, 0, -80\sin t)$, and the rotating center is at the point $(0.5, 2)$. On the time interval to be $[0, \pi]$ and $[0, 2\pi]$, the optimized channels are obtained as shown in Fig. 3.54a and b, respectively.

When the time interval is $[0, \pi]$, the direction of Coriolis force is counter-clockwise, so that the inclined channel in Fig. 3.54a is obtained. When the time interval is $[0, 2\pi]$, the Coriolis force has the same magnitude and inverse directions in the time intervals $[0, \pi]$ and $(\pi, 2\pi]$. Therefore, the straight channel shown in Fig. 3.54b is obtained. The cross comparison of the objective values for the two

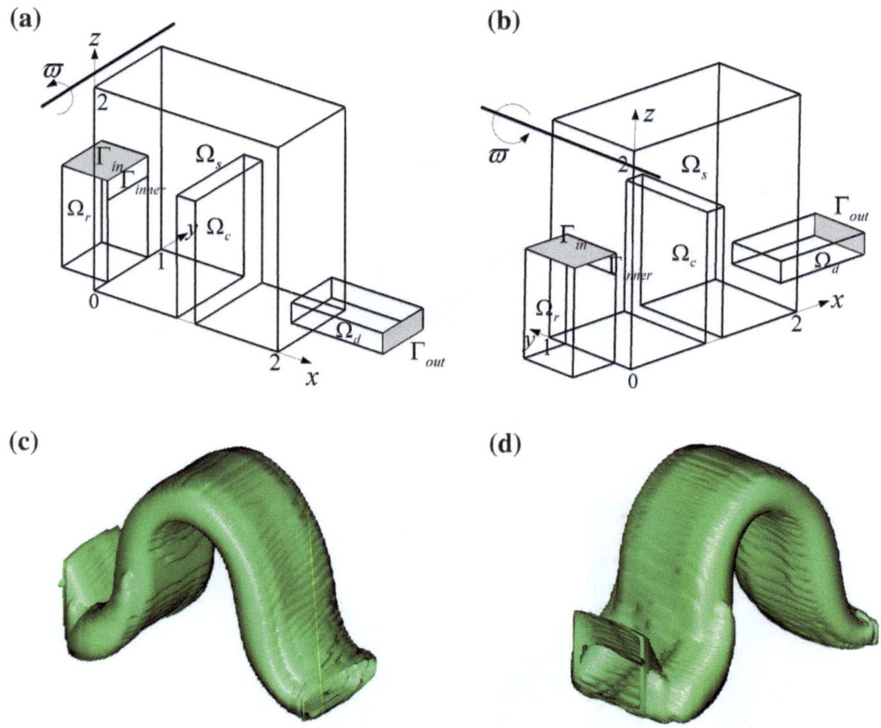

Fig. 3.52 Optimization domain for the siphon in the centrifugal and Coriolis force field from two different viewpoints (a) and (b). Ω_r is a reservoir to supply the liquid flowing in the siphon; Ω_d is a duct connected to the outlet of the siphon; $\Omega_s \cup \Omega_c$ is the optimization domain; the gap Ω_c is a solid domain, which is used to ensure the tiptop of the siphon higher than the inlet of the reservoir Ω_r; Γ_{in} and Γ_{out} are the inlet and outlet respectively; Γ_{inner} and the left exterior boundaries are set to be no slip type. The optimized 3D siphon in the centrifugal and Coriolis body force field is shown in (c) and (d) from two viewpoints corresponding to (a) and (b)

Table 3.12 Objective values corresponding to the optimized shape in Fig. 3.54 for the time intervals $[0, \pi]$ and $[0, 2\pi]$, respectively

	$[0, \pi]$	$[0, 2\pi]$
Figure 3.54a	1.683×10^5	4.127×10^5
Figure 3.54b	1.841×10^5	3.952×10^5

cases in Fig. 3.54 is in Table 3.12. When the initial distribution of the level set function is chosen as shown in Fig. 3.55a, the optimized shape and corresponding zero level set are obtained as shown in Fig. 3.55b and c. The agreement between the results in Figs. 3.54 and 3.55 demonstrates the robustness of the extended method for unsteady flows with body forces.

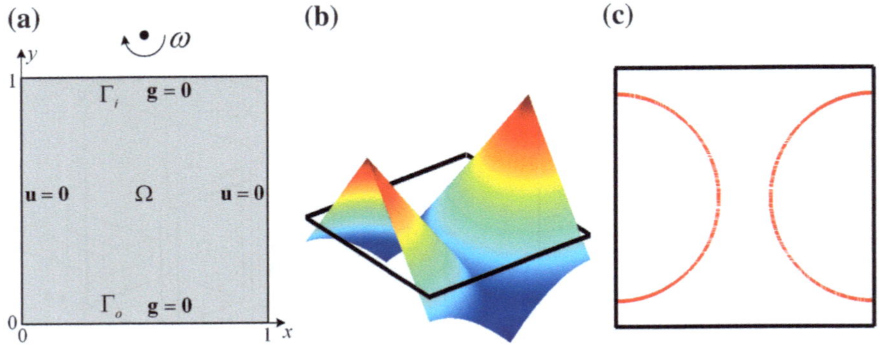

Fig. 3.53 **a** Design domain Ω and the schematic rotating center at (0.5, 2); **b** initial distribution of the level set function; **c** initial zero level set. $\boldsymbol{\omega}$ is the angular velocity; Γ_i is the inlet; Γ_o is the outlet

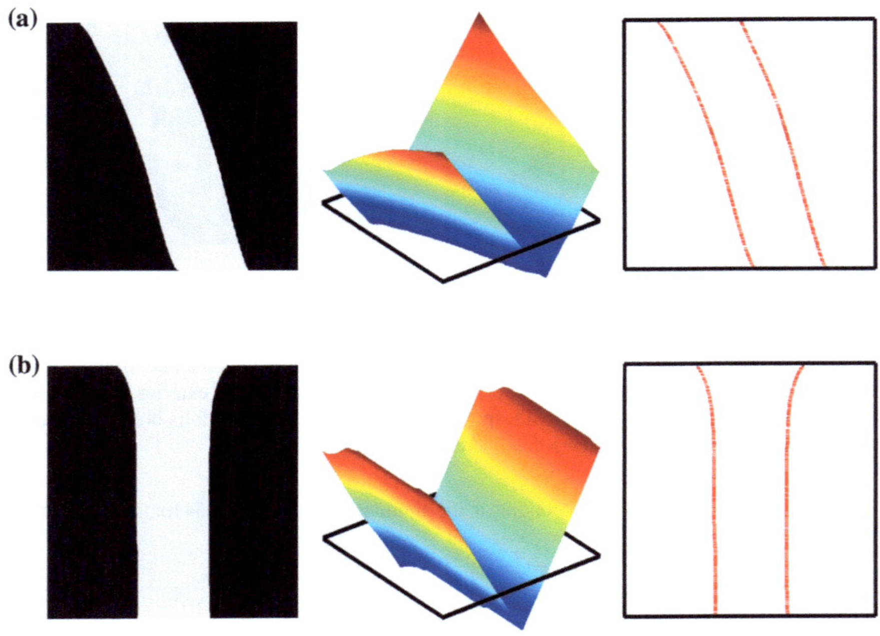

Fig. 3.54 Optimized shape, the corresponding distribution of the level set function and zero level set for the channel in centrifugal and Coriolis force field for the time interval $[0, \pi]$ **a** and $[0, 2\pi]$ **b**, respectively

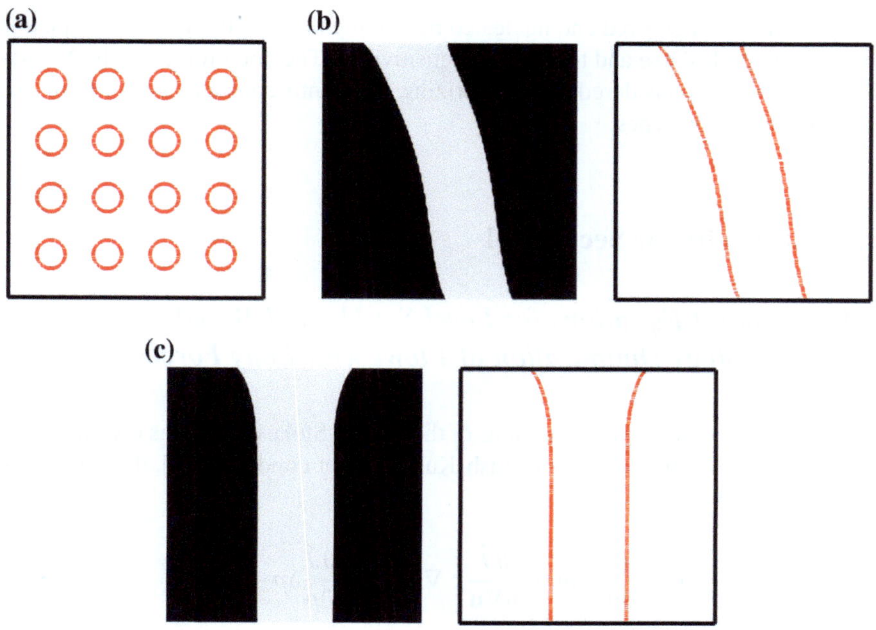

Fig. 3.55 **a** Initial distribution of the level set function; **b** optimized shape and the zero level set corresponding to the time interval $[0, \pi]$; **c** optimized shape and the zero level set corresponding to the time interval $[0, 2\pi]$

3.3 Summary

The topology optimization has been implemented for the steady and unsteady incompressible Navier-Stokes flows with body forces:

Based on the density method, the topology optimization of the steady and unsteady flows with body forces are performed by adding the artificial friction force into the Navier-Stokes equations and penalizing the physical body forces using the power-law approach. In the numerical examples, the topology optimization of the flows with the constant body force (e.g. the gravity), the centrifugal force and Coriolis force are implemented, respectively. The design of fluidic channels, where the objective is chosen to be the work done by the body forces, has demonstrated that the optimal topology of the steady and unsteady Navier-Stokes flows with body forces are influenced by the type of the body forces; the mass distribution of the flow driven by the body forces has been performed by adding flux constraint into the topology optimization problem.

Based on the local velocity control method, the topology optimization of the fluidic networks for the flows with body forces has also been implemented. Based on the level set method, topology optimization is implemented for the flows with general body forces, considering the shape and topological sensitivities. The optimization problem is analyzed by the continuous adjoint method; and the implicit boundary of the fluid flow is evolved and nucleated by solving the Hamilton-Jacobin equation considering both shape and topological sensitivities. The body forces of the Navier-Stokes flows are considered by categorizing them into constant, nonuniform and solution-dependent types.

3.4 Appendix for Section 3.1

3.4.1 Adjoint Equations for Level Set Method-Based Topology Optimization of Flows with Body Forces

The derivation of the adjoint equations of the Navier-Stokes equations is exhibited in this appendix. According to the Karush-Kuhn-Tucker conditions [6], the variational of $\hat{J}$ in Eq. 3.43

$$\delta \hat{J} = \frac{\partial \hat{J}}{\partial \mathbf{u}} \cdot \delta \mathbf{u} + \frac{\partial \hat{J}}{\partial \nabla \mathbf{u}} : \nabla (\delta \mathbf{u}) + \frac{\partial \hat{J}}{\partial p} \delta p + \frac{\partial \hat{J}}{\partial \phi} \delta \phi \tag{3.58}$$

should be zero corresponding to the optimal distribution of the level set function. This means that the variational of $\hat{J}$ to the velocity $\mathbf{u}$, the pressure p and the level set function ϕ are all zero

$$\frac{\partial \hat{J}}{\partial \mathbf{u}} \cdot \delta \mathbf{u} + \frac{\partial \hat{J}}{\partial \nabla \mathbf{u}} : \nabla (\delta \mathbf{u}) = 0, \quad \frac{\partial \hat{J}}{\partial p} \delta p = 0, \quad \frac{\partial \hat{J}}{\partial \phi} \delta \phi = 0 \tag{3.59}$$

Because $\mathbf{u} = \mathbf{0}$ in Ω_c, $(\mathbf{u}, \mathbf{u}_a)_{\Omega_c}$ in Eq. 3.43 is equal to zero. Based on the first two formulas of Eq. 3.59, one can obtain

$$\begin{aligned} & J(\delta \mathbf{u}, \nabla \mathbf{u}, p; \phi) + J(\mathbf{u}, \nabla (\delta \mathbf{u}), p; \phi) + a(\delta \mathbf{u}, \mathbf{u}_a)_\Omega + b(\delta \mathbf{u}; \mathbf{u}, \mathbf{u}_a)_\Omega \\ & + b(\mathbf{u}; \delta \mathbf{u}, \mathbf{u}_a)_\Omega - \left(\frac{\partial \mathbf{f}}{\partial \mathbf{u}} \cdot \delta \mathbf{u}, \mathbf{u}_a \right)_\Omega - (\nabla \cdot \delta \mathbf{u}, p_a)_\Omega + (H(\phi) \delta \mathbf{u}, \mathbf{u}_a)_\Omega = 0 \\ & J(\mathbf{u}, \nabla \mathbf{u}, \delta p; \phi) - (\delta p, \nabla \cdot \mathbf{u}_a)_\Omega - \left(\frac{\partial \mathbf{f}}{\partial p} \delta p, \mathbf{u}_a \right)_\Omega = 0 \end{aligned} \tag{3.60}$$

Note that

$$\begin{aligned} J(\delta \mathbf{u}, \nabla \mathbf{u}, p; \phi) &= \int_\Omega \left[H(-\phi) \frac{\partial A}{\partial \mathbf{u}} + \tau(\phi) \| \nabla \phi \| \frac{\partial B}{\partial \mathbf{u}} \right] \cdot \delta \mathbf{u} \, \mathrm{d}\Omega \\ &= \int_\Omega \frac{\partial \tilde{A}}{\partial \mathbf{u}} \cdot \delta \mathbf{u} \, \mathrm{d}\Omega \end{aligned} \tag{3.61}$$

$$
\begin{aligned}
J(\mathbf{u}, \nabla\delta\mathbf{u}, p; \phi) &= \int_\Omega \left[H(-\phi)\frac{\partial A}{\partial \nabla\mathbf{u}} + \tau(\phi)\|\nabla\phi\|\frac{\partial B}{\partial \nabla\mathbf{u}} \right] : \nabla(\delta\mathbf{u})\,\mathrm{d}\Omega \\
&= \int_\Omega \frac{\partial \tilde{A}}{\partial \nabla\mathbf{u}} : \nabla(\delta\mathbf{u})\,\mathrm{d}\Omega \\
&= \int_\Omega \nabla\cdot\left(\frac{\partial \tilde{A}}{\partial \nabla\mathbf{u}}\cdot\delta\mathbf{u}\right)\mathrm{d}\Omega - \int_\Omega \left(\nabla\cdot\frac{\partial \tilde{A}}{\partial \nabla\mathbf{u}}\right)\cdot\delta\mathbf{u}\,\mathrm{d}\Omega \\
&= \int_{\Gamma_N} \frac{\partial \tilde{A}}{\partial \nabla\mathbf{u}}\cdot\mathbf{n}\cdot\delta\mathbf{u}\,\mathrm{d}\Gamma - \int_\Omega \left(\nabla\cdot\frac{\partial \tilde{A}}{\partial \nabla\mathbf{u}}\right)\cdot\delta\mathbf{u}\,\mathrm{d}\Omega
\end{aligned}
\tag{3.62}
$$

$$
\begin{aligned}
J(\mathbf{u}, \nabla\mathbf{u}, \delta p; \phi) &= \int_\Omega \left[H(-\phi)\frac{\partial A}{\partial p} + \tau(\phi)\|\nabla\phi\|\frac{\partial B}{\partial p} \right]\delta p\,\mathrm{d}\Omega \\
&= \int_\Omega \frac{\partial \tilde{A}}{\partial p}\delta p\,\mathrm{d}\Omega
\end{aligned}
\tag{3.63}
$$

where $\tilde{A} = H(-\phi)A + \tau(\phi)\|\nabla\phi\|B$.

$$a(\delta\mathbf{u}, \mathbf{u}_a)_\Omega = -\eta\big(\Delta\mathbf{u}_a, \delta\mathbf{u}\big)_\Omega + \eta\big((\nabla\mathbf{u}_a + \nabla\mathbf{u}_a^{\mathrm{T}})\cdot\mathbf{n}, \delta\mathbf{u}\big)_{\Gamma_N} \tag{3.64}$$

$$b(\delta\mathbf{u}; \mathbf{u}, \mathbf{u}_a)_\Omega = \big(\rho\mathbf{u}_a\cdot(\nabla\mathbf{u}), \delta\mathbf{u}\big)_\Omega \tag{3.65}$$

$$b(\mathbf{u}; \delta\mathbf{u}, \mathbf{u}_a)_\Omega = -\big(\rho(\mathbf{u}\cdot\nabla)\mathbf{u}_a, \delta\mathbf{u}\big)_\Omega + \big(\rho(\mathbf{u}\cdot\mathbf{n})\mathbf{u}_a, \delta\mathbf{u}\big)_{\Gamma_N} \tag{3.66}$$

$$\big(\nabla\cdot\delta\mathbf{u}, p_a\big)_\Omega = -\big(\nabla p_a, \delta\mathbf{u}\big)_\Omega + \big(p_a\mathbf{n}, \delta\mathbf{u}\big)_{\Gamma_N} \tag{3.67}$$

where $\Gamma_N = \partial\Omega\backslash\Gamma_D$ is the Neumann boundary. According to the third equation in Eq. 3.29, one can obtain

$$\begin{cases} \mathbf{u} = \mathbf{0} \Rightarrow \delta\mathbf{u} = \mathbf{0}, & \phi > 0 \\ H(\phi) = 0, & \phi < 0 \end{cases} \tag{3.68}$$

In addition, the measure of the implicit boundary Γ in Ω is zero. Therefore,

$$\big(H(\phi)\delta\mathbf{u}, \mathbf{u}_a\big)_\Omega = 0 \tag{3.69}$$

By combining Eq. 3.60 with the above equations, one can obtain

$$
\begin{aligned}
&\int_\Omega \left[\frac{\partial \tilde{A}}{\partial \mathbf{u}} - \nabla\cdot\frac{\partial \tilde{A}}{\partial \nabla\mathbf{u}} - \frac{\partial \mathbf{f}}{\partial \mathbf{u}}\cdot\mathbf{u}_a - \eta\Delta\mathbf{u}_a + \rho\mathbf{u}_a\cdot(\nabla\mathbf{u}) - \rho(\mathbf{u}\cdot\nabla)\mathbf{u}_a + \nabla p_a\right]\cdot\delta\mathbf{u}\,\mathrm{d}\Omega \\
&\quad + \int_{\Gamma_N}\left[\eta(\nabla\mathbf{u}_a + \nabla\mathbf{u}_a^{\mathrm{T}})\cdot\mathbf{n} + \rho(\mathbf{u}\cdot\mathbf{n})\mathbf{u}_a - p_a\mathbf{n} + \frac{\partial \tilde{A}}{\partial \nabla\mathbf{u}}\cdot\mathbf{n}\right]\cdot\delta\mathbf{u}\,\mathrm{d}\Gamma = 0 \\
&\int_\Omega \left[\frac{\partial \tilde{A}}{\partial p} - \frac{\partial \mathbf{f}}{\partial p}\cdot\mathbf{u}_a - \nabla\cdot\mathbf{u}_a\right]\delta p\,\mathrm{d}\Omega = 0
\end{aligned}
\tag{3.70}
$$

For the variational of $\mathbf{u} \in \mathscr{H}_E^1(\Omega)$ and $p \in \mathscr{L}^2(\Omega)$ have arbitrariness, the adjoint equations of the Navier-Stokes equations with body force for the topology optimization problem 3.39 can be obtained as

$$\begin{split} -\eta \Delta \mathbf{u}_a + \rho \mathbf{u}_a \cdot (\nabla \mathbf{u}) - \rho (\mathbf{u} \cdot \nabla) \mathbf{u}_a + \nabla p_a &= -\frac{\partial \tilde{A}}{\partial \mathbf{u}} + \nabla \cdot \frac{\partial \tilde{A}}{\partial \nabla \mathbf{u}} + \frac{\partial \mathbf{f}}{\partial \mathbf{u}} \cdot \mathbf{u}_a, \text{ in } \Omega \\ -\nabla \cdot \mathbf{u}_a &= -\frac{\partial \tilde{A}}{\partial p} + \frac{\partial \mathbf{f}}{\partial p} \cdot \mathbf{u}_a, \text{ in } \Omega \\ \mathbf{u}_a &= \mathbf{0}, \text{ on } \Gamma_D \\ \left[-p_a \mathbf{I} + \eta (\nabla \mathbf{u}_a + \nabla \mathbf{u}_a^{\mathrm{T}}) \right] \cdot \mathbf{n} &= -\rho (\mathbf{u} \cdot \mathbf{n}) \mathbf{u}_a - \frac{\partial \tilde{A}}{\partial \nabla \mathbf{u}} \cdot \mathbf{n}, \text{ on } \Gamma_N \end{split} \tag{3.71}$$

3.4.2 *Calculus of Shape Sensitivity*

The shape sensitivity for the topology optimization problem in Eq. 3.39 can be obtained by the adjoint analysis of the augmented functional in Eq. 3.43:

$$\begin{split} \delta \hat{J} &= \frac{\partial \hat{J}}{\partial \mathbf{u}} \cdot \frac{\partial \mathbf{u}}{\partial \phi} \delta \phi + \frac{\partial \hat{J}}{\partial \nabla \mathbf{u}} : \frac{\partial (\nabla \mathbf{u})}{\partial \phi} \delta \phi + \frac{\partial \hat{J}}{\partial p} \frac{\partial p}{\partial \phi} \delta \phi + \frac{\partial \hat{J}}{\partial \phi} \delta \phi \\ &= \frac{\partial \hat{J}}{\partial \mathbf{u}} \cdot \delta \mathbf{u} + \frac{\partial \hat{J}}{\partial \nabla \mathbf{u}} : \nabla (\delta \mathbf{u}) + \frac{\partial \hat{J}}{\partial p} \delta p + \frac{\partial \hat{J}}{\partial \phi} \delta \phi \\ &= \frac{\partial \hat{J}}{\partial \phi} \delta \phi \end{split} \tag{3.72}$$

Based on the first two formulas of Eq. 3.59, the shape sensitivity can be reduced to

$$\begin{split} \delta \hat{J} =& \frac{\partial J}{\partial \phi} \delta \phi + \lambda \int_\Omega \tau(\phi) \, \delta \phi \, \mathrm{d}\Omega + \int_\Omega \tau(\phi) \mathbf{u} \cdot \mathbf{u}_a \delta \phi \, \mathrm{d}\Omega \\ & - \Lambda \int_\Omega \tau(\phi) \, \delta \phi \, \mathrm{d}\Omega \left(\int_\Omega H(-\phi) \, \mathrm{d}\Omega - V_r V_\Omega \right) \end{split} \tag{3.73}$$

In addition,

$$\begin{split} \delta \|\nabla \phi\| &= \delta \left(\frac{(\nabla \phi)^2}{\|\nabla \phi\|} \right) \\ &= \frac{2 \nabla \phi \cdot \nabla (\delta \phi)}{\|\nabla \phi\|} - \frac{(\nabla \phi)^2 \cdot \delta \|\nabla \phi\|}{\|\nabla \phi\|^2} \\ &= \frac{2 \nabla \phi \cdot \nabla (\delta \phi)}{\|\nabla \phi\|} - \delta \|\nabla \phi\| \end{split} \tag{3.74}$$

Therefore,

$$\delta\|\nabla\phi\| = \frac{\nabla\phi\cdot\nabla(\delta\phi)}{\|\nabla\phi\|} \tag{3.75}$$

Based on the similar transformation in [22] and Eq. 3.75, the following transformation can be obtained:

$$\begin{aligned}
\frac{\partial J}{\partial\phi}\delta\phi =& -\int_{\Omega}\tau(\phi)A\,\delta\phi\,\mathrm{d}\Omega + \int_{\Omega}\delta\big(\tau(\phi)\|\nabla\phi\|\big)B\,\mathrm{d}\Omega \\
=& -\int_{\Omega}\tau(\phi)A\,\delta\phi\,\mathrm{d}\Omega + \int_{\Omega}\big[\delta\big(\tau(\phi)\big)\|\nabla\phi\| + \tau(\phi)\,\delta\|\nabla\phi\|\big]B\,\mathrm{d}\Omega \\
=& -\int_{\Omega}\tau(\phi)A\,\delta\phi\,\mathrm{d}\Omega + \int_{\Omega}\left[\delta\big(\tau(\phi)\big)\|\nabla\phi\| + \tau(\phi)\frac{\nabla\phi\cdot\nabla(\delta\phi)}{\|\nabla\phi\|}\right]B\,\mathrm{d}\Omega \\
=& -\int_{\Omega}\tau(\phi)A\,\delta\phi\,\mathrm{d}\Omega + \int_{\Omega}\left[\delta\big(\tau(\phi)\big)\|\nabla\phi\| + \nabla\cdot\left(\tau(\phi)\frac{\nabla\phi}{\|\nabla\phi\|}\delta\phi\right)\right. \\
&\left. - \nabla\cdot\left(\tau(\phi)\frac{\nabla\phi}{\|\nabla\phi\|}\right)\delta\phi\right]B\,\mathrm{d}\Omega \\
=& -\int_{\Omega}\tau(\phi)A\,\delta\phi\,\mathrm{d}\Omega + \int_{\Omega}\Big[\delta\big(\tau(\phi)\big)\|\nabla\phi\| + \nabla\cdot\big(\tau(\phi)\mathbf{n}_{\Gamma}\,\delta\phi\big) \\
& - \delta\phi\tau(\phi)\nabla\cdot\mathbf{n}_{\Gamma} - \delta\big(\tau(\phi)\big)\|\nabla\phi\|\Big]B\,\mathrm{d}\Omega \\
=& -\int_{\Omega}\tau(\phi)A\,\delta\phi\,\mathrm{d}\Omega + \int_{\Omega}\big[\nabla\cdot\big(\tau(\phi)\mathbf{n}_{\Gamma}\,\delta\phi\big) - \delta\phi\tau(\phi)\nabla\cdot\mathbf{n}_{\Gamma}\big]B\,\mathrm{d}\Omega \\
=& -\int_{\Omega}\tau(\phi)A\,\delta\phi\,\mathrm{d}\Omega - \int_{\Omega}\tau(\phi)B\nabla\cdot\mathbf{n}_{\Gamma}\,\delta\phi\,\mathrm{d}\Omega \\
& + \int_{\Omega}\big[\nabla\cdot\big(\tau(\phi)\mathbf{n}_{\Gamma}B\,\delta\phi\big) - \nabla B\cdot\mathbf{n}_{\Gamma}\tau(\phi)\,\delta\phi\big]\,\mathrm{d}\Omega \\
=& -\int_{\Omega}\tau(\phi)A\,\delta\phi\,\mathrm{d}\Omega + \int_{\partial\Omega}\tau(\phi)B\,\delta\phi\,\mathbf{n}_{\Gamma}\cdot\mathbf{n}\,\mathrm{d}\Gamma - \int_{\Omega}\nabla B\cdot\mathbf{n}_{\Gamma}\tau(\phi)\,\delta\phi\,\mathrm{d}\Omega \\
& - \int_{\Omega}\tau(\phi)B\nabla\cdot\mathbf{n}_{\Gamma}\,\delta\phi\,\mathrm{d}\Omega \\
=& -\int_{\Omega}\tau(\phi)A\,\delta\phi\,\mathrm{d}\Omega - \int_{\Omega}\nabla B\cdot\mathbf{n}_{\Gamma}\tau(\phi)\,\delta\phi\,\mathrm{d}\Omega - \int_{\Omega}\tau(\phi)B\kappa\,\delta\phi\,\mathrm{d}\Omega
\end{aligned} \tag{3.76}$$

where $\mathbf{n}_{\Gamma} = \nabla\phi/\|\nabla\phi\|$ and $\kappa = \nabla\cdot\mathbf{n}_{\Gamma}$. Substituting Eq. 3.76 into Eq. 3.73, the shape sensitivity for the topology optimization problem in Eq. 3.39 can be obtained as

$$\begin{aligned}\delta\hat{J} =& -\int_{\Omega} \tau(\phi) A\,\delta\phi\,\mathrm{d}\Omega - \int_{\Omega} \nabla B \cdot \mathbf{n}_{\Gamma} \tau(\phi)\,\delta\phi\,\mathrm{d}\Omega - \int_{\Omega} \tau(\phi) B\kappa\,\delta\phi\,\mathrm{d}\Omega \\ &+ \int_{\Omega} \tau(\phi)\mathbf{u}\cdot\mathbf{u}_a \delta\phi\,\mathrm{d}\Omega + \lambda \int_{\Omega} \tau(\phi)\,\delta\phi\,\mathrm{d}\Omega \\ &- \Lambda\left(\int_{\Omega} H(-\phi)\,\mathrm{d}\Omega - V_r V_{\Omega}\right)\int_{\Omega} \tau(\phi)\,\delta\phi\,\mathrm{d}\Omega \\ =& -\int_{\Omega}\Bigg[A + \nabla B\cdot\mathbf{n}_{\Gamma} + B\kappa - \mathbf{u}\cdot\mathbf{u}_a - \lambda \\ &+ \Lambda\left(\int_{\Omega} H(-\phi)\,\mathrm{d}\Omega - V_r V_{\Omega}\right)\Bigg]\tau(\phi)\delta\phi\,\mathrm{d}\Omega\end{aligned} \tag{3.77}$$

References

1. R.L. Panton, *Incompressible Flow* (Wiley, 1984)
2. T. Borrvall, J. Petersson, Topology optimization of fluid in Stokes flow. Int. J. Numer. Meth. Fluids **41**, 77–107 (2003)
3. A. Gersborg-Hansen, O. Sigmund, R.B. Haber, Topology optimization of channel flow problems. Struct. Multidisc. Optim. **29**, 1–12 (2005)
4. L.H. Olesen, F. Okkels, H. Bruus, A high-level programming-language implementation of topology optimization applied to steady-state Navier-Stokes flow. Int. J. Numer. Meth. Eng. **65**, 975–1001 (2006)
5. D.N. Srinath, S. Mittal, An adjoint method for shape optimization in unsteady viscous flows. J. Comput. Phys. **229**, 1994–2008 (2010)
6. M. Hinze, R. Pinnau, M. Ulbrich, S. Ulbrich, *Optimization with PDE Constraints* (Springer, 2009)
7. M.B. Giles, N.A. Pierce, An introduction to the adjoint approach to design. Flow, Turbul. Combust. **65**, 393–415 (2000)
8. B. Mohammadi, O. Pironneau, *Applied Shape Optimization for Fluids*, (Oxford, 2010)
9. J. Nocedal, S.J. Wright, *Numerical Optimization* (Springer, 1998)
10. K. Svanberg, The method of moving asymptotes: a new method for structural optimization. Int. J. Numer. Meth. Eng. **24**, 359–373 (1987)
11. http://www.comsol.com
12. H. C. Elman, D.J. Silvester, A.J. Wathen, *Finite Elements and Fast Iterative Solvers: with Applications in Incompressible Fluid Dynamics*, (Oxford, 2006)
13. G.K. Batchelor, *An Introduction to Fluid Dynamics* (Cambridge University Press, Cambridge, England, 1967)
14. A. Gersborg-Hansen, M. Berggren, B. Dammann, Topology optimizatiopn of mass distribution problems in Stokes flow. Solid Mech. Appl. **137**, 365–374 (2006)
15. Z. Liu, Q. Gao, P. Zhang, M. Xuan, Y. Wu, Topology optimization of fluid channels with flow rate equality constraints. Struct. Multidisc. Optim. **44**, 31–37 (2011)
16. S. Osher, J.A. Sethian, Front propagating with curvature dependent speed: algorithms based on Hamilton-Jacobi formulations. J. Comput. Phys. **78**, 12–49 (1988)
17. S. Zahedi, A.K. Tornberg, Delta function approximations in level set methods by distance function extension. J. Comput. Phys. **229**, 2199C2219 (2010)
18. B. Engquist, A.K. Tornberg, R. Tsai, Discretization of Dirac delta functions in level set methods. J. Comput. Phys. **207**, 28–51 (2005)

19. A.K. Tornberg, B. Engquist, Regularization techniques for numerical approximation of PDEs with singularities. J. Scientif. Comput. **19**, 1–3 (2003)
20. M.Y. Wang, X. Wang, D. Guo, A level set method for structural optimization. Comput. Meth. Appl. Mech. Eng. **192**, 227–246 (2003)
21. G. Allaire, F. Jouve, A. Toader, Structural optimization using sensitivity analysis and a level-set method. J. Comput. Phys. **194**, 363–393 (2004)
22. S. Zhou, Q. Li, A variational level set method for the topology optimization of steady-state Navier-Stokes flow. J. Comput. Phys. **227**, 10178–10195 (2008)
23. M. Burger, B. Hackl, W. Ring, Incorporating topological derivatives into level set methods. J. Comput. Phys. **194**, 344–362 (2004)
24. L.H. Olessen, F. Okkels, H. Bruus, A high-level programming-language implementation of optimization applied to steady-state Navier-Stokes flow. Int. J. Numer. Meth. Eng. **65**, 975–1001 (2006)
25. V.J. Challis, A discrete levle-set topology optimization code written in Matlab. Struc. Multidisc. Optim. **41**, 453–464 (2009)
26. Y. Jung, K.T. Chu, S. Torquato, A variational level set approach for surface area minimization of triply-periodic surfaces. J. Comput. Phys. **223**, 711–730 (2007)
27. P. Michaleris, D.A. Tortorelli, C.A. Vidal, Tangent operators and design sensitivity formulations for transient nonlinear coupled problems with applications to elasto-plasticity. Int. J. Numer. Meth. Eng. **37**, 2471–2499 (1994)
28. J. Nocedal, S.J. Wright, *Numerical Optimization* (Springer, New York, 2000)
29. K. Ito, K. Kunisch, *Lagrange Multiplier Approach to Variational Problems and Applications*, SIAM, 2008
30. T. Belytschko, S.P. Xiao, C. Parimi, Topology optimization with implictly function and regularization. Int. J. Numer. Meth. Eng. **57**, 1177–1196 (2003)
31. T. Slawig, PDE-constrained control using FEMLAB-Control of the Navier-Stokes equations. Numer. Algor. **42**, 107–126 (2006)
32. H. Maatoug, Shape optimization for the Stokes equations using topological sensitivity analysis. ARIMA **5**, 216–229 (2006)
33. S. Amstutz, The topological asymptotic for the Navier-Stokes equations. ESAIM Control Optim. Calcul. Var. **11**, 401–425 (2005)
34. S. Amstutz, Topological sensitivity analysis for some nonlinear PDE systems. J. Math. Pures Appl. **85**, 540–557 (2006)
35. G. Hauke, T.J.R. Hughes, A unified approach to compressible and incompressible flows. Comp. Meth. Appl. Mech. Eng. **113**, 389–395 (1994)
36. S. Osher, R. Fedkiw, *Level set methods and dynamic implicit surfaces* (Springer, New York, 2003)
37. C. Min, On reinitializing level set functions. J. Comput. Phys. **229**, 2764–2772 (2010)
38. J. Ducrée, S. Haeberle, S. Lutz, S. Pausch, F. Stetten, R. Zengerle, Design and fabrication of a centrifugally driven microfluidic disk for fully integrated metabolic assays on whole blood. J. Micromech. Microeng. **17**, 103–115 (2007)
39. J. Steigert, T. Brenner, M. Grumann, L. Riegger, R. Zengerle, J. Ducrée, Design and fabrication of a centrifugally driven microfluidic disk for fully integrated metabolic assays on whole blood. IEEE MEMS Conf. **229**, 2764–2772 (2006)
40. http://scienceworld.wolfram.com/physics/Navier-StokesEquationsRotational.html

Chapter 4
Topology Optimization for Two-Phase Flows

This chapter presents topology optimization of two-phase flow with immiscible fluids, where the level set method and diffuse-interface model are combined to implement the proposed method. The two-phase flow is described by the diffuse-interface model with essential no slip condition imposed on the wall, where the singularity at the contact line is regularized by the molecular diffusion at the interface between two immiscible fluids. The level set method is utilized to express the fluid and solid phases in the flows and the wall energy at the implicit fluid-solid interface, where the no slip condition on the implicit fluid-solid interface is achieved by the fluid velocity penalization method. Based on the variational procedure for the total free energy of two-phase flow, the Cahn-Hilliard equations for the diffuse-interface model are modified for the two-phase flow with implicit boundary expressed by the level set method. Then the topology optimization problem for the two-phase flow is constructed for the cost functional with general formulation. The sensitivity analysis is implemented by using the continuous adjoint method. The level set function is evolved by solving the Hamilton-Jacobian equation, and numerical test is carried out for capillary to demonstrate the robustness of the proposed topology optimization method for the two-phase flow with two immiscible fluids.

4.1 Level Set Method-Based Implicit Expression of Two-Phase Flow

The solid and fluid phases in the two-phase flow is distinguished implicitly using the sign of the level set function, which is expressed to be a signed distance function defined in the optimization domain [1, 2]. Then, the solid-fluid boundary is represented by the zero level set of the level set function. The implicit expression is (Fig. 4.1)

Y. Deng et al., *Topology Optimization Theory for Laminar Flow*,
https://doi.org/10.1007/978-981-10-4687-2_4

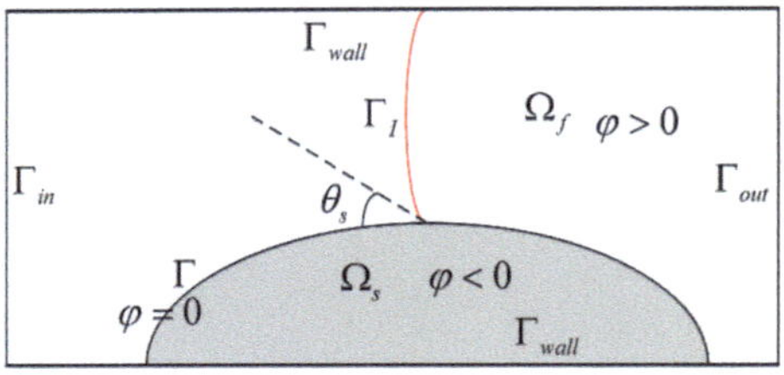

Fig. 4.1 Schematic for the level set based implicit expression of the two-phase flow. Ω_s is the solid phase; Ω_f is the fluid phase; Γ is the implicit solid-fluid boundary; Γ_I is the interface between the two immiscible fluids; θ_s is the contact angle formed by the tangents of the solid surface and immiscible interface; φ is the level set function; Γ_{in}, Γ_{wall} and Γ_{out} are the inlet, wall and outlet boundaries of the optimization domain, respectively

$$\begin{cases} \phi(\mathbf{x}) < 0, & \forall \mathbf{x} \in \Omega_s \\ \phi(\mathbf{x}) > 0, & \forall \mathbf{x} \in \Omega_l \\ \phi(\mathbf{x}) = 0, & \forall \mathbf{x} \in \Gamma \end{cases} \tag{4.1}$$

where ϕ is the level set function satisfying $|\nabla\phi| = 1$; Ω_s, Ω_l and Γ are the solid, fluid and implicit boundary, respectively; Ω_s and Ω_l are open, and they satisfy $\overline{\Omega}_s \cup \overline{\Omega}_l = \overline{\Omega}$, with Ω representing the optimization domain. Based on the implicit expression in Eq. 4.1, the solid is approximated by the fluid velocity penalization method, i.e. adding the artificial Darcy force into the body force term of the Navier-Stokes equations [3], with the artificial Darcy force expressed as

$$\mathbf{f}_r = -\rho\alpha\left(\phi\right)\mathbf{u} \tag{4.2}$$

where ρ is the fluid density; $\mathbf{u}$ is the fluid velocity; $\alpha\left(\phi\right)$ is the impermeability, and it is interpolated by the level set function

$$\alpha\left(\phi\right) = \alpha_s + \left(\alpha_f - \alpha_s\right) H\left(\phi\right) \tag{4.3}$$

with α_f and α_s are the impermeability of the fluid and artificial solid, respectively. The impermeability of fluid is zero, i.e. $\alpha_f = 0$. Theoretically, the impermeability of solid is infinity. However, α_s is chosen to be a high but finite positive value to simultaneously approximate the solid and ensure the numerical stability. In Eq. 4.3, $H\left(\phi\right)$ is the Heaviside function of the level set function

$$H\left(\phi\right) = \begin{cases} 1, & \phi \geq 0 \\ 0, & \phi < 0 \end{cases} \tag{4.4}$$

The derivative of the Heaviside function to ϕ is the Dirac function $\tau\left(\phi\right)$. In numerical computation, the Heaviside function and the Dirac function need to be regularized and approximated by the smoothed Heaviside function and Dirac function [4]

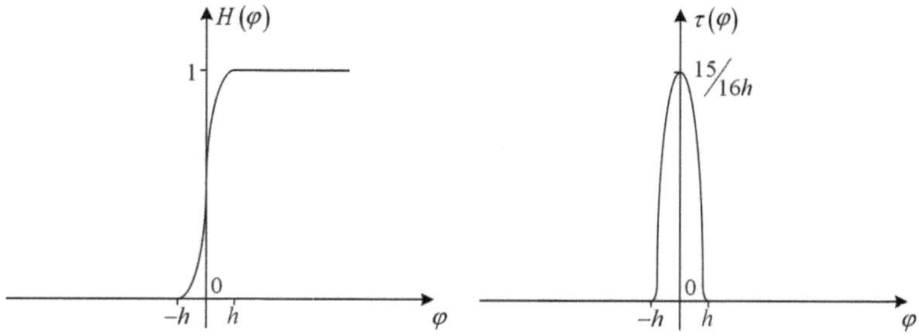

(a) Heaviside function (b) Dirac function

Fig. 4.2 Schematic for the regularized Heaviside function and Dirac function of the level set function, where h is the support size

$$H(\phi) = \begin{cases} 0, \ \phi \leq -h \\ \dfrac{1}{2} + \dfrac{15\phi}{16h} - \dfrac{5\phi^3}{8h^3} + \dfrac{3\phi^5}{16h^5}, & -h < \phi \leq h \\ 1, \ \phi > h \end{cases} \tag{4.5}$$

$$\tau(\phi) = \begin{cases} \dfrac{15}{16h}\left(1 - \dfrac{\phi^2}{h^2}\right)^2, & |\phi| \leq h \\ 0, \ |\phi| > h \end{cases} \tag{4.6}$$

where h is the support size (Fig. 4.2).

4.2 Governing Equations

The diffuse-interface model is used to describe the two-phase flow, in which the singularity at the contact line is regularized by the molecular diffusion in the two-phase interface. And the no slip boundary condition is imposed on the wall without generating infinite dissipation and stress singularity. Therefore, the fluid velocity penalization method is utilized to implicitly express the no slip boundary condition at the interface between solid and fluid.

In the diffuse-interface model, the phase field is defined to represent and evolve the two immiscible fluids of two-phase flow. The evolution of the phase field is governed by the Cahn-Hilliard equations [5, 6], derived based on the surface free energy minimum principle [5]. In topology optimization, the interface between solid and fluid is expressed implicitly. This causes the change of the expression of the total free energy. According to [5], the total free energy of two-phase flow in the optimization domain shown in Fig. 4.1 is written as follows, by introducing the phase field variable φ satisfying $\varphi = \pm 1$ in the two-fluid bulks and $\varphi = 0$ at the immiscible interface

$$\begin{aligned} F &= \int_{\Omega} f_m(\varphi, \nabla\varphi)\, \mathrm{d}\Omega + \int_{\Gamma} f_w(\varphi)\, \mathrm{d}\Gamma + \int_{\Gamma_{wall}} f_w(\varphi)\, \mathrm{d}\Gamma \\ &= \int_{\Omega} \left[f_m(\varphi, \nabla\varphi) + f_w(\varphi)\, \tau(\phi)\, |\nabla\phi| \right] \mathrm{d}\Omega + \int_{\Gamma_{wall}} f_w(\varphi)\, \mathrm{d}\Gamma \end{aligned} \tag{4.7}$$

where $f_m(\varphi, \nabla\varphi)$ and $f_w(\varphi)$ are respectively the mixing energy in the two immiscible fluids [5] and wall energy at the immiscible interface Γ_{wall} [7, 8]. Based on the implicit expression described in Sect. 4.1, the mixing energy and wall energy are

$$\begin{aligned} f_m(\varphi, \nabla\varphi) &= \frac{\lambda(\phi)}{2} |\nabla\varphi|^2 + \frac{\lambda(\phi)}{4\varepsilon_c^2} \left(\varphi^2 - 1\right)^2 \\ f_w(\varphi) &= \begin{cases} -\sigma_0 \dfrac{\varphi\left(3 - \varphi^2\right)}{4} \cos\theta_s + \dfrac{\sigma_{w1} + \sigma_{w2}}{2}, & \text{on } \Gamma \\ -\sigma(\phi) \dfrac{\varphi\left(3 - \varphi^2\right)}{4} \cos\theta_s + \dfrac{\sigma_{w1} + \sigma_{w2}}{2} H(\phi), & \text{on } \Gamma_{wall} \end{cases} \end{aligned} \tag{4.8}$$

where $\lambda(\phi)$ is the mixing energy density interpolated by the level set function; ε_c is the capillary width; θ_s is the contact angle formed by the tangents of the solid surface and immiscible interface; σ_0 is the surface tension at the immiscible interface; σ_{w1} and σ_{w2} are the fluid-solid interfacial tension for the two fluids, respectively; in equilibrium, the surface tension at the immiscible interface is determined by the mixing energy density $\lambda(\phi)$ and capillary width ε_c

$$\sigma = \frac{2\sqrt{2}}{3} \frac{\lambda(\phi)}{\varepsilon_c} \tag{4.9}$$

In principle, the interpolated mixing energy density $\lambda(\phi)$ should keep its value in the fluid Ω_f and vanish to be zero in the solid Ω_s. To ensure the numerical stability, $\lambda(\phi)$ is set to be a nonzero but small enough positive value. Then the mixing energy density is interpolated to be

$$\lambda(\phi) = \lambda_s + \left(\lambda_f - \lambda_s\right) H(\phi) \tag{4.10}$$

where λ_f and λ_s are the mixing energy density for the fluid and approximated solid, respectively. This implicit expression of mixing energy density is also consistent with the surface tension in Eq. 4.9.

Based on the variational procedure in [7, 9], the chemical potential and the boundary condition for the phase field are derived to be

$$
\begin{aligned}
\psi =& \frac{\delta F}{\delta \varphi} \\
=& \left[-\nabla \cdot \left(\lambda\left(\phi\right) \nabla \varphi\right) + \frac{\lambda\left(\phi\right)}{\varepsilon_c^2} \varphi \left(\varphi^2 - 1\right)\right] - \frac{3}{4} \sigma_0 \left(1 - \varphi^2\right) \cos\theta_s \tau\left(\phi\right) \left|\nabla \phi\right|, \text{ in } \Omega \\
& \lambda\left(\phi\right) \mathbf{n} \cdot \nabla \varphi - \frac{3}{4} \sigma\left(\phi\right) \left(1 - \varphi^2\right) \cos\theta_s = 0, \text{ on } \Gamma_{wall}
\end{aligned}
\tag{4.11}
$$

and the body force that is the diffuse-interface equivalent of the interfacial tension [10]

$$
\mathbf{f}_s = H\left(\phi\right) \psi \nabla \varphi \tag{4.12}
$$

is added to the Navier-Stokes equations, where $H\left(\phi\right)$ is used to penalize the interfacial tension and avoid the fluid flow driven by the interfacial tension in the artificial solid phase with $\phi < 0$ [11]. The evolution of the phase filed variable is governed by the convection-diffusion equation [12]

$$
\frac{\partial \varphi}{\partial t} + \mathbf{u} \cdot \nabla \varphi = \nabla \cdot \left[M\left(\phi\right) \nabla \psi\right] \tag{4.13}
$$

where $M\left(\phi\right) > 0$ is the mobility or Onsager coefficient. Equation 4.13 models the creation, evolution, and dissolution of diffusively controlled phase-field interfaces. In the implicitly expressed two-phase flow, the mobility keeps its value in the fluid and is equal to infinity in the solid. Therefore, the mobility is also interpolated by the level set function. Being similar with the impermeability, the mobility is assigned to be a high but finite value in the approximated solid, to ensure the numerical stability. Then the interpolation of the mobility is

$$
M\left(\phi\right) = M_s + \left(M_f - M_s\right) H\left(\phi\right) \tag{4.14}
$$

where M_f and M_s are the mobility for the fluid and approximated solid, respectively.

Based on the above procedure, the governing equations for the implicitly expressed two-phase flow are

$$
\begin{aligned}
&\text{Navier-Stokes equations:} \\
&\rho \left(\frac{\partial \mathbf{u}}{\partial t} + \mathbf{u} \cdot \nabla \mathbf{u}\right) = \nabla \cdot \left[-p\mathbf{I} + \eta \left(\nabla \mathbf{u} + \nabla \mathbf{u}^{\mathrm{T}}\right)\right] + H\left(\phi\right) \psi \nabla \varphi - \rho \alpha\left(\phi\right) \mathbf{u} \\
&\qquad -\nabla \cdot \mathbf{u} = 0 \\
&\text{Cahn-Hilliard equations:} \\
&\frac{\partial \varphi}{\partial t} + \mathbf{u} \cdot \nabla \varphi = \nabla \cdot \left[M\left(\phi\right) \nabla \psi\right] \\
&\qquad \psi = \left[-\nabla \cdot \left(\lambda\left(\phi\right) \nabla \varphi\right) + \frac{\lambda\left(\phi\right)}{\varepsilon_c^2} \varphi \left(\varphi^2 - 1\right)\right] \\
&\qquad\qquad - \frac{3}{4} \sigma_0 \left(1 - \varphi^2\right) \cos\theta_s \tau\left(\phi\right) \left|\nabla \phi\right|
\end{aligned}
\tag{4.15}
$$

where ρ is the fluid density; η is the dynamic viscosity; p is the pressure; $\mathbf{I}$ is the unit tensor. The density and dynamic viscosity are the functions of the phase field variable

$$\begin{aligned} \rho(\varphi) &= \rho_2 + (\rho_1 - \rho_2) H(\varphi) \\ \eta(\varphi) &= \eta_2 + (\eta_1 - \eta_2) H(\varphi) \end{aligned} \tag{4.16}$$

where ρ_1 and ρ_2 are the density of the two immiscible fluids, respectively; η_1 and η_2 are the dynamic viscosity of the two immiscible fluids, respectively; $H(\varphi)$ is the Heaviside function of the phase field variable φ; $H(\varphi)$ and its derivative $\tau(\varphi)$ are smoothed similar to Eqs. 4.5 and 4.6. The initial condition and boundary conditions for the governing Eq. 4.15 are:
initial condition

$$\begin{aligned} \mathbf{u}(t = 0) &= \mathbf{u}_0 \\ \varphi(t = 0) &= \varphi_0 \end{aligned} \tag{4.17}$$

with initial interface represented by $\varphi_0 = 0$; boundary condition at the inlet

$$\begin{aligned} \left[-p\mathbf{I} + \eta\left(\nabla\mathbf{u} + \nabla\mathbf{u}^{\mathrm{T}}\right)\right] \cdot \mathbf{n} &= \mathbf{0} \\ \mathbf{n} \cdot [M(\phi)\nabla\psi] &= 0 \\ \varphi &= \varphi_{in} \end{aligned} \tag{4.18}$$

boundary condition at the wall

$$\begin{aligned} \mathbf{u} &= \mathbf{0} \\ \mathbf{n} \cdot [M(\phi)\nabla\psi] &= 0 \\ \lambda(\phi)\,\mathbf{n} \cdot \nabla\varphi - \frac{3}{4}\sigma(\phi)\left(1 - \varphi^2\right)\cos\theta_s &= 0 \end{aligned} \tag{4.19}$$

and boundary condition at the outlet

$$\begin{aligned} \left[-p\mathbf{I} + \eta\left(\nabla\mathbf{u} + \nabla\mathbf{u}^{\mathrm{T}}\right)\right] \cdot \mathbf{n} &= \mathbf{0} \\ \mathbf{n} \cdot [M(\phi)\nabla\psi] &= 0 \\ \lambda(\phi)\,\mathbf{n} \cdot \nabla\varphi &= 0 \end{aligned} \tag{4.20}$$

where the inlet and outlet boundaries are set to be open type and the wall is set to be no slip type.

4.3 Topology Optimization Problem and Sensitivity Analysis

Based on the level set method-based implicit expression of two-phase flow in Sect. 4.1 and the modification of its governing equations in Sect. 4.2, the corresponding topology optimization problem is constructed and analyzed as follows.

4.3.1 Topology Optimization Problem

The topology optimization problem for the two-phase flow is constructed to be

$$\text{Find } \phi \text{ to maximize or minimize } J\left(\mathbf{u}, p, \psi, \varphi; \phi\right), \text{ subject to}$$
$$\left\{\begin{aligned} & \rho\left(\frac{\partial \mathbf{u}}{\partial t} + \mathbf{u}\cdot\nabla\mathbf{u}\right) = \nabla\cdot\left[-p\mathbf{I} + \eta\left(\nabla\mathbf{u} + \nabla\mathbf{u}^{\mathrm{T}}\right)\right] + H\left(\phi\right)\psi\nabla\varphi \\ & \qquad\qquad\qquad - \rho\alpha\left(\phi\right)\mathbf{u}, \text{ in } (0, T)\times\Omega \\ & -\nabla\cdot\mathbf{u} = 0, \text{ in } (0, T)\times\Omega \\ & \frac{\partial\varphi}{\partial t} + \mathbf{u}\cdot\nabla\varphi = \nabla\cdot\left[M\left(\phi\right)\nabla\psi\right], \text{ in } (0, T)\times\Omega \\ & \psi = -\nabla\cdot\left[\lambda\left(\phi\right)\nabla\varphi\right] + \frac{\lambda\left(\phi\right)}{\varepsilon_c^2}\varphi\left(\varphi^2 - 1\right) \\ & \qquad - \frac{3}{4}\sigma_0\left(1-\varphi^2\right)\cos\theta_s\tau\left(\phi\right)\left|\nabla\phi\right|, \text{ in } (0, T)\times\Omega \\ & \int_\Omega H\left(\phi\right)\,\mathrm{d}\Omega = V_r V_0 \end{aligned}\right. \tag{4.21}$$

where the optimization problem is regularized by a volume constraint; V_r is the volume fraction of the fluids; V_0 is the volume of the optimization domain; $(0, T)$ is the time interval. In this optimization problem, the level set function ϕ is the design variable, which will be iteratively evolved to derive the optimal structural topology used to control the two-phase flow. The objective functional is set to be the general form, including both the domain and boundary integrations about the unknowns of the governing equations

$$\begin{aligned} J\left(\mathbf{u}, p, \psi, \varphi; \phi\right) = & \int_0^T \int_\Omega \beta_1 A\left(\mathbf{u}, \nabla\mathbf{u}, p, \psi, \nabla\psi, \varphi, \nabla\varphi; H\left(\phi\right)\right)\mathrm{d}\Omega\mathrm{d}t \\ & + \int_0^T \int_{\partial\Omega} \beta_2 B\left(\mathbf{u}, p, \psi, \varphi; H\left(\phi\right)\right)\mathrm{d}\Gamma\mathrm{d}t \\ & + \int_0^T \int_\Gamma \omega_3 C\left(\mathbf{u}, p, \psi, \varphi\right)\mathrm{d}\Gamma\mathrm{d}t \end{aligned} \tag{4.22}$$

where A, B and C are respectively the integral functional defined on the optimization domain, domain boundary and zero level set; β_1, β_2 and β_3 are the weights of the integrations in the objective. The objective functional can be further transformed into

$$\begin{aligned} J(\mathbf{u}, p, \psi, \varphi; \phi) = & \int_0^T \int_\Omega \beta_1 A(\mathbf{u}, \nabla\mathbf{u}, p, \psi, \nabla\psi, \varphi, \nabla\varphi; H(\phi)) \\ & + \beta_3 \tau(\phi) |\nabla\phi| C(\mathbf{u}, p, \psi, \varphi) \, \mathrm{d}\Omega \mathrm{d}t \\ & + \int_0^T \int_{\partial\Omega} \beta_2 B(\mathbf{u}, p, \psi, \varphi; H(\phi)) \, \mathrm{d}\Gamma \mathrm{d}t \end{aligned} \tag{4.23}$$

In the level set method, the level set function is evolved by the Hamilton-Jacobian equation [13]

$$\frac{\partial\phi}{\partial\theta} + \mathbf{v} \cdot \nabla\phi + v_n |\nabla\phi| = b\kappa |\nabla\phi| \tag{4.24}$$

where θ is the evolving time of the level set function; $\mathbf{v}$ is the evolving velocity vector; v_n is the normal evolving velocity of the level set function; b is the mean curvature evolving velocity; and κ is the mean curvature of the level set function. For the positive variational of the evolving time ($\delta\theta > 0$), Eq. 4.24 is equivalent to

$$\delta\phi = -(\mathbf{v} \cdot \nabla\phi + v_n |\nabla\phi| - b\kappa |\nabla\phi|) \delta\theta \tag{4.25}$$

To solve the topology optimization problem, the level set function is evolved using the Hamilton-Jacobian equation with reasonably chosen $\mathbf{v}$, v_n and b, derived from the gradient information of the optimization problem.

4.3.2 Sensitivity Analysis

Using the Lagrangian multiplier based adjoint method, the sensitivity analysis of the topology optimization problem in Eq. 4.21 is implemented as that detailed in Appendix 4.7. Based on the derived sensitivity information, $\mathbf{v}$, v_n and b in the Hamilton-Jacobian Eq. 4.24 are chosen to be

$$\begin{aligned} \mathbf{v} &= \int_0^T \nabla \left[\frac{3}{4}\sigma_0 \left(1 - \varphi^2\right) \cos\theta_s \psi_a + \beta_3 C \right] \mathrm{d}t \\ v_n &= -\int_0^T \beta_1 \frac{\partial A}{\partial H(\phi)} - \psi\nabla\varphi \cdot \mathbf{u}_a + \rho\left(\alpha_f - \alpha_s\right) \mathbf{u} \cdot \mathbf{u}_a \\ & \quad + \left(M_f - M_s\right) \nabla\psi \cdot \nabla\varphi_a - \left(\lambda_f - \lambda_s\right) \nabla\varphi \cdot \nabla\psi_a \\ & \quad - \left(\lambda_f - \lambda_s\right) \frac{1}{\varepsilon_c^2} \varphi \left(\varphi^2 - 1\right) \mu_a \, \mathrm{d}t + \xi \\ b &= -\int_0^T \frac{3}{4}\sigma_0 \left(1 - \varphi^2\right) \cos\theta_s \psi_a + \beta_3 C \, \mathrm{d}t \end{aligned} \tag{4.26}$$

for maximizing the objective J; and the sign of Eq. 4.26 is reversed, for minimizing the objective J. In Eq. 4.26, $\mathbf{u}_a$, p_a, φ_a and ψ_a are the adjoint variables of $\mathbf{u}$, p, φ and ψ, respectively; and the adjoint variables are derived by solving the adjoint equations of the Navier-Stokes equations

$$\begin{aligned}
& -\rho\frac{\partial \mathbf{u}_a}{\partial t} + \rho\mathbf{u}_a \cdot (\nabla\mathbf{u}) - \rho(\mathbf{u}\cdot\nabla)\mathbf{u}_a + \nabla\cdot\left[-\eta\left(\nabla\mathbf{u}_a + \nabla\mathbf{u}_a^{\mathrm{T}}\right) + p_a\mathbf{I}\right] + \rho\alpha(\phi)\mathbf{u}_a \\
& \quad + \varphi_a\nabla\varphi + \beta_1\frac{\partial A}{\partial \mathbf{u}} + \beta_3\tau(\phi)|\nabla\phi|\frac{\partial C}{\partial \mathbf{u}} - \beta_1\nabla\cdot\frac{\partial A}{\partial \nabla\mathbf{u}} = 0, \text{ in } (0,T)\times\Omega \\
& -\nabla\cdot\mathbf{u}_a + \left(\beta_1\frac{\partial A}{\partial p} + \beta_3\tau(\phi)|\nabla\phi|\frac{\partial C}{\partial p}\right) = 0, \text{ in } (0,T)\times\Omega \\
& \mathbf{u}_a(t=T) = \mathbf{0}, \text{ in } \Omega \\
& \mathbf{u}_a = -\beta_2\frac{\partial B}{\partial p}\mathbf{n}, \text{ on } \Gamma_{in}\cup\Gamma_{wall} \\
& \left[-p_a\mathbf{I} + \eta\left(\nabla u_a + \nabla u_a^{\mathrm{T}}\right)\right]\cdot\mathbf{n} = -\rho(\mathbf{u}\cdot\mathbf{n})\mathbf{u}_a - \beta_1\frac{\partial A}{\partial\nabla\mathbf{u}}\cdot\mathbf{n}, \text{ on } \Gamma_{out}
\end{aligned} \tag{4.27}$$

and the adjoint equations of the Cahn-Hilliard equations

$$\begin{aligned}
& -\frac{\partial\varphi_a}{\partial t} - (\mathbf{u}\cdot\nabla)\varphi_a + \nabla\cdot(\lambda\nabla\psi_a) - \frac{\lambda}{\varepsilon_c^2}\left(3\varphi^2 - 1\right)\psi_a - \frac{3}{2}\sigma_0\varphi\cos\theta_s\tau(\phi)|\nabla\phi|\psi_a - \\
& \frac{\partial\rho}{\partial\varphi}\mathbf{u}\cdot\frac{\partial\mathbf{u}_a}{\partial t} + \frac{\partial\rho}{\partial\varphi}\mathbf{u}\cdot\nabla\mathbf{u}\cdot\mathbf{u}_a + \frac{\partial\eta}{\partial\varphi}\left(\nabla\mathbf{u} + \nabla\mathbf{u}^{\mathrm{T}}\right):\nabla\mathbf{u}_a + \nabla\cdot[H(\phi)\psi\mathbf{u}_a] - \\
& \beta_1\nabla\cdot\frac{\partial A}{\partial\nabla\varphi} + \beta_1\frac{\partial A}{\partial\varphi} + \beta_3\tau(\phi)|\nabla\phi|\frac{\partial C}{\partial\varphi} + \frac{\partial\rho}{\partial\varphi}\alpha(\phi)\mathbf{u}\cdot\mathbf{u}_a = 0, \text{ in } (0,T)\times\Omega \\
& -\nabla\cdot(M\nabla\varphi_a) - \nabla\cdot\left(\beta_1\frac{\partial A}{\partial\nabla\psi}\right) + \psi_a + \beta_1\frac{\partial A}{\partial\psi} + \beta_3\tau(\phi)|\nabla\phi|\frac{\partial C}{\partial\psi} - \\
& H(\phi)\nabla\varphi\cdot\mathbf{u}_a = 0, \text{ in } (0,T)\times\Omega \\
& \varphi_a = 0, \text{ on } \Gamma_{in} \\
& -\lambda\nabla\psi_a\cdot\mathbf{n} - H(\phi)\psi\mathbf{u}_a\cdot\mathbf{n} + \beta_1\frac{\partial A}{\partial\nabla\varphi}\cdot\mathbf{n} - \frac{3}{2}\sigma\varphi\cos\theta_s\psi_a - \\
& \frac{\partial\eta}{\partial\varphi}\left(\nabla\mathbf{u} + \nabla\mathbf{u}^{\mathrm{T}}\right)\cdot\mathbf{u}_a\cdot\mathbf{n} + \beta_2\frac{\partial B}{\partial\varphi} = 0, \text{ on } \Gamma_{wall} \\
& -\lambda\nabla\psi_a\cdot\mathbf{n} + (\mathbf{u}\cdot\mathbf{n})\varphi_a - H(\phi)\psi\mathbf{u}_a\cdot\mathbf{n} + \beta_1\frac{\partial A}{\partial\nabla\varphi}\cdot\mathbf{n} + \beta_2\frac{\partial B}{\partial\varphi} = 0, \text{ on } \Gamma_{out} \\
& M\nabla\varphi_a\cdot\mathbf{n} + \beta_1\frac{\partial A}{\partial\nabla\psi}\cdot\mathbf{n} + \beta_2\frac{\partial B}{\partial\psi} = 0, \text{ on } \partial\Omega \\
& \varphi_a(t=T) = -\frac{\partial\rho}{\partial\varphi}\mathbf{u}\cdot\mathbf{u}_a, \text{ in } \Omega
\end{aligned} \tag{4.28}$$

With the evolving velocity vector, normal evolving velocity and mean curvature evolving velocity in Eq. 4.26 or its reverse, the level set function can be iteratively evolved by solving the Hamilton-Jacobian Eq. 4.24.

4.4 Numerical Implementation

The topology optimization problem of two-phase flow is solved by iteratively evolving the level set function. The procedure for the iterative evolution process includes the following steps (Table 4.1):

- the coupled system of the Navier-Stokes equations and Cahn-Hilliard equations (Eq. 4.15) are solved with the current distribution of the level set function;
- the adjoint equations (Eqs. 4.27 and 4.28) are solved based on the numerical solution of the coupled system;
- the evolving velocity vector, normal evolving velocity and mean curvature evolving velocity of the level set function are computed according to Eq. 4.26 or its reverse;
- the level set function is evolved by the Hamilton-Jacobian equation, which is solved by the upwind finite difference method. The above steps are iterated until the maximal iteration number is reached.

In the iterations, the Lagrangian multiplier of the volume constraint is determined using the bisection method to ensure the given volume of the fluids [14].

In the above procedure, the coupled equation system and the corresponding adjoint equations are solved by the finite element method implemented in the finite element package COMSOL Multiphysics [4]. For the details on the numerical setting of the optimization procedure implemented in COMSOL, one can refer to [15, 16]. During the optimization procedure, the Navier-Stokes equations and the adjoint equations are solved using Taylor-Hood elements, which respectively interpolate the fluidic velocity quadratically and the pressure linearly [17]; the Cahn-Hilliard equations and the corresponding adjoint equations are solved using the quadratical elements; the level set function is interpolated linearly based on the corner nodes of the elements (Fig. 4.3).

To evolve the level set function, the Hamilton-Jacobin equation is solved by the upwind finite difference method on a regular mesh with ghost elements (Fig. 4.4). The support size h of the smoothed Heaviside function and Dirac function is set to be the size of the regular mesh. The time step for the finite difference scheme is chosen based on the CFL stability condition [18]

Table 4.1 Procedure of the iterative approach for solving the topology optimization problem

1.	Give the initial distribution of the level set function ϕ;
2.	Solve the coupled system of the Navier-Stokes equations and Cahn-Hilliard equations by the finite element method;
3.	Solve the adjoint equations; compute the adjoint sensitivity and the corresponding objective value;
4.	Evolve the level set function by solving the Hamilton-Jacobian equation;
5.	Check for convergence; if the stopping conditions are not satisfied, go to 2;
6.	Post processing

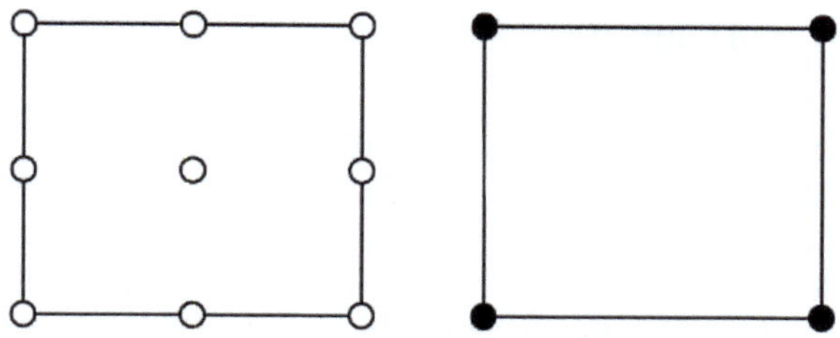

Fig. 4.3 Schematic for the quadratical and linear elements used in the finite element discretization. The fluid velocity **u**, the phase field variable ϕ, the chemical potential ψ and their adjoints are discretized using the quadratical elements; the fluid pressure p and level set function φ are discretized using the linear elements

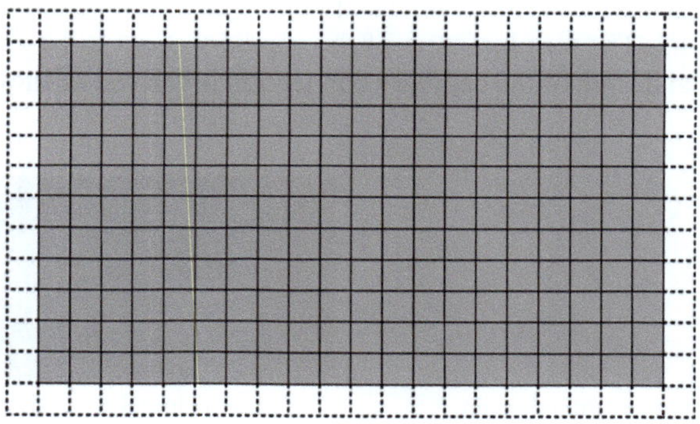

Fig. 4.4 Schematic for the *grid mesh* with ghost elements. The *solid mesh* is used to solve the Navier-Stokes equations and the Cahn-Hilliard equations; the combination of the *solid* and *dash meshes* is used to solve the Hamilton-Jacobian equation for evolving the level set function

$$\Delta\theta \leq \frac{\beta h_E}{\max\left\{\left|\mathbf{v}\cdot\mathbf{n}_\phi + v_n - b\kappa\right|\right\}}, \text{ where } \beta \in (0, 1) \tag{4.29}$$

where $\mathbf{n}_\phi = \frac{\nabla\phi}{|\nabla\phi|}$ is the unit normal vector of the level set function; h_E is the size of the elements and β is chosen to be 0.1. The evolved level set function is reinitialized after several time steps. The reinitialization of the level set function is performed by solving the Hamilton-Jacobin equation without explicitly determining the zero level set [18]

$$\frac{\partial\phi}{\partial\theta} = sign\left(\phi_0\right)\left(1 - |\nabla\phi|\right) \tag{4.30}$$

where $sign\left(\phi_0\right)$, with value either 1, -1 or 0, denotes the sign value of the un-reinitialized level set funtion ϕ_0. Equation 4.30 is also solved using the upwind finite difference method on the regular mesh with ghost elements as shown in Fig. 4.4.

4.5 Numerical Examples

To demonstrate the topology optimization method for two-phase flow, the connection of two capillaries is optimized. The cost functional is set to maximize the kinetic energy of the fluids in the capillary, i.e. A is set to be $-\frac{1}{2}\rho\mathbf{u}^2 H(\phi)$ with the coefficient β_1 equal to 1. The computational domain, the initial distribution of the phase field variable and the initial distribution of the zero level set are respectively set as shown in Fig. 4.5a–c. The size of the rectangular computational domain is set to be $200 \times 100\,\mu\mathrm{m}^2$, with square optimization domain sized $100 \times 100\,\mu\mathrm{m}^2$. The level set function is evolved 10 time steps in every iteration, and it is reinitialized after every 10 iterations. The maximal iteration number is 600. The two fluids in the optimization domain are respectively set to be water and air with density ratio 1×10^3 and dynamic viscosity ratio 1×10^2. The physical and optimization parameters are listed in Table 4.2. The terminal time of the capillary is set to be $T = 1.2 \times 10^{-4}$. Then the optimal design of the capillary connection is derived as shown in Fig. 4.5d.

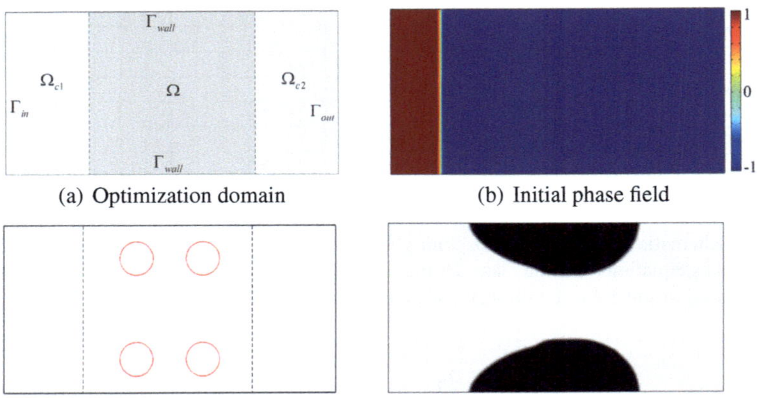

(a) Optimization domain (b) Initial phase field

(c) Initial zero level set (d) Optimal topology

Fig. 4.5 **a** Optimization domain of the capillary connection, where Ω is the optimization domain, Ω_{c1} and Ω_{c2} are capillaries connected to the optimization domain; **b** initial distribution of the phase field variable; **c** initial distribution of the zero level set in the design domain Ω; **d** optimal topology of the capillary connection obtained by the outline topology optimization method for two-phase flow

Table 4.2 Parameter setting for the topology optimization of the capillary connection

ρ_1 (kg/m^3)	ρ_2 (kg/m^3)	η_1 (Pa s)	η_2 (Pa s)	M_f	M_s
1×10^3	1×10^0	1×10^{-3}	1×10^{-5}	1×10^0	1×10^3
λ_f (N/m^2)	λ_s (N/m^2)	ε_c (m)	σ_0 (N/m)	θ_s (rad)	α_s (kg/s)
1×10^0	1×10^{-3}	2.5×10^{-6}	7.2×10^{-2}	$\pi/4$	1×10^7

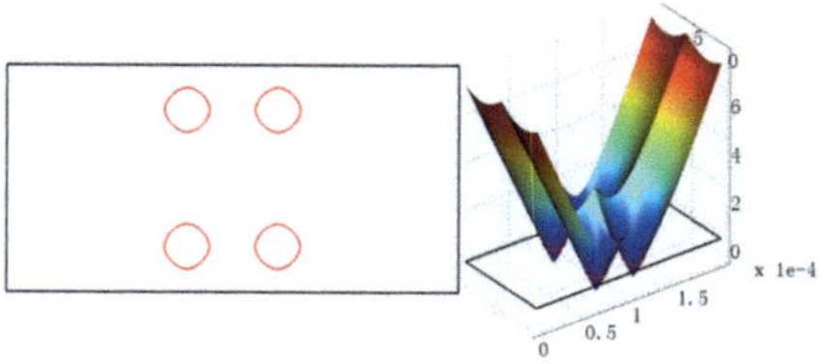

(a) Initial

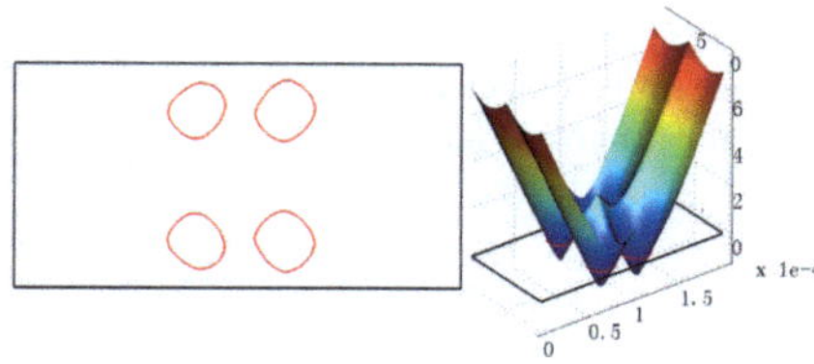

(b) Iteration 10

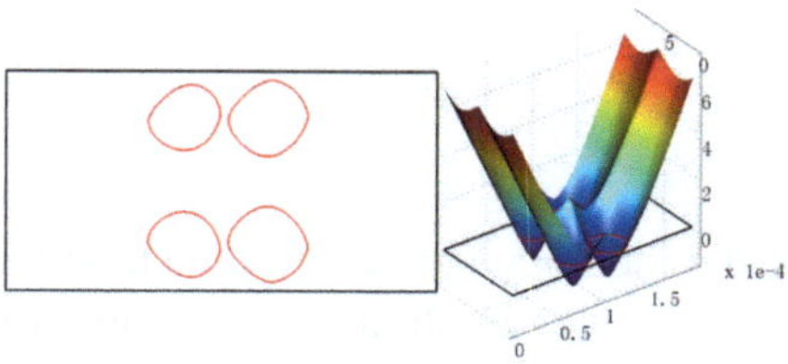

(c) Iteration 17

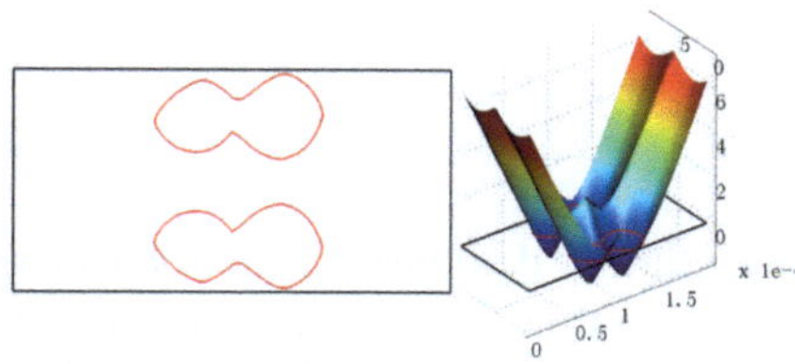

(d) Iteration 20

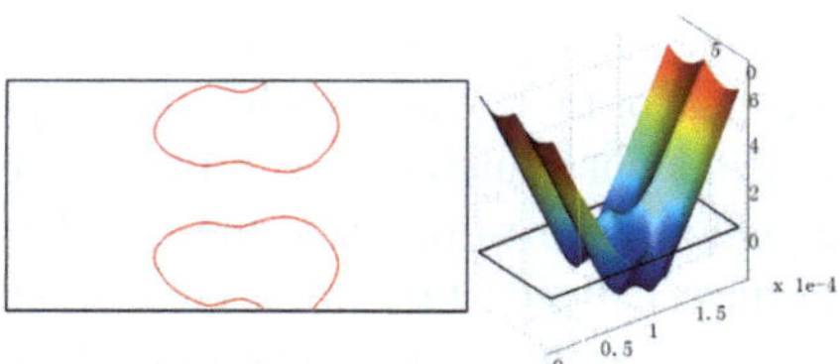

(e) Iteration 30

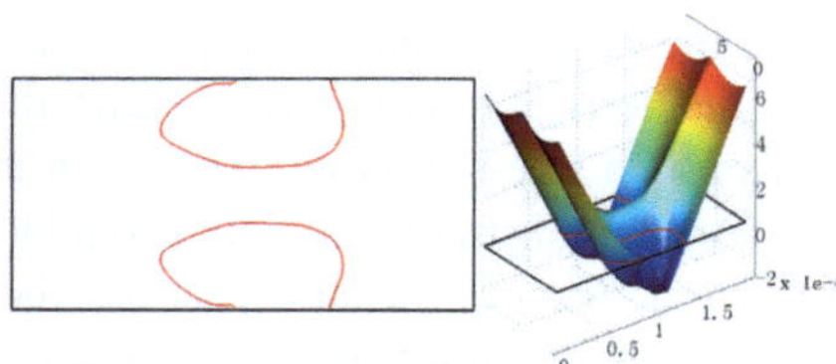

(f) Iteration 60

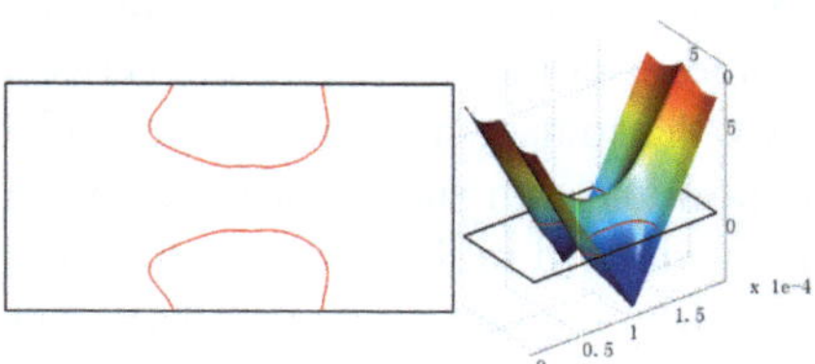

(g) Iteration 100

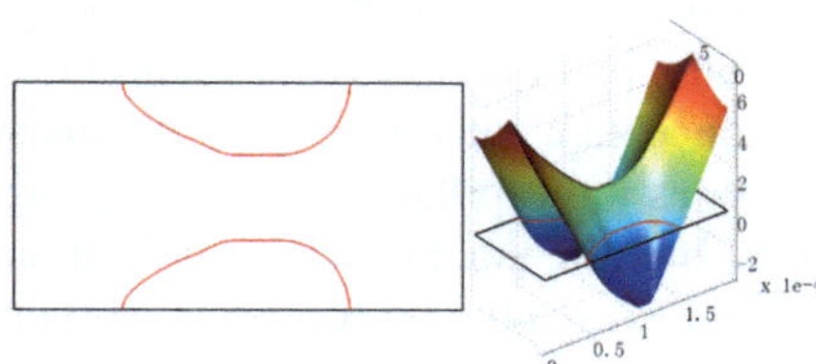

(h) Iteration 200

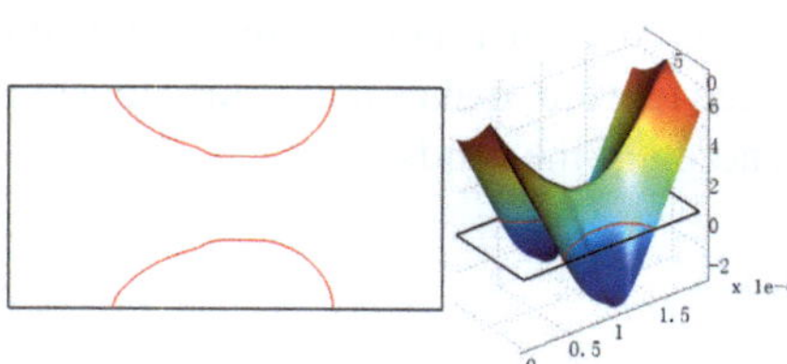

(i) Iteration 580

Fig. 4.6 Snapshots for the evolution of the zero level set and the corresponding level set function

Snapshots for the evolution of the zero level set and level set function are shown in Fig. 4.6. And the convergent histories of the objective value and volume fraction are shown in Fig. 4.7.

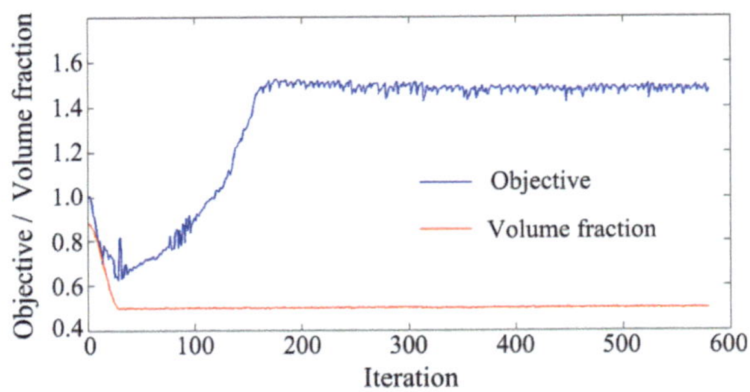

Fig. 4.7 Convergence histories of the objective value and volume fraction

To further demonstrate the robustness of the topology optimization method for two-phase flow, the initial zero level set for the topology optimization of capillary connection is changed to be the one shown in Fig. 4.8a. Correspondingly, snapshots for the evolution of the zero level set and level set function are shown in Fig. 4.8, where the same optimal topology is derived for the capillary connection.

In the derived topology (Fig. 4.5d), the inlet part of the connection has the shape with smooth size-shrinking, which can increase the curvature of the capillary interface. As demonstrated in Young-Laplace equation used to describe the capillary pressure difference sustained across the interface between two immiscible fluids [19, 20], the capillary pressure difference is proportional to the curvature of the capillary interface. Then the shape of the inlet part is to increase the capillary pressure difference and enhance the driving effect of the surface tension. Meanwhile, the outlet part of the derived connection smoothes over the size expansion between the connection and right capillary. This removes the contact angle anchoring, which occurs as the capillary interface passing the position with sudden size expansion. Therefore, the kinetic energy of the fluids is maximized in the derived connection with the sequential performance on increasing the curvature of the capillary interface and removing the contact angle anchoring. This analysis can be confirmed by the snapshots for the evolution of the capillary interface in the optimized connection shown in Fig. 4.9, which demonstrates that the optimized connection keeps the driving effect of the surface tension, makes the surface tension do positive work, and maximizes the kinetic energy of the fluids.

4.6 Summary

This chapter has presented the topology optimization method for the two-phase flow with two immiscible fluids. In the developed method, the diffuse-interface model with essential no slip boundary condition is chosen to describe the two-phase flow, where the applicability of the fluid velocity penalization method is ensured, because the singularity at the contact line of the two-phase flow is regularized by the molecular diffusion instead of slip boundary condition with velocity slip. To avoid the

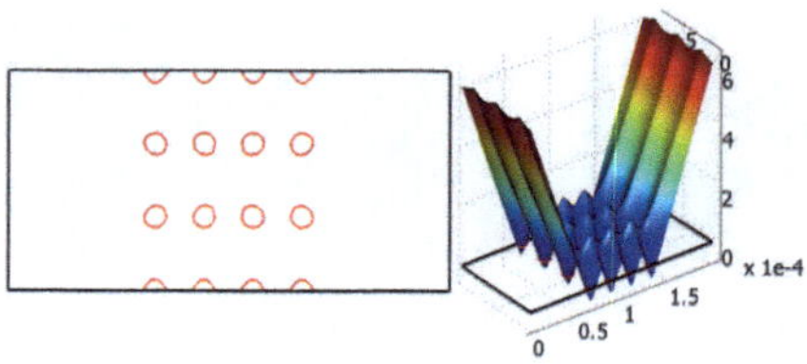

(a) Initial

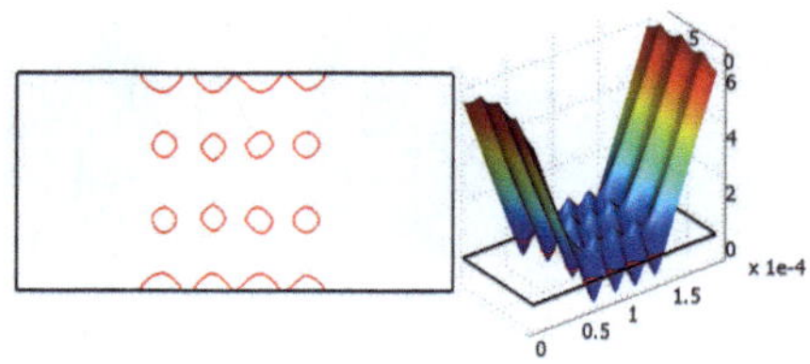

(b) Iteration 10

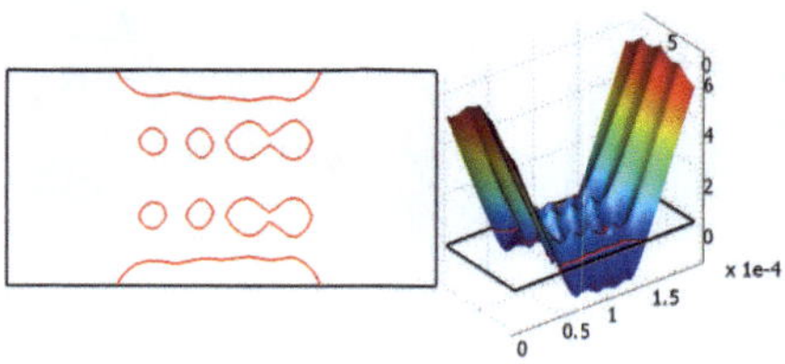

(c) Iteration 20

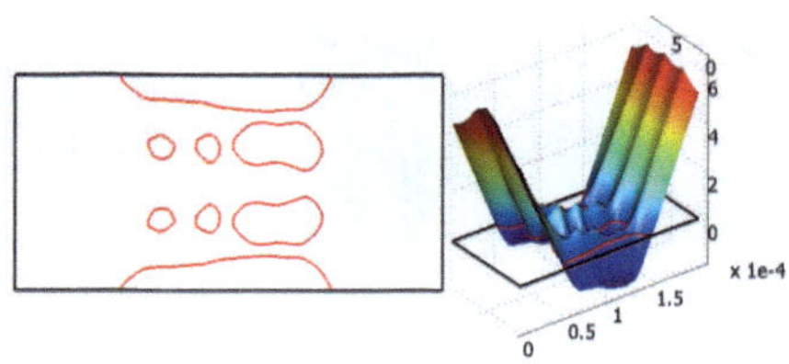

(d) Iteration 25

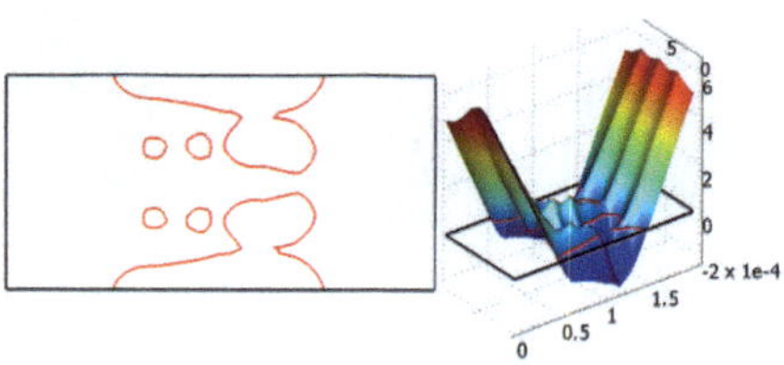

(e) Iteration 30

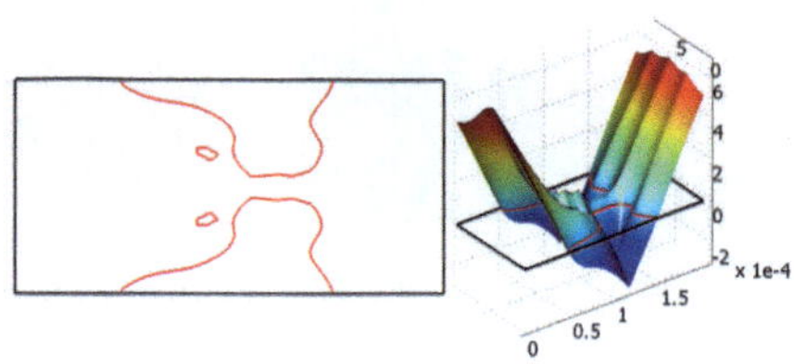

(f) Iteration 48

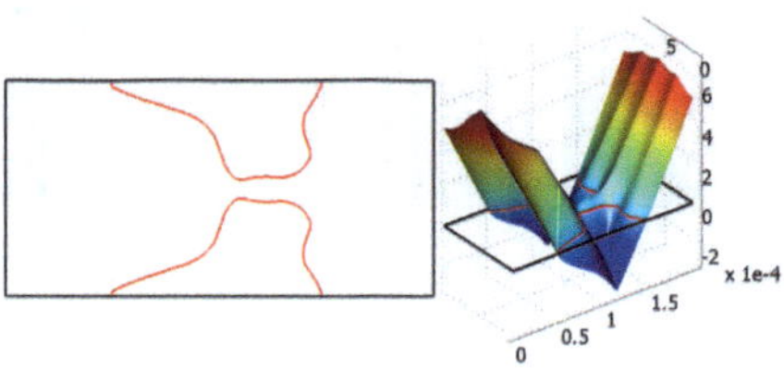

(g) Iteration 55

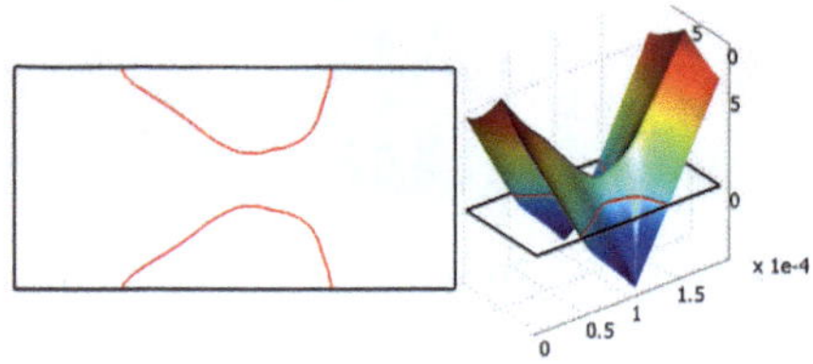

(h) Iteration 140

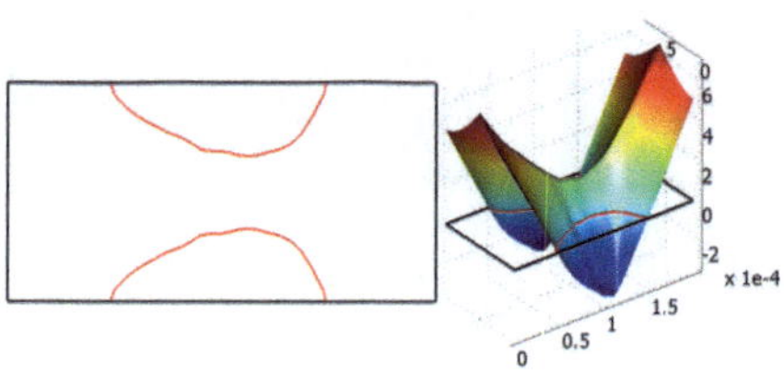

(i) Iteration 180

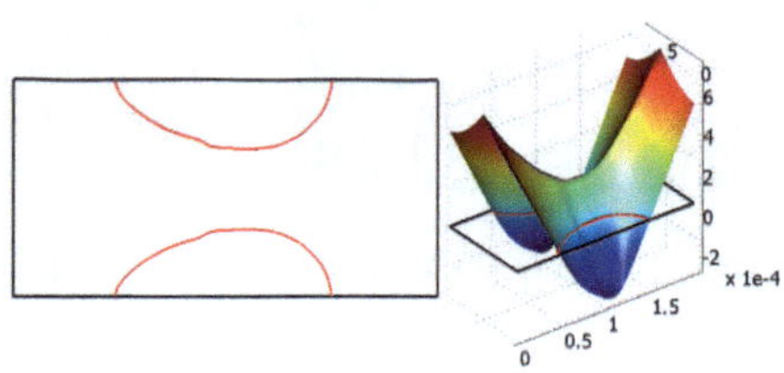

(j) Iteration 255

Fig. 4.8 Snapshots for the evolution of the zero level set and the corresponding level set function when the initial level set function is set to be the one shown in Fig. 4.8a

Fig. 4.9 Snapshots for the evolution of the capillary interface in the optimized capillary connection

artificial fluid flow driven by the interfacial tension in the implicit solid phase and keep the interfacial tension in the fluids, the surface tension term in the Navier-Stokes equations is penalized by the Heaviside of the level set function. To express the wall energy at the implicit solid-fluid interface, the level set method, which can exactly express the normal vector and curvature of the implicit fluid-solid interface, is utilized to implement the topology optimization of two-phase flow. Based on the total free energy of the two-phase flow, the Cahn-Hilliard equations are modified to govern the phase field of the diffuse-interface model defined on the fluids and solid implicitly expressed by the level set function. Then the level set method based variational problem, constrained by the coupled system of the Navier-Stokes equations and the Cahn-Hilliard equations, is constructed and analyzed for the topology optimization of two-phase flow. Numerical examples, with topology optimization of the connection between two capillaries, demonstrate the robustness of the proposed topology optimization method for two-phase flow.

4.7 Appendix

The Lagrangian multiplier based adjoint method are used to perform the sensitivity analysis of the topology optimization problem in Eq. 4.21 [21–23]. According to the Lagrangian multiplier based adjoint method, the augmented Lagrangian corresponding to the optimization problem in Eq. 4.21 is introduced to be

$$\hat{J} = J(\mathbf{u}, p, \psi, \varphi; \phi) + e(\mathbf{u}, p, \psi, \varphi; \phi) - \xi \left(\int_\Omega H(\phi)\, \mathrm{d}\Omega - V_r V_0 \right) \tag{4.31}$$

where

$$\begin{aligned} e(\mathbf{u}, p, \psi, \varphi; \phi) &= \int_\Omega [\rho \mathbf{u} \cdot \mathbf{u}_a]_{t=0}^{t=T}\, \mathrm{d}\Omega + \int_0^T \int_\Omega -\rho \mathbf{u} \cdot \frac{\partial \mathbf{u}_a}{\partial t} + (\rho \mathbf{u} \cdot \nabla \mathbf{u}) \cdot \mathbf{u}_a \\ &+ \left[-p\mathbf{I} + \eta \left(\nabla \mathbf{u} + \nabla \mathbf{u}^\mathrm{T}\right)\right] : \nabla \mathbf{u}_a - H(\phi)\, \psi \nabla \varphi \cdot \mathbf{u}_a + \rho \alpha(\phi)\, \mathbf{u} \cdot \mathbf{u}_a \\ &- p_a \nabla \cdot \mathbf{u}\, \mathrm{d}\Omega \mathrm{d}t + \int_0^T \int_{\partial\Omega} -\left[-p\mathbf{I} + \eta \left(\nabla \mathbf{u} + \nabla \mathbf{u}^\mathrm{T}\right)\right] \cdot \mathbf{u}_a \cdot \mathbf{n}\, \mathrm{d}\Gamma \mathrm{d}t \\ &+ \int_\Omega [\varphi \varphi_a]_{t=0}^{t=T}\, \mathrm{d}\Omega + \int_0^T \int_\Omega -\varphi \frac{\partial \varphi_a}{\partial t} + (\mathbf{u} \cdot \nabla \varphi)\, \varphi_a + M(\phi)\, \nabla \psi \cdot \nabla \varphi_a\, \mathrm{d}\Omega \mathrm{d}t \\ &+ \int_0^T \int_\Omega \psi \psi_a - \lambda(\phi)\, \nabla \varphi \cdot \nabla \psi_a - \frac{\lambda(\phi)}{\varepsilon_c^2} \varphi \left(\varphi^2 - 1\right) \psi_a \\ &+ \frac{3}{4} \sigma_0 \left(1 - \varphi^2\right) \cos\theta_s \tau(\phi)\, |\nabla \phi|\, \psi_a\, \mathrm{d}\Omega \mathrm{d}t + \int_0^T \int_{\Gamma_{in} \bigcup \Gamma_{wall}} \lambda(\phi)\, \nabla \varphi \cdot \mathbf{n} \psi_a\, \mathrm{d}\Gamma \mathrm{d}t \end{aligned} \tag{4.32}$$

$\mathbf{u}_a$, p_a, φ_a and ψ_a are the adjoint variables of $\mathbf{u}$, p, φ and ψ, respectively; φ, ψ, φ_a, ψ_a, and the components of $\mathbf{u}$ and $\mathbf{u}_a$ are valued in the functional space $\mathscr{L}^2\left((0,T);\mathscr{H}(\Omega)\right)$; p and p_a are valued in $\mathscr{L}^2\left((0,T);\mathscr{L}^2(\Omega)\right)$; $\mathscr{L}^2$ is the second-order integrable Lebesgue space; $\mathscr{H}$ is the first-order Sobolev space; the fluid volume constraint is dealt with the Lagrangian multiplier method, and ξ is the corresponding Lagrangian multiplier. According to Karush-Kuhn-Tucker conditions of partial differential equation constrained optimization problem [22], the variational of $\hat{J}$

$$\delta\hat{J} = \left\langle \frac{\partial J}{\partial \mathbf{u}}, \delta\mathbf{u} \right\rangle + \left\langle \frac{\partial J}{\partial p}, \delta p \right\rangle + \left\langle \frac{\partial J}{\partial \varphi}, \delta\varphi \right\rangle + \left\langle \frac{\partial J}{\partial \psi}, \delta\psi \right\rangle + \left\langle \frac{\partial J}{\partial \phi}, \delta\phi \right\rangle \tag{4.33}$$

should be zero at the optimal distribution of the level set function. Therefore, the adjoint equations are derived based on the following equations

$$\left\langle \frac{\partial J}{\partial \mathbf{u}}, \delta\mathbf{u} \right\rangle = 0;\ \left\langle \frac{\partial J}{\partial p}, \delta p \right\rangle = 0;\ \left\langle \frac{\partial J}{\partial \varphi}, \delta\varphi \right\rangle = 0;\ \left\langle \frac{\partial J}{\partial \psi}, \delta\psi \right\rangle = 0 \tag{4.34}$$

The variational of the augmented Lagrangian is

$$\begin{aligned}
\delta\hat{J} = & \int_0^T \int_\Omega \left[\beta_1 \left(\frac{\partial A}{\partial \mathbf{u}} - \nabla\cdot\frac{\partial A}{\partial \nabla\mathbf{u}} \right) + \beta_3 \tau(\phi) |\nabla\phi| \frac{\partial C}{\partial \mathbf{u}} \right] \cdot \delta\mathbf{u}\, \mathrm{d}\Omega \mathrm{d}t \\
& + \int_0^T \int_{\partial\Omega} \beta_1 \frac{\partial A}{\partial \nabla\mathbf{u}} \cdot \mathbf{n} \cdot \delta\mathbf{u}\, \mathrm{d}\Gamma \mathrm{d}t + \int_0^T \int_\Omega \left[\beta_1 \frac{\partial A}{\partial p} + \beta_3 \tau(\phi) |\nabla\phi| \frac{\partial C}{\partial p} \right] \delta p\, \mathrm{d}\Omega \mathrm{d}t \\
& + \int_0^T \int_\Omega \left[\beta_1 \left(\frac{\partial A}{\partial \psi} - \nabla\cdot\frac{\partial A}{\partial \nabla\psi} \right) + \beta_3 \tau(\phi) |\nabla\phi| \frac{\partial C}{\partial \psi} \right] \delta\psi\, \mathrm{d}\Omega \mathrm{d}t \\
& + \int_0^T \int_{\partial\Omega} \beta_1 \frac{\partial A}{\partial \nabla\psi} \cdot \mathbf{n}\delta\psi\, \mathrm{d}\Gamma \mathrm{d}t + \int_0^T \int_\Omega \left[\beta_1 \left(\frac{\partial A}{\partial \varphi} - \nabla\cdot\frac{\partial A}{\partial \nabla\varphi} \right) \right. \\
& \left. + \beta_3 \tau(\phi) |\nabla\phi| \frac{\partial C}{\partial \varphi} \right] \delta\varphi\, \mathrm{d}\Omega \mathrm{d}t + \int_0^T \int_{\partial\Omega} \beta_1 \frac{\partial A}{\partial \nabla\varphi} \cdot \mathbf{n}\delta\varphi\, \mathrm{d}\Gamma \mathrm{d}t \\
& + \int_0^T \int_\Omega \left[\beta_1 \frac{\partial A}{\partial H(\phi)} \tau(\phi) + \beta_3 \frac{\partial(\tau(\phi)|\nabla\phi|)}{\partial\phi} C \right] \delta\phi\, \mathrm{d}\Omega \mathrm{d}t \\
& + \int_0^T \int_{\partial\Omega} \beta_2 \left[\frac{\partial B}{\partial \mathbf{u}} \delta\mathbf{u} + \frac{\partial B}{\partial p} \delta p + \frac{\partial B}{\partial \psi} \delta\psi + \frac{\partial B}{\partial \varphi} \delta\varphi + \frac{\partial B}{\partial H(\phi)} \tau(\phi) \delta\phi \right] \mathrm{d}\Gamma \mathrm{d}t
\end{aligned}$$

$$\begin{aligned}
& + \int_\Omega \left[\rho\delta\mathbf{u}\cdot\mathbf{u}_a + \frac{\partial\rho}{\partial\varphi} \mathbf{u}\cdot\mathbf{u}_a\delta\varphi \right]_{t=T} \mathrm{d}\Omega + \int_0^T \int_\Omega -\rho\delta\mathbf{u}\cdot\frac{\partial\mathbf{u}_a}{\partial t} - \frac{\partial\rho}{\partial\varphi}\mathbf{u}\cdot\frac{\partial\mathbf{u}_a}{\partial t}\delta\varphi \\
& + (\rho\delta\mathbf{u}\cdot\nabla\mathbf{u})\cdot\mathbf{u}_a + (\rho\mathbf{u}\cdot\nabla\delta\mathbf{u})\cdot\mathbf{u}_a + \left(\frac{\partial\rho}{\partial\varphi}\mathbf{u}\cdot\nabla\mathbf{u} \right)\cdot\mathbf{u}_a\delta\varphi
\end{aligned}$$

$$
\begin{aligned}
&+\left[-\delta p\mathbf{I}+\eta\left(\nabla\delta\mathbf{u}+\nabla\delta\mathbf{u}^{\mathrm{T}}\right)\right]:\nabla\mathbf{u}_a+\left[\frac{\partial\eta}{\partial\varphi}\left(\nabla\mathbf{u}+\nabla\mathbf{u}^{\mathrm{T}}\right)\right]:\nabla\mathbf{u}_a\delta\varphi\\
&-H\left(\phi\right)\psi\nabla\delta\varphi\cdot\mathbf{u}_a-H\left(\phi\right)\delta\psi\nabla\varphi\cdot\mathbf{u}_a-\tau\left(\phi\right)\psi\nabla\varphi\cdot\mathbf{u}_a\delta\phi\\
&+\rho\alpha\left(\phi\right)\delta\mathbf{u}\cdot\mathbf{u}_a+\frac{\partial\rho}{\partial\varphi}\alpha\left(\phi\right)\mathbf{u}\cdot\mathbf{u}_a\delta\varphi+\rho\frac{\partial\alpha\left(\phi\right)}{\partial\phi}\mathbf{u}\cdot\mathbf{u}_a\delta\phi-p_a\nabla\cdot\delta\mathbf{u}\,\mathrm{d}\Omega\mathrm{d}t\\
&+\int_0^T\int_{\Gamma_{in}\bigcup\Gamma_{wall}}\mathbf{u}_a\cdot\mathbf{n}\delta p-\frac{\partial\eta}{\partial\varphi}\left(\nabla\mathbf{u}+\nabla\mathbf{u}^{\mathrm{T}}\right)\cdot\mathbf{u}_a\cdot\mathbf{n}\delta\varphi\,\mathrm{d}\Gamma\mathrm{d}t+\int_\Omega\left[\phi_a\delta\varphi\right]_{t=T}\mathrm{d}\Omega\\
&+\int_0^T\int_\Omega-\delta\varphi\frac{\partial\varphi_a}{\partial t}+\left(\delta\mathbf{u}\cdot\nabla\varphi\right)\varphi_a+\left(\mathbf{u}\cdot\nabla\delta\varphi\right)\varphi_a+\frac{\partial M}{\partial\phi}\nabla\psi\cdot\nabla\varphi_a\delta\phi\\
&+M\nabla\delta\psi\cdot\nabla\varphi_a\,\mathrm{d}\Omega\mathrm{d}t+\int_0^T\int_\Omega\delta\psi\psi_a-\lambda\nabla\delta\varphi\cdot\nabla\psi_a-\frac{\partial\lambda}{\partial\phi}\nabla\varphi\cdot\nabla\psi_a\delta\phi\\
&-\frac{\lambda}{\varepsilon_c^2}\left(3\varphi^2-1\right)\psi_a\delta\varphi-\frac{\partial\lambda}{\partial\phi}\frac{1}{\varepsilon_c^2}\varphi\left(\varphi^2-1\right)\psi_a\delta\phi-\frac{3}{2}\sigma_0\varphi\cos\theta_s\tau\left(\phi\right)\left|\nabla\phi\right|\psi_a\delta\varphi\\
&+\frac{3}{4}\sigma_0\left(1-\varphi^2\right)\cos\theta_s\frac{\partial(\tau\left(\phi\right)\left|\nabla\phi\right|)}{\partial\phi}\psi_a\delta\phi\,\mathrm{d}\Omega\mathrm{d}t+\int_0^T\int_{\Gamma_{in}}\lambda\nabla\delta\varphi\cdot\mathbf{n}\psi_a\\
&+\frac{\partial\lambda}{\partial\phi}\nabla\varphi\cdot\mathbf{n}\psi_a\delta\phi\,\mathrm{d}\Gamma\mathrm{d}t+\int_0^T\int_{\Gamma_{wall}}-\frac{3}{2}\sigma\varphi\cos\theta_s\psi_a\delta\varphi+\\
&\frac{\sqrt{2}}{2}\frac{\lambda_h-\lambda_l}{\varepsilon_c}\left(1-\varphi^2\right)\cos\theta_s\tau\left(\phi\right)\psi_a\delta\phi\,\mathrm{d}\Gamma\mathrm{d}t+\int_\Omega-\xi\tau\left(\phi\right)\delta\phi\,\mathrm{d}\Omega
\end{aligned}
\tag{4.35}
$$

Then based on the first two formulas in Eq. 4.34, the weak form of the adjoint equations for the Navier-Stokes equations is derived as

$$
\begin{aligned}
&\int_0^T\int_\Omega-\rho\delta\mathbf{u}\cdot\frac{\partial\mathbf{u}_a}{\partial t}+\left(\rho\delta\mathbf{u}\cdot\nabla\mathbf{u}\right)\cdot\mathbf{u}_a\\
&+\left(\rho\mathbf{u}\cdot\nabla\delta\mathbf{u}\right)\cdot\mathbf{u}_a+\left[-\delta p\mathbf{I}+\eta\left(\nabla\delta\mathbf{u}+\nabla\delta\mathbf{u}^{\mathrm{T}}\right)\right]:\nabla\mathbf{u}_a\\
&+\rho\alpha\left(\phi\right)\delta\mathbf{u}\cdot\mathbf{u}_a-p_a\nabla\cdot\delta\mathbf{u}+\left(\delta\mathbf{u}\cdot\nabla\varphi\right)\varphi_a\\
&+\left(\beta_1\frac{\partial A}{\partial p}+\beta_3\tau\left(\phi\right)\left|\nabla\phi\right|\frac{\partial C}{\partial p}\right)\cdot\delta p+\left(\beta_1\frac{\partial A}{\partial\mathbf{u}}+\beta_3\tau\left(\phi\right)\left|\nabla\phi\right|\frac{\partial C}{\partial\mathbf{u}}\right)\cdot\delta\mathbf{u}\\
&+\beta_1\frac{\partial A}{\partial\nabla\mathbf{u}}:\nabla\delta\mathbf{u}\,\mathrm{d}\Omega\mathrm{d}t+\int_0^T\int_{\Gamma_{in}\bigcup\Gamma_{wall}}\left(\mathbf{u}_a\cdot\mathbf{n}+\beta_2\frac{\partial B}{\partial p}\right)\delta p\,\mathrm{d}\Gamma\mathrm{d}t\\
&+\int_0^T\int_{\Gamma_{out}}\beta_2\frac{\partial B}{\partial\mathbf{u}}\cdot\delta\mathbf{u}\,\mathrm{d}\Gamma\mathrm{d}t=0;\\
&\mathbf{u}_a\left(t=T\right)=\mathbf{0},\ \text{in }\Omega
\end{aligned}
\tag{4.36}
$$

And the corresponding strong form of the adjoint equations for the Navier-Stokes equations is

$$
\begin{aligned}
&-\rho\frac{\partial \mathbf{u}_a}{\partial t}+\rho\mathbf{u}_a\cdot(\nabla\mathbf{u})-\rho(\mathbf{u}\cdot\nabla)\mathbf{u}_a+\nabla\cdot\left[-\eta\left(\nabla\mathbf{u}_a+\nabla\mathbf{u}_a^{\mathrm{T}}\right)+p_a\mathbf{I}\right]\\
&+\rho\alpha(\phi)\mathbf{u}_a+\varphi_a\nabla\varphi+\beta_1\frac{\partial A}{\partial\mathbf{u}}+\beta_3\tau(\phi)|\nabla\phi|\frac{\partial C}{\partial\mathbf{u}}\\
&-\beta_1\nabla\cdot\frac{\partial A}{\partial\nabla\mathbf{u}}=0,\ \text{in}\ (0,T)\times\Omega\\
&-\nabla\cdot\mathbf{u}_a+\left(\beta_1\frac{\partial A}{\partial p}+\beta_3\tau(\phi)|\nabla\phi|\frac{\partial C}{\partial p}\right)=0,\ \text{in}\ (0,T)\times\Omega\\
&\mathbf{u}_a(t=T)=\mathbf{0},\ \text{in}\ \Omega\\
&\mathbf{u}_a=-\beta_2\frac{\partial B}{\partial p}\mathbf{n},\ \text{on}\ (0,T)\times(\Gamma_{in}\cup\Gamma_{wall})\\
&\left[-p_a\mathbf{I}+\eta\left(\nabla u_a+\nabla u_a^{\mathrm{T}}\right)\right]\cdot\mathbf{n}=-\rho(\mathbf{u}\cdot\mathbf{n})\mathbf{u}_a-\beta_1\frac{\partial A}{\partial\nabla\mathbf{u}}\cdot\mathbf{n},\ \text{on}\ (0,T)\times\Gamma_{out}
\end{aligned}
\tag{4.37}
$$

Based on the third and fourth formulas in Eq. 4.34, the weak form of the adjoint equation for the Cahn-Hilliard equations is derived as

$$
\begin{aligned}
&\int_0^T\int_\Omega -\delta\varphi\frac{\partial\varphi_a}{\partial t}+(\mathbf{u}\cdot\nabla\delta\varphi)\varphi_a-\lambda\nabla\delta\varphi\cdot\nabla\psi_a-\frac{\lambda}{\varepsilon_c^2}\left(3\varphi^2-1\right)\psi_a\delta\varphi\\
&-\frac{3}{2}\sigma_0\varphi\cos\theta_s\tau(\phi)|\nabla\phi|\psi_a\delta\varphi-\frac{\partial\rho}{\partial\varphi}\mathbf{u}\cdot\frac{\partial\mathbf{u}_a}{\partial t}\delta\varphi+\frac{\partial\rho}{\partial\varphi}\mathbf{u}\cdot\nabla\mathbf{u}\cdot\mathbf{u}_a\delta\varphi\\
&+\frac{\partial\eta}{\partial\varphi}\left(\nabla\mathbf{u}+\nabla\mathbf{u}^{\mathrm{T}}\right):\nabla\mathbf{u}_a\delta\varphi-H(\phi)\psi\nabla\delta\varphi\cdot\mathbf{u}_a+\beta_1\frac{\partial A}{\partial\nabla\varphi}\cdot\nabla\delta\varphi\\
&+\left(\beta_1\frac{\partial A}{\partial\varphi}+\beta_3\tau(\phi)|\nabla\phi|\frac{\partial C}{\partial\varphi}\right)\delta\varphi+\frac{\partial\rho}{\partial\varphi}\alpha(\phi)\mathbf{u}\cdot\mathbf{u}_a\delta\varphi\,\mathrm{d}\Omega\mathrm{d}t\\
&+\int_0^T\int_{\Gamma_{in}}\lambda\nabla\delta\varphi\cdot\mathbf{n}\psi_a-\frac{\partial\eta}{\partial\varphi}\left(\nabla\mathbf{u}+\nabla\mathbf{u}^{\mathrm{T}}\right)\cdot\mathbf{u}_a\cdot\mathbf{n}\delta\varphi\,\mathrm{d}\Gamma\mathrm{d}t\\
&+\int_0^T\int_{\Gamma_{wall}}-\frac{3}{2}\sigma\varphi\cos\theta_s\psi_a\delta\varphi-\frac{\partial\eta}{\partial\varphi}\left(\nabla\mathbf{u}+\nabla\mathbf{u}^{\mathrm{T}}\right)\cdot\mathbf{u}_a\cdot\mathbf{n}\delta\varphi+\beta_2\frac{\partial B}{\partial\varphi}\delta\varphi\,\mathrm{d}\Gamma\mathrm{d}t\\
&+\int_0^T\int_{\Gamma_{out}}\beta_2\frac{\partial B}{\partial\varphi}\delta\varphi\,\mathrm{d}\Gamma\mathrm{d}t+\int_0^T\int_\Omega M\nabla\delta\psi\cdot\nabla\varphi_a+\beta_1\frac{\partial A}{\partial\nabla\psi}\cdot\nabla\delta\psi\\
&+\left(\psi_a+\beta_1\frac{\partial A}{\partial\psi}+\beta_3\tau(\phi)|\nabla\phi|\frac{\partial C}{\partial\psi}-H(\phi)\nabla\varphi\cdot\mathbf{u}_a\right)\delta\psi\,\mathrm{d}\Omega\mathrm{d}t\\
&+\int_0^T\int_{\partial\Omega}\beta_2\frac{\partial B}{\partial\psi}\delta\psi\,\mathrm{d}\Gamma\mathrm{d}t=0;\\
&\varphi_a(t=T)=-\frac{\partial\rho}{\partial\varphi}\mathbf{u}\cdot\mathbf{u}_a,\ \text{in}\ \Omega
\end{aligned}
\tag{4.38}
$$

And the corresponding strong form is

$$
\begin{aligned}
& -\frac{\partial \varphi_a}{\partial t}-(\mathbf{u}\cdot\nabla)\varphi_a+\nabla\cdot(\lambda\nabla\psi_a)-\frac{\lambda}{\varepsilon_c^2}\left(3\varphi^2-1\right)\psi_a-\frac{3}{2}\sigma_0\varphi\cos\theta_s\tau(\phi)|\nabla\phi|\psi_a- \\
& \frac{\partial\rho}{\partial\varphi}\mathbf{u}\cdot\frac{\partial\mathbf{u}_a}{\partial t}+\frac{\partial\rho}{\partial\varphi}\mathbf{u}\cdot\nabla\mathbf{u}\cdot\mathbf{u}_a+\frac{\partial\eta}{\partial\varphi}\left(\nabla\mathbf{u}+\nabla\mathbf{u}^{\mathrm{T}}\right):\nabla\mathbf{u}_a+\nabla\cdot[H(\phi)\psi\mathbf{u}_a]- \\
& \beta_1\nabla\cdot\frac{\partial A}{\partial\nabla\varphi}+\beta_1\frac{\partial A}{\partial\varphi}+\beta_3\tau(\phi)|\nabla\phi|\frac{\partial C}{\partial\varphi}+\frac{\partial\rho}{\partial\varphi}\alpha(\phi)\mathbf{u}\cdot\mathbf{u}_a=0,\text{ in }(0,T)\times\Omega \\
& -\nabla\cdot(M\nabla\varphi_a)-\nabla\cdot\left(\beta_1\frac{\partial A}{\partial\nabla\psi}\right)+\psi_a+\beta_1\frac{\partial A}{\partial\psi}+\beta_3\tau(\phi)|\nabla\phi|\frac{\partial C}{\partial\psi}- \\
& H(\phi)\nabla\varphi\cdot\mathbf{u}_a=0,\text{ in }(0,T)\times\Omega \\
& \varphi_a=0,\text{ on }(0,T)\times\Gamma_{in} \\
& -\lambda\nabla\psi_a\cdot\mathbf{n}-H(\phi)\psi\mathbf{u}_a\cdot\mathbf{n}+\beta_1\frac{\partial A}{\partial\nabla\varphi}\cdot\mathbf{n}-\frac{3}{2}\sigma\varphi\cos\theta_s\psi_a- \\
& \frac{\partial\eta}{\partial\varphi}\left(\nabla\mathbf{u}+\nabla\mathbf{u}^{\mathrm{T}}\right)\cdot\mathbf{u}_a\cdot\mathbf{n}+\beta_2\frac{\partial B}{\partial\varphi}=0,\text{ on }(0,T)\times\Gamma_{wall} \\
& -\lambda\nabla\psi_a\cdot\mathbf{n}+(\mathbf{u}\cdot\mathbf{n})\varphi_a-H(\phi)\psi\mathbf{u}_a\cdot\mathbf{n}+\beta_1\frac{\partial A}{\partial\nabla\varphi}\cdot\mathbf{n}+\beta_2\frac{\partial B}{\partial\varphi}=0,\text{ on }(0,T)\times\Gamma_{out} \\
& M\nabla\varphi_a\cdot\mathbf{n}+\beta_1\frac{\partial A}{\partial\nabla\psi}\cdot\mathbf{n}+\beta_2\frac{\partial B}{\partial\psi}=0,\text{ on }(0,T)\times\partial\Omega \\
& \varphi_a(t=T)=-\frac{\partial\rho}{\partial\varphi}\mathbf{u}\cdot\mathbf{u}_a,\text{ in }\Omega
\end{aligned}
\tag{4.39}
$$

Then, the adjoint sensitivity of the optimization problem in the optimization domain Ω is derived to be

$$
\begin{aligned}
& \delta\hat{J}(\Omega)= \\
& \int_0^T\int_\Omega\left\{\beta_1\frac{\partial A}{\partial H(\phi)}-\psi\nabla\varphi\cdot\mathbf{u}_a+\rho\left(\alpha_f-\alpha_s\right)\mathbf{u}\cdot\mathbf{u}_a+\left(M_f-M_s\right)\nabla\psi\cdot\nabla\varphi_a\right. \\
& -\left(\lambda_f-\lambda_s\right)\nabla\varphi\cdot\nabla\psi_a-\left(\lambda_f-\lambda_s\right)\frac{1}{\varepsilon_c^2}\varphi\left(\varphi^2-1\right)\psi_a \\
& -\nabla\left[\frac{3}{4}\sigma_0\left(1-\varphi^2\right)\cos\theta_s\psi_a+\beta_3 C\right]\cdot\mathbf{n}_\phi \\
& \left.-\left[\frac{3}{4}\sigma_0\left(1-\varphi^2\right)\cos\theta_s\psi_a+\beta_3 C\right]\kappa\right\}\tau(\phi)\,\delta\phi\,\mathrm{d}\Omega\mathrm{d}t \\
& +\int_\Omega-\xi\tau(\phi)\,\delta\phi\,\mathrm{d}\Omega
\end{aligned}
\tag{4.40}
$$

where $\mathbf{n}_\phi=\frac{\nabla\phi}{|\nabla\phi|}$ and $\kappa=\nabla\cdot\mathbf{n}_\phi$ are the unitary normal vector and mean curvature of the level set function, respectively. In the optimization domain Ω, $\delta\hat{J}$ should satisfy $\delta\hat{J}>0$ for maximizing J, then $\mathbf{v}$, v_n and b are respectively chosen as

$$
\begin{aligned}
\mathbf{v} &= \int_0^T \nabla \left[\frac{3}{4}\sigma_0 \left(1-\varphi^2\right) \cos\theta_s \psi_a + \beta_3 C \right] \mathrm{d}t \\
v_n &= -\int_0^T \beta_1 \frac{\partial A}{\partial H(\phi)} - \psi \nabla \varphi \cdot \mathbf{u}_a + \rho \left(\alpha_f - \alpha_s\right) \mathbf{u} \cdot \mathbf{u}_a \\
&\quad + \left(M_f - M_s\right) \nabla \psi \cdot \nabla \varphi_a - \left(\lambda_f - \lambda_s\right) \nabla \varphi \cdot \nabla \psi_a \\
&\quad - \left(\lambda_f - \lambda_s\right) \frac{1}{\varepsilon_c^2} \varphi \left(\varphi^2 - 1\right) \psi_a \, \mathrm{d}t + \xi \\
b &= -\int_0^T \frac{3}{4}\sigma_0 \left(1-\varphi^2\right) \cos\theta_s \psi_a + \beta_3 C \, \mathrm{d}t
\end{aligned}
\tag{4.41}
$$

For minimizing J, $\delta \hat{J}$ should satisfy $\delta \hat{J} < 0$, then the sign of $\mathbf{v}$, v_n and b in Eq. 4.41 is reversed.

References

1. S. Zhou, Q. Li, A variational level set method for the topology optimization of steady-state Navier-Stokes flow. J. Comput. Phys. **227**, 10178–10195 (2008)
2. X. Duan, Y. Ma, R. Zhang, Shape-topology optimization for Navier-Stokes problem using variational level set method. J. Comput. Appl. Math. **222**, 487–499 (2008)
3. T. Borrvall, J. Petersson, Topology optimization of fluid in Stokes flow. Int. J. Numer. Meth. Fluids **41**, 77–107 (2003)
4. http://www.comsol.com
5. J.W. Cahn, J. Hilliard, Free energy of a nonuniform system. I. Interfacial free energy. J. Chem. Phys. **28**, 258–267 (1958)
6. J.W. Cahn, On spinodal decomposition. Acta Metall. **9**, 795–801 (1961)
7. D. Jacqmin, Contact-line dynamics of a diffuse fluid interface. J. Fluid Mech. **402**, 57–88 (2000)
8. J.W. Cahn, Critical-point wetting. J. Chem. Phys. **66**, 3667–3672 (1977)
9. T. Qian, X.P. Wang, P. Sheng, Molecular hydrodynamics of the moving contact line in two-phase immiscible flows. Comm. Comput. Phys. **1**, 1–52 (2006)
10. P. Yue, C. Zhou, J.J. Feng, C.F. Ollivier-Gooch, H.H. Hu, Phase-field simulations of interfacial dynamics in viscoelastic fluids using finite elements with adaptive meshing. J. Comput. Phys. **219**, 47–67 (2006)
11. Y. Deng, Z. Liu, Y. Wu, Topology optimization of steady and unsteady incompressible Navier-Stokes flows driven by body forces. Struct. Multidisc. Optim. **47**, 555–570 (2013)
12. P. Yue, J.J. Feng, C. Liu, J. Shen, A diffuse-interface method for simulating two-phase flows of complex fluids. J. Fluid Mech. **515**, 293–317 (2004)
13. S. Osher, R.P. Fedkiw, *Level Set Methods and Dynamic Implicit Surfaces* (Springer, New York, 2002)
14. J. Nocedal, S. Wright, *Numerical Optimization*, 2nd edn. (Springer, 2000)
15. L.H. Olesen, F. Okkels, H. Bruus, A high-level programming-language implementation of topology optimization applied to steady-state Navier-Stokes flow. Int. J. Numer. Methods Eng. **65**, 975–1001 (2006)
16. Y. Deng, Z. Liu, P. Zhang, Y. Liu, Y. Wu, Topology optimization of unsteady incompressible Navier-Stokes flow. J. Comput. Phys. **230**, 6688–6708 (2011)
17. H.C. Elman, D.J. Silvester, A.J. Wathen, *Finite Elements and Fast Iterative: Solvers With Applications in Incompressible Fluid Dynamics* (Oxford University Press, 2006)

18. S. Osher, R. Fedkiw, *Level Set Methods and Dynamic Implicit Surfaces* (Springer, New York, 2003)
19. R. Defay, I. Prigogine, *Surface Tension and Adsorption, Longmans* (Green & Co Ltd, London, 1966)
20. L.D. Landau, E.M. Lifshitz, *Fluid Mechanics* (Pergamon Press, Oxford, 1987)
21. M.B. Giles, N.A. Pierce, An introduction to the adjoint approach to design. Flow, Turbulence and Combustion **65**, 393–415 (2000)
22. M. Hinze, R. Pinnau, M. Ulbrich, S. Ulbrich, *Optimization with PDE Constraints* (Springer, Berlin, 2009)
23. B. Mohammadi, O. Pironneau, *Applied Shape Optimization for Fluids* (Oxford University Press, USA, Oxford, 2010)

Chapter 5
Combination of Topology Optimization and Optimal Control Method

This chapter presents the combination of topology optimization and optimal control method to find the optimal match between the material topology and control. In this method, the material topology is determined using the density method. The design variable is relaxed and bounded in the interval [0, 1], and the evolution of the design variable is implemented by the method of moving asymptotes (MMA), which can deal with optimization problem with multiple integral constraints and bound constraint of the design variable. In the combination of topology optimization and optimal control method, the control variable appears along with the design variable. In order to evolve the control variable and design variable using MMA simultaneously, the control variable is regularized using a bound constraint and the corresponding bound constraint is projected onto the interval [0, 1], which is the same as the bound constraint of the design variable. The optimization problem is analyzed using the adjoint method to obtain the adjoint sensitivity. During the optimization procedure, the design and control variables are filtered by the Helmholtz filters to ensure the smoothness of the distribution. To remove the gray area in the material topology, the filtered design variable is projected by the threshold method. The feasibility and robustness of the combination of these two methods are demonstrated by several test problems, including fluid flow, heat transfer and compliance minimization.

5.1 Variational Problem and Regularization

The variational problem for the combination of topology optimization and optimal control method are typical optimization problem constrained by partial differential equations, and it can be written in the following abstract form

Y. Deng et al., *Topology Optimization Theory for Laminar Flow*,
https://doi.org/10.1007/978-981-10-4687-2_5

$$\begin{aligned}&\text{Find } (\gamma, u) \text{ to minimize } J(y; \gamma, u)\\&\text{Subject to } e(y; \gamma, u) = 0;\ \gamma \in \mathrm{K}_\gamma,\ u \in \mathrm{K}_u\end{aligned} \tag{5.1}$$

where $y \in \mathrm{Y}$ is the state variable which can be a scalar or vector; $\gamma \in \mathrm{V}$ is the design variable; $u \in \mathrm{U}$ is the control variable; y, γ and u are functionals defined on the open and bounded domain Ω with Lipschitz boundary; $\mathrm{K}_\gamma \subset \mathrm{V}$ and $\mathrm{K}_u \subset \mathrm{U}$ are sets of the feasible values of γ and u; $J : \mathrm{Y} \times \mathrm{V} \times \mathrm{U} \rightarrow \mathbb{R}$ and $e : \mathrm{Y} \times \mathrm{V} \times \mathrm{U} \rightarrow \mathrm{Z}$ are bounded continuous mapping operator; Y, V, U and Z are Banach spaces; K_γ and K_u are convex sets. A quite common situation is that $e : \mathrm{Y} \times \mathrm{V} \times \mathrm{U} \rightarrow Z$ is continuously Fréchet differentiable and $e_y(y; \gamma, u) \in \mathscr{L}(\mathrm{Y}, Z)$ has the bounded inverse. Then according to the implicit function theorem [1], the partial differential equation $e(y; \gamma, u) = 0$ locally defines a continuously Fréchet differentiable map $(\gamma, u) \mapsto y(\gamma, u)$, which has the following Fréchet derivative

$$y'(\gamma, u) = -e_y^{-1}(y; \gamma, u)\, e_{(\gamma,u)}(y; \gamma, u) \tag{5.2}$$

Therefore, the optimization problem in Eq. 5.1 is transformed into

$$\min \hat{J}(\gamma, u) = J(y(\gamma, u); \gamma, u);\ \gamma \in \mathrm{K}_\gamma, u \in \mathrm{K}_u \tag{5.3}$$

In topology optimization based on the density method, the design variable is relaxed and bounded in a given interval, typically [0, 1], where the values 0 and 1 represent two different materials respectively. Then the set of the feasible values of the design variable in Eq. 5.1 is $\mathscr{K}_\gamma = [0, 1]$. Based on the gradient information of the optimization problem, the design variable is penalized and evolved to the value 0 or 1 at a fixed point in the design domain. Currently, the evolution of the design variable in the SIMP optimization is usually achieved using the method of moving asymptotes (MMA) [2], which can be used to deal with the optimization problem with multiple integral constraints and bound constraint of the design variable. In the combination of topology optimization and optimal control method, the control variable appears along with the design variable. In order to evolve the control and design variables using MMA simultaneously, the control variable is regularized using a bound constraint and the bound constraint is projected onto the interval [0, 1], which is the same as the bound constraint of the design variable in topology optimization. The bound constraint of the control variable is

$$u_l \le u \le u_h \tag{5.4}$$

where u_l and u_h are the lower and upper bounds of the control variable, and the bounds can be determined based on the engineering requirement. Then the feasible set of the control variable in Eq. 5.1 is $\mathscr{K}_u = [u_l, u_h]$. The projection of the bound constraint of the control variable is carried out using the following substitution:

$$u(u_{01}) = u_l + u_{01}(u_h - u_l),\ u_{01} \in [0, 1] \tag{5.5}$$

where u_{01} is the projected control variable. Because the mapping between u_{01} and u is continuous and linear, u_{01} satisfies $u_{01} \in \mathrm{U}$. By Eq. 5.5, the bound constraint of the control variable $u \in [u_l, u_h]$ is projected onto $u_{01} \in [0, 1]$.

Based on the above regularization, the optimization problem in Eq. 5.3 can be transformed furthermore

$$\min \hat{J}(\gamma, u(u_{01})) = J(y(\gamma, u(u_{01})); \gamma, u(u_{01})); \ \gamma \in [0,1], u_{01} \in [0,1] \quad (5.6)$$

The gradient information of γ and u_{01} in Eq. 5.6 can be obtained after the adjoint analysis. Based on the gradient information, the variational problem for the combination of topology optimization and optimal control method can be solved using the iterative approach.

5.2 Adjoint Analysis

In this section, the variational problem is analyzed to obtain the gradient information. According to [3–5], the adjoint method is an efficient approach to derive the derivative of the objective in the partial differential equation constrained optimization problem. From the Gâteaux derivative of $\hat{J}$, one can obtain

$$\begin{aligned}
&\left\langle \hat{J}', s \right\rangle_{(\mathrm{V}\times\mathrm{U})^*, \mathrm{V}\times\mathrm{U}} \\
&= \langle J_y, y's \rangle_{\mathrm{Y}^*, \mathrm{Y}} + \langle (J_\gamma, J_u u_{u_{01}}), s \rangle_{(\mathrm{V}\times\mathrm{U})^*, \mathrm{V}\times\mathrm{U}} \\
&= \langle y'^* J_y, s \rangle_{(\mathrm{V}\times\mathrm{U})^*, \mathrm{V}\times\mathrm{U}} + \langle (J_\gamma, J_u u_{u_{01}}), s \rangle_{(\mathrm{V}\times\mathrm{U})^*, \mathrm{V}\times\mathrm{U}} \\
&= \langle y'^* J_y + (J_\gamma, J_u u_{u_{01}}), s \rangle_{(\mathrm{V}\times\mathrm{U})^*, \mathrm{V}\times\mathrm{U}} \\
&= \left\langle (e_\gamma, e_u u_{u_{01}})^* \left(-e_y^{-1}\right)^* J_y + (J_\gamma, J_u u_{u_{01}}), s \right\rangle_{(\mathrm{V}\times\mathrm{U})^*, \mathrm{V}\times\mathrm{U}}; \ \forall s \in \mathrm{V}\times\mathrm{U}
\end{aligned} \quad (5.7)$$

where $(\cdot)^*$ is the adjoint operator of $(\cdot)$; $\langle \cdot, \cdot \rangle$ represents the dual pairing of $(\mathrm{V}\times\mathrm{U})^*$ and $\mathrm{V}\times\mathrm{U}$. By setting

$$\lambda = \left(-e_y^{-1}\right)^* J_y = -\left(e_y^*\right)^{-1} J_y \quad (5.8)$$

one can obtain

$$\langle e_y^* \lambda, v \rangle_{\mathrm{Y}^*, \mathrm{Y}} = \langle -J_y, v \rangle_{\mathrm{Y}^*, \mathrm{Y}}, \ \forall v \in \mathrm{Y} \quad (5.9)$$

where $\lambda \in Z^*$ is the adjoint variable of y; e_y^* is the adjoint operator of e_y. And the Gâteaux derivative of $\hat{J}$ can be reduced to be:

$$\left\langle \hat{J}', s \right\rangle_{(\mathrm{V}\times\mathrm{U})^*, \mathrm{V}\times\mathrm{U}} = \langle (e_\gamma, e_u u_{u_{01}})^* \lambda + (J_\gamma, J_u u_{u_{01}}), s \rangle_{(\mathrm{V}\times\mathrm{U})^*, \mathrm{V}\times\mathrm{U}}, \ \forall s \in \mathrm{V}\times\mathrm{U} \quad (5.10)$$

If $\hat{J}$ is Fréchet differentiable, the adjoint equation

$$e_y^* \lambda = -J_y \tag{5.11}$$

and the adjoint derivative

$$\hat{J}' = \left(e_\gamma, e_u u_{u_{01}}\right)^* \lambda + \left(J_\gamma, J_u u_{u_{01}}\right) \tag{5.12}$$

can be obtained. According to the Kurash-Kuhn-Tucker condition [5], the solution of the variational problem in Eq. 5.1 satisfies the following system:

$$\begin{aligned} e\left(y; \gamma, u\left(u_{01}\right)\right) &= 0 \\ e_y^* \lambda &= -J_y \\ \frac{DJ}{D\gamma} &= e_\gamma^* \lambda + J_\gamma \begin{cases} = 0, & \text{if } 0 < \gamma < 1 \\ \geq 0, & \text{if } \gamma = 0 \\ \leq 0, & \text{if } \gamma = 1 \end{cases} \\ \frac{DJ}{Du_{01}} &= \left(e_u u_{u_{01}}\right)^* \lambda + J_u u_{u_{01}} \begin{cases} = 0, & \text{if } 0 < u_{01} < 1 \\ \geq 0, & \text{if } u_{01} = 0 \\ \leq 0, & \text{if } u_{01} = 1 \end{cases} \end{aligned} \tag{5.13}$$

In engineering, the variational problem may need to be regularized furthermore by some integral constraints, e.g. volume constraint. The adjoint analysis of these integral constraints is the same as that for the objective. The equation system in Eq. 5.13 usually has strong nonlinearity and it is difficult to solve the system directly. Therefore, the iterative approach is widely adopted. In the iterative approach, the descent direction can be defined based on the adjoint derivative, where the state and adjoint variables are obtained by solving the partial differential equations and corresponding adjoint equations respectively.

5.3 Numerical Implementation

After the adjoint analysis, the variational problem for the combination of topology optimization and optimal control method can be solved using the iterative approach based on the gradient information. The procedure for the iterative approach includes the following steps (Table 5.1): (a) the partial differential equations are solved with the given values of the design and control variables; (b) the adjoint equations are solved based on the numerical solution of the partial differential equations; (c) the adjoint derivatives of the objective functional and integral constraint are computed; (d) the design and control variables are updated using MMA. The above steps are

Table 5.1 Procedure of the iterative approach

1.	Give the initial values of the design and control variables
2.	Solve the partial differential equations
3.	Solve the adjoint equations; compute the adjoint derivatives and corresponding objective and integral constraint values
4.	Update the design and control variables using MMA
5.	Check for convergence; if the stopping conditions are not satisfied, go to 2
6.	Post processing

implemented iteratively until the stopping criteria are satisfied. The stopping criteria are specified as the change of the objective values in 5 consecutive iterations satisfying

$$\sum_{i=0}^{4} \left| J_{k-i} - J_{k-i-1} \right| \Big/ \left| J_k \right| < \varepsilon \tag{5.14}$$

in the k-th iteration, where J_k and γ_k are the objective value and the distribution of the design variable in the k-th iteration, respectively; ε is the tolerance chosen to be 1×10^{-3}. The maximal iterative number is chosen to be 400. If there is an integral constraint, the relative residual of the integral constraint is required to be less than 1×10^{-3} when the iterative procedure converges. In the above iterative procedure, the partial differential equations and corresponding adjoint equations are solved by the finite element method using the finite element package *COMSOL Multiphysics* [6]. For the details on the setting of the numerical integrations in the optimization procedure, one can refer to [7, 8].

During the optimization procedure, the design and control variables are filtered by the Helmholtz filters [9, 10] to ensure the distribution smoothness, where the Helmholtz filters are implemented by solving the following Helmholtz equations:

$$\begin{aligned} -r^2 \nabla \cdot \nabla \widetilde{\gamma} + \widetilde{\gamma} = \gamma, \text{ in } \Omega \\ \nabla \widetilde{\gamma} \cdot \mathbf{n} = 0, \text{ on } \partial \Omega \end{aligned} \tag{5.15}$$

$$\begin{aligned} -r^2 \nabla \cdot \nabla \widetilde{u} + \widetilde{u} = u, \text{ in } \Omega \\ \nabla \widetilde{u} \cdot \mathbf{n} = 0, \text{ on } \partial \Omega \end{aligned} \tag{5.16}$$

where r is the filter radius chosen based on numerical experiments [10]; $\widetilde{\gamma}$ and $\widetilde{u}$ are the filtered design and control variables, respectively. The filtered control variable $\widetilde{u}$ is called physical control. To ensure the minimum scale length and remove the gray area in the obtained material topology, the filtered design variable is projected by the threshold method [11–13]:

$$\overline{\widetilde{\gamma}} \left(\widetilde{\gamma} \right) = \frac{\tanh \left(\beta \xi \right) + \tanh \left(\beta \left(\widetilde{\gamma} - \xi \right) \right)}{\tanh \left(\beta \xi \right) + \tanh \left(\beta \left(1 - \xi \right) \right)} \tag{5.17}$$

where $\overline{\widetilde{\gamma}}$ is the projected design variable called physical density representing the material topology [13]; $\xi \in [0, 1]$ and β are the threshold and projection parameters for the threshold projection, respectively. On the choice of the values of ξ and β, one can refer to [11, 14]. Because the design and control variables are filtered by Helmholtz filters, the Helmholtz equations are included in the partial differential equation constraints of the variational problem (Eq. 5.1). Then $\widetilde{\gamma}$ and $\widetilde{u}$ are regarded as two of the components of the state vector y. And the adjoint equations of the Helmholtz equations are included in the adjoint equations.

5.4 Test Problems

The combination of topology optimization and optimal distribution control method is presented in Sect. 5.4.1; the combination of topology optimization and optimal boundary control method is demonstrated in Sect. 5.4.2; the combination of topology optimization and optimal control method is exhibited for the transient problem in Sect. 5.4.3. Several problems are defined and tested to verify the feasibility and robustness of the combination of topology optimization. n and optimal control method. The test problems include the heat transfer, fluid flow and compliance minimization. In these test problems, the partial differential equations are solved using the standard Galerkin finite element method. In the heat transfer problem, the temperature and its adjoint variable are interpolated quadratically; in the fluid flow problem, the Navier-Stokes equations and corresponding adjoint equations are solved using Taylor-Hood elements [15], where the fluid velocity and its adjoint are interpolated quadratically, and the pressure and its adjoint are interpolated linearly; in the compliance minimization, the displacement and its adjoint are interpolated quadratically; in the Helmholtz filters, the filtered design variable, filtered control variable and the corresponding adjoint variables are solved using linear elements; in addition, the design and control variables are interpolated linearly. All the computation are performed on a DELL workstation (DELL PRECISION T5500, two Intel Xeon Quad X5550 CPUs).

5.4.1 Combination of Topology Optimization and Optimal Distribution Control

5.4.1.1 Fluid Flow

In this section, the combination of topology optimization and optimal distribution control method is tested for the fluid flow problem. The flow of the Newtonian fluid is described by the Navier-Stokes equations widely [16]. Topology optimization using the SIMP method for fluid problems was first researched for Stokes flows [17–19] and Darcy-Stokes flows [20, 21]. It was later extended to Navier-Stokes flows [7, 22–25] and non-Newtonian flows [26]. Additionally, the topology optimization method has been applied to design fluid devices [27–32]. In the topology optimization of fluid flow, the artificial Darcy force is added to the Navier-Stokes equations, where the solid and fluid phases are approximated by the porous media with high and low impermeability respectively [17]. In the optimal distribution control of the fluid flow, the control variable is the distribution of the body force in a specified domain, and this has been investigated for several engineering applications, i.e. velocity tracking problem [33, 34]. At the same time, the reasonable match between the topology of the fluid domain and distribution control can be more effective for achieving the desired performance of the fluid flow. Therefore, the following variational problem is constructed for the fluid flow to find the optimal match:

$$
\begin{aligned}
&\text{Find } (\gamma, \mathbf{f}) \text{ to minimize } J(\mathbf{u}, p; \gamma, \mathbf{f}) = \int_\Omega \omega_1 A\left(\mathbf{u}, \nabla\mathbf{u}, p; \overline{\tilde{\gamma}}, \tilde{\mathbf{f}}\right) \mathrm{d}\Omega \\
&\qquad + \int_{\partial\Omega} \omega_2 B\left(\mathbf{u}, p; \overline{\tilde{\gamma}}, \tilde{\mathbf{f}}\right) \mathrm{d}\Gamma \\
&\text{subject to } \left\{
\begin{array}{l}
\left\{
\begin{array}{l}
\rho\mathbf{u}\cdot\nabla\mathbf{u} - \nabla\cdot\left[\eta\left(\nabla\mathbf{u}+\nabla\mathbf{u}^{\mathrm{T}}\right)\right] + \nabla p = -\alpha\mathbf{u} + \tilde{\mathbf{f}}\,\overline{\tilde{\gamma}}^m, \text{ in } \Omega \\
-\nabla\cdot\mathbf{u} = 0, \text{ in } \Omega \\
\mathbf{u} = \mathbf{u}_D, \text{ on } \Gamma_D \\
\left[p\mathbf{I} - \eta\left(\nabla\mathbf{u}+\nabla\mathbf{u}^{\mathrm{T}}\right)\right]\cdot\mathbf{n} = \mathbf{g}, \text{ on } \Gamma_N
\end{array}
\right. \\
\left\{
\begin{array}{l}
-r^2\nabla\cdot\nabla\tilde{\gamma} + \tilde{\gamma} = \gamma, \text{ in } \Omega \\
\nabla\tilde{\gamma}\cdot\mathbf{n} = 0, \text{ on } \partial\Omega
\end{array}
\right. \quad \text{(Filter of } \gamma\text{)} \\
\left\{
\begin{array}{l}
-r^2\nabla\cdot\nabla\tilde{\mathbf{f}} + \tilde{\mathbf{f}} = \mathbf{f}, \text{ in } \Omega \\
\nabla\tilde{\mathbf{f}}\cdot\mathbf{n} = \mathbf{0}, \text{ on } \partial\Omega
\end{array}
\right. \quad \text{(Filter of } \mathbf{f}\text{)} \\
\int_\Omega \overline{\tilde{\gamma}}\,\mathrm{d}\Omega \le \theta_{Vol} V_0 \quad \text{(Volume constraint)} \\
0 \le \gamma \le 1 \quad \text{(Bound of } \gamma\text{)} \\
f_{l_i}(\mathbf{x}) \le f_i \le f_{h_i}(\mathbf{x}), \ \mathbf{x}\in\Omega, \ i = 1, 2 \text{ or } 3 \quad \text{(Bound of } \mathbf{f}\text{)}
\end{array}
\right.
\end{aligned} \tag{5.18}
$$

where ρ and η are the density and dynamic viscosity of the fluid, respectively; $\mathbf{u}$ is the fluid velocity; p is the pressure; $-\alpha\mathbf{u}$ is the artificial Darcy force; $\mathbf{u}_D$ is the known velocity distribution on the boundary Γ_D; $\mathbf{g}$ is the known stress distribution on the boundary Γ_N; Ω is the design domain; ω_1 and ω_2 are the weights of the two integrations of A and B in the objective; θ_{Vol} is the upper bound of the volume fraction of the fluid; f_{l_i} and f_{h_i}, chosen based on the engineering requirement, are the lower and upper bounds of the i-th component of the control variable $\mathbf{f}$, respectively; the control variable $\mathbf{f}$ is filtered by the Helmholtz filter; $\tilde{\mathbf{f}}$ is the filtered control variable; the filtered design variable is projected by the threshold method (Eq. 5.17) and $\overline{\tilde{\gamma}}$ is the projected design variable. In Eq. 5.18, the control variable $\mathbf{f}$ corresponds to the variable u in Eq. 5.1; the velocity $\mathbf{u}$, the pressure p, the filtered variables $\widetilde{\gamma}$ and $\widetilde{\mathbf{f}}$ correspond to the components of the state variable y in Eq. 5.1.

In order to avoid the fluid flow driven by the body force in the approximated solid region, the physical control is penalized using the power-law approach, i.e. $\widetilde{\mathbf{f}}\,\overline{\overline{\gamma}}^m$, where m is the penalty chosen to be 3 [35]. The penalized filtered control variable $\widetilde{\mathbf{f}}\,\overline{\overline{\gamma}}^m$ is called physical control representing the body force loaded on the fluid. In the design domain, the interpolation of the impermeability α is [17]

$$\alpha\left(\overline{\overline{\gamma}}\right) = \alpha_S\left[1 - \overline{\overline{\gamma}}\,(1+q)\Big/\left(q+\overline{\overline{\gamma}}\right)\right] \tag{5.19}$$

where α_S, chosen as a high but finite number to approximate the solid and ensure the numerical stability, is the impermeability of the solid phase; q is a positive number used to tune the convexity of the interpolation. According to Eq. 5.5, the following substitution is performed for the control variable:

$$f_i = f_{li} + f_{01i}\left(f_{hi} - f_{li}\right),\ f_{01i} \in [0,1],\ i = 1, 2 \text{ or } 3 \tag{5.20}$$

where f_{01i} is the i-th component of the projected control variable. When the iterative procedure is converged, the physical density $\overline{\overline{\gamma}}$ and filtered control $\widetilde{\mathbf{f}}$ can be obtained based on the obtained distributions of γ and $\mathbf{f}_{01}$, respectively.

Based on the adjoint method in Sect. 5.2, the adjoint equations of the Navier-Stokes equations are obtained as:

$$\begin{aligned} \rho\left(\nabla\mathbf{u}\cdot\mathbf{u}_a\right) - \rho\mathbf{u}\cdot\nabla\mathbf{u}_a - \nabla\cdot\left[\eta\left(\nabla\mathbf{u}_a + \nabla\mathbf{u}_a^{\mathrm{T}}\right)\right] & \\ +\nabla p_a = & -\omega_1\left(\frac{\partial A}{\partial\mathbf{u}} - \nabla\cdot\frac{\partial A}{\partial\nabla\mathbf{u}}\right) - \alpha\mathbf{u}_a,\ \text{in } \Omega \\ -\nabla\cdot\mathbf{u}_a = & -\omega_1\frac{\partial A}{\partial p},\ \text{in } \Omega \\ \mathbf{u}_a = & -\omega_2\frac{\partial B}{\partial p}\mathbf{n},\ \text{on } \Gamma_D \\ \left[p_a\mathbf{I} - \eta\left(\nabla\mathbf{u}_a + \nabla\mathbf{u}_a^{\mathrm{T}}\right)\right]\cdot\mathbf{n} = \omega_1\frac{\partial A}{\partial\nabla\mathbf{u}}\cdot\mathbf{n} & + \omega_2\frac{\partial B}{\partial\mathbf{u}} + \rho\left(\mathbf{u}\cdot\mathbf{n}\right)\mathbf{u}_a,\ \text{on } \Gamma_N \end{aligned} \tag{5.21}$$

where $\mathbf{u}_a$ and p_a are the adjoint variables of $\mathbf{u}$ and p, respectively. The adjoint equations of the Helmholtz filters for the design and control variables are obtained as:

$$\begin{aligned} -r^2\nabla\cdot\nabla\widetilde{\gamma}_a + \widetilde{\gamma}_a = & -\left(\omega_1\frac{\partial A}{\partial\overline{\overline{\gamma}}} + \frac{\partial\alpha}{\partial\overline{\overline{\gamma}}}\mathbf{u}\cdot\mathbf{u}_a - m\overline{\overline{\gamma}}^{m-1}\widetilde{\mathbf{f}}\cdot\mathbf{u}_a\right)\frac{\partial\overline{\overline{\gamma}}}{\partial\widetilde{\gamma}},\ \text{in } \Omega \\ r^2\nabla\widetilde{\gamma}_a\cdot\mathbf{n} = & -\omega_2\frac{\partial B}{\partial\overline{\overline{\gamma}}}\frac{\partial\overline{\overline{\gamma}}}{\partial\widetilde{\gamma}},\ \text{on } \partial\Omega \end{aligned} \tag{5.22}$$

$$
\begin{aligned}
-r^2\nabla\cdot\nabla\widetilde{\mathbf{f}}_a+\widetilde{\mathbf{f}}_a &= -\omega_1\frac{\partial A}{\partial\widetilde{\mathbf{f}}}+\overline{\widetilde{\gamma}}^m\mathbf{u}_a,\ \text{in}\ \Omega \\
r^2\nabla\widetilde{\mathbf{f}}_a\cdot\mathbf{n} &= -\omega_2\frac{\partial B}{\partial\widetilde{\mathbf{f}}},\ \text{on}\ \partial\Omega
\end{aligned}
\tag{5.23}
$$

where $\widetilde{\gamma}_a$ and $\widetilde{\mathbf{f}}_a$ are the adjoint variables of $\widetilde{\gamma}$ and $\widetilde{\mathbf{f}}$, respectively. And the adjoint derivatives are obtained as:

$$
\begin{aligned}
\left.\frac{DJ}{D\gamma}\right|_{\Omega} &= -\,\widetilde{\gamma}_a \\
\left.\frac{DJ}{D\mathbf{f}_{01}}\right|_{\Omega} &= -\,\widetilde{\mathbf{f}}_a\cdot\frac{\partial\mathbf{f}}{\partial\mathbf{f}_{01}}
\end{aligned}
\tag{5.24}
$$

In the following, one four-terminal fluid channel is optimized to demonstrate the combination of topology optimization and optimal distribution control method for the fluid flow problem. The design domain, discretized by 200×100 rectangular elements, is shown in Fig. 5.1a. The objective is to maximize the kinetic energy of the fluid and the flux of the flow by setting $A=-1/2\rho\mathbf{u}^2$ in Ω and $B=\mathbf{u}\cdot\mathbf{n}$ on the inlet boundaries. The parameter choices are listed in Table 5.2. The optimal match is obtained with the fluid topology and distribution of the physical control variable shown in Fig. 5.1b, c respectively. The convergent histories of the objective values

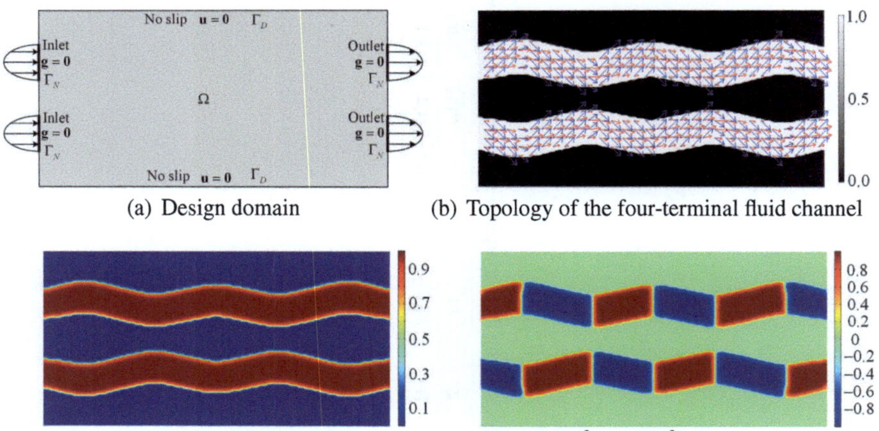

(a) Design domain (b) Topology of the four-terminal fluid channel

(c) Distribution of the physical control $\widetilde{f_1}\overline{\widetilde{\gamma}}^3$ and $\widetilde{f_2}\overline{\widetilde{\gamma}}^3$

Fig. 5.1 **a** Design domain of the fluid flow problem, where the size is 2×1; **b** topology of the four-terminal fluid channel in the optimal match, where the *blue arrows* and *red arrows* represent the physical control vectors and velocity vectors respectively; **c** distribution of the physical control in the optimal match

Table 5.2 Parameter setting for solving the variational problem of fluid flow in the four-terminal channel. s is the reference coordinate on the inlet boundary

ρ	η	$\mathbf{u}_D$	$\mathbf{g}$	m	r	θ_{Vol}	α_S	q	ξ	ω_1	ω_2	f_{l1}	f_{h1}	f_{l2}	f_{h2}
1	1	$(4s(1-s), 0)$	$\mathbf{0}$	3	1×10^{-2}	0.4	1×10^4	1	0.5	0.9	0.1	-1	1	-1	1

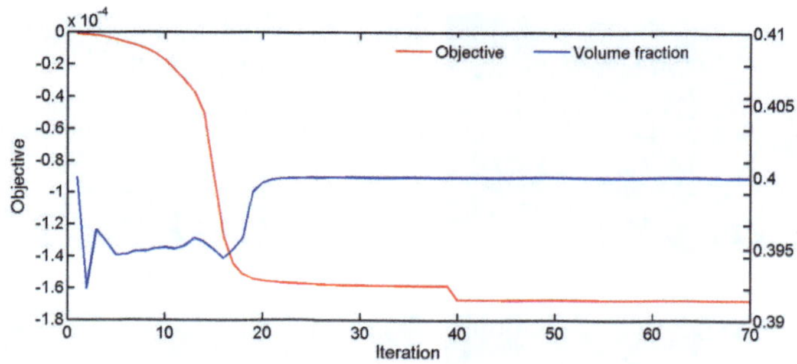

Fig. 5.2 Convergent histories of the objective values and volume constraint. The cost of CPU time is 1.55 h

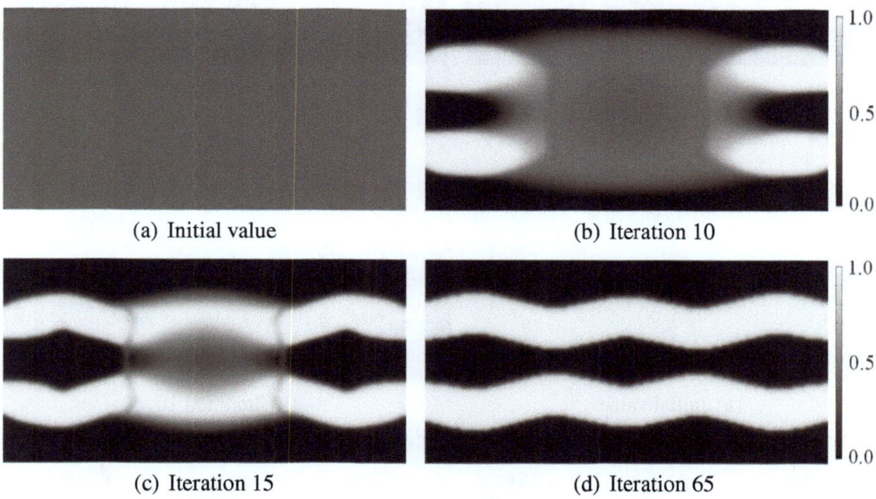

Fig. 5.3 Snapshots for the evolution of the physical density

and volume fraction are plotted in Fig. 5.2. Snapshots for the evolution of the physical density and physical control are shown in Figs. 5.3, 5.4 and 5.5, respectively.

From the distribution of the physical control vectors representing the body force (blue arrows in Fig. 5.1b) and the velocity vectors (red arrows in Fig. 5.1b) in the fluid topology, one can see that the physical control tries to reach the maximum value of the bound constraint of the control and the fluid topology makes the direction of the velocity vectors trend to be consistent with the direction of the physical control to maximize the kinetic energy of the fluid and the flux at the inlet boundaries. On the physical realizability, the obtained physical control as a body force can be realized by imposing electromagnetic force on the ions in the liquid solution [36–38].

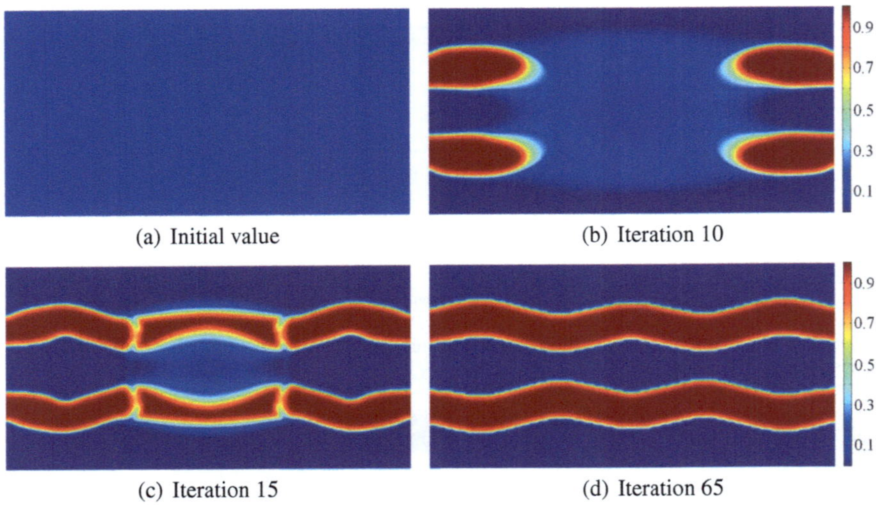

(a) Initial value (b) Iteration 10

(c) Iteration 15 (d) Iteration 65

Fig. 5.4 Snapshots for the evolution of $\widetilde{f}_1\overline{\overline{\gamma}}^3$

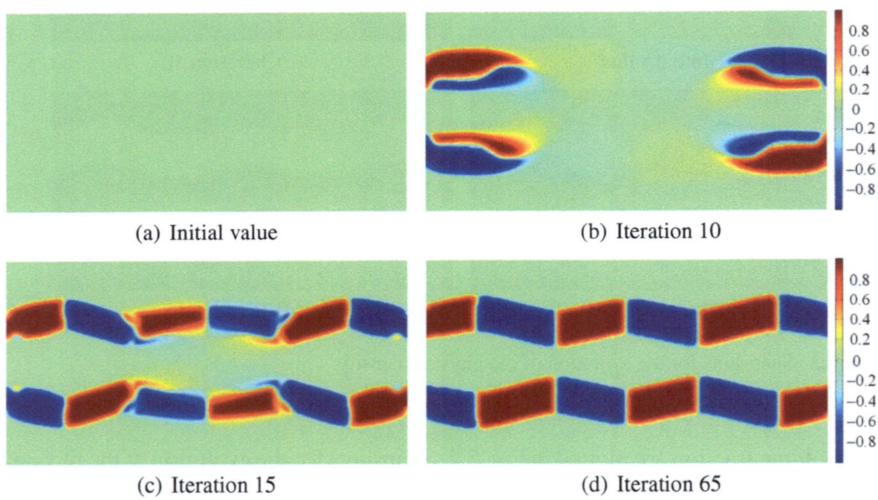

(a) Initial value (b) Iteration 10

(c) Iteration 15 (d) Iteration 65

Fig. 5.5 Snapshots for the evolution of $\widetilde{f}_2\overline{\overline{\gamma}}^3$

To check the optimality of the obtained match of the topology and control in Fig. 5.1, the similar method is adopted as that in Sect. 5.4.1.2. The optimal topology in Fig. 5.6a is obtained using the topology optimization method by fixing the physical control as shown in Fig. 5.1c; the optimal control in Fig. 5.6b is obtained using the optimal distribution control method by fixing the physical density as shown in Fig. 5.1b; and the objective values corresponding to Figs. 5.1b, and 5.6a, c are listed

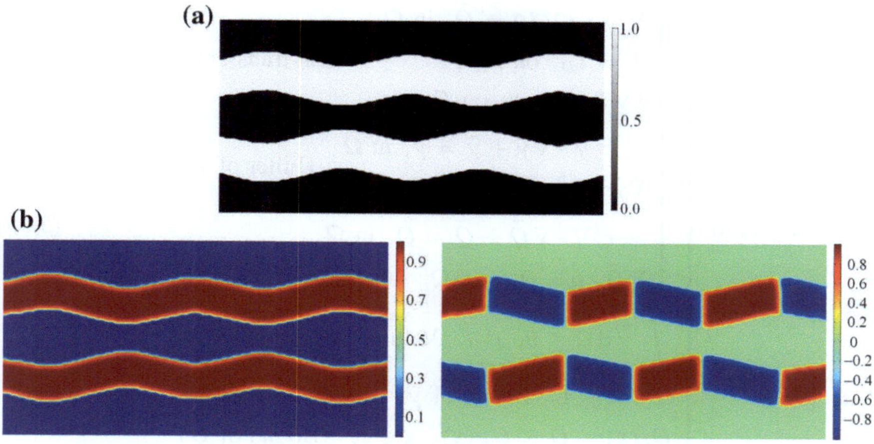

Fig. 5.6 **a** Optimal topology computed by fixing the physical distribution control as shown in Fig. 5.1d; **b** optimal distribution of the physical control computed by fixing the topology as shown in Fig. 5.1b

Table 5.3 Objective values corresponding to Figs. 5.1b and 5.6a, b

Figure 5.1b	Figure 5.6a	Figure 5.6b
1.6802×10^{-4}	1.6815×10^{-4}	1.6745×10^{-4}

in Table 5.3. Based on the consistency of the values in Table 5.3, the optimality is confirmed for the obtained match between the topology and control.

5.4.1.2 Heat Transfer

The heat transfer problem has been investigated using the topology optimization method [39–42] and optimal control method [43] respectively, where the heat source is fixed in topology optimization and the layout of the heat conductive materials are fixed in optimal control. Based on the intuition, one can conclude that a reasonable match between the layout of the heat conductive materials and the distribution of the heat source can be more effective. Then the combination of topology optimization and optimal distribution control is used to achieve this by solving the following variational problem:

$$\text{Find } (\gamma, Q), \text{ to minimize } J(T; \gamma, Q) = \int_{\Omega} \omega_1 A\left(T, \nabla T; \overline{\tilde{\gamma}}, \tilde{Q}\right) \mathrm{d}\Omega + \int_{\partial\Omega} \omega_2 B\left(T; \overline{\tilde{\gamma}}, \tilde{Q}\right) \mathrm{d}\Gamma$$

$$
\text{subject to} \left\{ \begin{array}{l} \left\{ \begin{array}{l} \nabla \cdot (-k \nabla T) = \tilde{Q}, \text{ in } \Omega \\ T = T_D, \text{ on } \Gamma_D \\ k \nabla T \cdot \mathbf{n} = g, \text{ on } \Gamma_N \end{array} \right. \quad \text{(Heat transfer)} \\ \left\{ \begin{array}{l} -r^2 \nabla \cdot \nabla \tilde{\gamma} + \tilde{\gamma} = \gamma, \text{ in } \Omega \\ \nabla \tilde{\gamma} \cdot \mathbf{n} = 0, \text{ on } \partial \Omega \end{array} \right. \quad \text{(Filter of } \gamma) \\ \left\{ \begin{array}{l} -r^2 \nabla \cdot \nabla \tilde{Q} + \tilde{Q} = Q, \text{ in } \Omega \\ \nabla \tilde{Q} \cdot \mathbf{n} = 0, \text{ on } \partial \Omega \end{array} \right. \quad \text{(Filter of } Q) \\ \int_\Omega \bar{\tilde{\gamma}} \, \mathrm{d}\Omega \le \theta_{Vol} V_0 \quad \text{(Volume constraint)} \\ 0 \le \gamma \le 1 \quad \text{(Bound of } \gamma) \\ Q_l(\mathbf{x}) \le Q \le Q_h(\mathbf{x}), \ \mathbf{x} \in \Omega \quad \text{(Bound of } Q) \end{array} \right. \tag{5.25}
$$

where T is the temperature; k is the thermal conductivity; T_D is the known temperature on the boundary Γ_D; g is the heat flux on the boundary Γ_N, which is the insulative boundary as $g = 0$; Ω is the design domain; ω_1 and ω_2 are the weights of the two integrations of A and B in the objective; Q_l and Q_h, chosen based on the engineering requirement, are the lower and upper bounds of the control variable Q, respectively; the design variable γ and control variable Q are filtered by the Helmholtz filters; $\tilde{\gamma}$ and $\tilde{Q}$ are the filtered design variable and filtered control variable, respectively; the filtered control variable $\tilde{Q}$ is called physical control representing the heating source; $\bar{\tilde{\gamma}}$ is the projected design variable (Eq. 5.17); the optimization problem is regularized by a volume constraint; θ_{Vol} is the upper bound of the volume fraction of the thermal conductive material; V_0 is the volume of the design domain Ω. In the above described variational problem, the control variable Q corresponds to the variable u in Eq. 5.1; the temperature T, the filtered variables $\tilde{\gamma}$ and $\tilde{Q}$ correspond to the components of the state vector y in Eq. 5.1.

In the above variational problem (Eq. 5.25), the material interpolation of the thermal conductivity k is the same as that in the topology optimization of heat transfer problem [39]:

$$
k\left(\bar{\tilde{\gamma}}\right) = k_{min} + \left(k_{max} - k_{min}\right) \bar{\tilde{\gamma}}^m \tag{5.26}
$$

where k_{min} and k_{max} are the thermal conductivity of the thermal insulative and conductive materials, respectively; m is the penalty, chosen to be 3. The bound constraint of the control variable is projected onto [0, 1] using the following substitution:

$$
Q = Q_l + Q_{01}\left(Q_h - Q_l\right), \ Q_{01} \in [0, 1] \tag{5.27}
$$

where Q_{01} is the projected control variable. When the iterative procedure is converged, the physical density $\bar{\tilde{\gamma}}$ and physical control $\tilde{Q}$ can be obtained based on the obtained distributions of γ and Q_{01}, respectively.

The analysis of the variational problem in Eq. 5.25 can be implemented based on the adjoint method in Sect. 5.2. Then the adjoint equation of the heat transfer equation is obtained as:

$$\begin{aligned} \nabla \cdot (-k\nabla T_a) &= -\omega_1 \left(\frac{\partial A}{\partial T} - \nabla \cdot \frac{\partial A}{\partial \nabla T} \right), \text{ in } \Omega \\ T_a &= 0, \text{ on } \Gamma_D \\ k\nabla T_a \cdot \mathbf{n} &= -\left(\omega_1 \frac{\partial A}{\partial \nabla T} \cdot \mathbf{n} + \omega_2 \frac{\partial B}{\partial T} \right), \text{ on } \Gamma_N \end{aligned} \tag{5.28}$$

where T_a is the adjoint variable of T. The adjoint equations of the Helmholtz filters for the design and control variables are obtained as:

$$\begin{aligned} -r^2\nabla \cdot \nabla \widetilde{\gamma}_a + \widetilde{\gamma}_a &= -\left(\omega_1 \frac{\partial A}{\partial \overline{\widetilde{\gamma}}} + \frac{\partial k}{\partial \overline{\widetilde{\gamma}}} \nabla T \cdot \nabla T_a \right) \frac{\partial \overline{\widetilde{\gamma}}}{\partial \widetilde{\gamma}}, \text{ in } \Omega \\ r^2\nabla \widetilde{\gamma}_a \cdot \mathbf{n} &= -\omega_2 \frac{\partial B}{\partial \overline{\widetilde{\gamma}}} \frac{\partial \overline{\widetilde{\gamma}}}{\partial \widetilde{\gamma}}, \text{ on } \partial\Omega \end{aligned} \tag{5.29}$$

$$\begin{aligned} -r^2\nabla \cdot \nabla \widetilde{Q}_a + \widetilde{Q}_a &= -\omega_1 \frac{\partial A}{\partial \widetilde{Q}} + T_a, \text{ in } \Omega \\ r^2\nabla \widetilde{Q}_a \cdot \mathbf{n} &= -\omega_2 \frac{\partial B}{\partial \widetilde{Q}}, \text{ on } \partial\Omega \end{aligned} \tag{5.30}$$

where $\widetilde{\gamma}_a$ and $\widetilde{Q}_a$ are the adjoint variables of $\widetilde{\gamma}$ and $\widetilde{Q}$, respectively. And the adjoint derivatives are obtained as:

$$\begin{aligned} \left. \frac{DJ}{D\gamma} \right|_{\Omega} &= -\widetilde{\gamma}_a \\ \left. \frac{DJ}{DQ_{01}} \right|_{\Omega} &= -\widetilde{Q}_a \frac{\partial Q}{\partial Q_{01}} \end{aligned} \tag{5.31}$$

It is valuable to notice that the partial differential equations and their corresponding adjoint equations have the same global stiffness matrix, when the finite element method is used. Therefore, the corresponding global stiffness matrix needs to be assembled only once in the whole iterative procedure.

One numerical example is presented to demonstrate the combination of topology optimization and optimal distribution control for the heat transfer problem. By setting the design domain as shown in Fig. 5.7a, the thermal compliance as the objective with $A = k\nabla T \cdot \nabla T$ and $B = 0$, and the parameters as listed in Table 5.4, the variational problem is solved, where the design domain is discretized by 200×200 rectangular elements. The projection parameter β with the initial value 1 is doubled every 40 iterations before reaching the maximum value 1024, and the same method for updating β is used in the following numerical examples. The intention

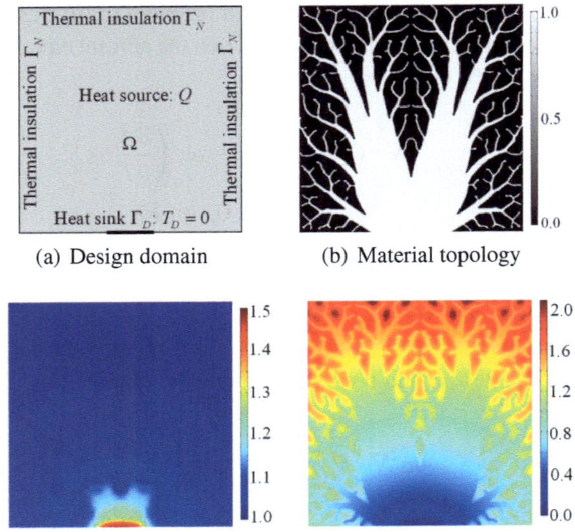

(a) Design domain (b) Material topology

(c) Heating source distribution (d) Temperature

Fig. 5.7 a Design domain of the heat transfer problem, where the size is 1×1; **b** material topology of the optimal match; **c** heating source distribution of the optimal match; **d** distribution of the temperature corresponding to the optimal match

Table 5.4 Parameter setting for solving the variational problem of heat transfer

T_D	r	θ_{Vol}	m	k_{min}	k_{max}	ξ	ω_1	ω_2	Q_l	Q_h
0	1×10^{-2}	0.4	3	1×10^{-3}	1×10^{0}	0.5	1	0	1	2

of these setting is to find the optimal match of the material topology and distribution of the heating source to minimize the thermal compliance. The obtained material topology, distribution of the heating source and the corresponding distribution of the temperature are shown in Fig. 5.7b–d, respectively. The convergent histories of the objective values and volume fraction are plotted in Fig. 5.8. Snapshots for the evolution of the material topology and distribution of the heating source during the iterative procedure are plotted in Figs. 5.9 and 5.10, respectively.

The material topology in Fig. 5.9 has the multi-branch shape which increases the area of the interface between the thermal conductive and insulative materials. From the distribution of the heating source and temperature in Fig. 5.7c, d, one can see that the region with low temperature corresponds to higher power of heating source, and the power of heating source in the conductive material is higher than that in the insulative material. This is propitious to descend the temperature gradient and decrease the thermal compliance.

By fixing the heating source distribution as shown in Fig. 5.7c, the corresponding optimal material topology shown in Fig. 5.11a is computed using the topology optimization method. On the other hand, the optimal distribution of the heating source is computed (Fig. 5.11b) by fixing the material topology as shown in Fig. 5.7b. The objective values corresponding to Figs. 5.7b and 5.11a, b are listed in Table 5.5.

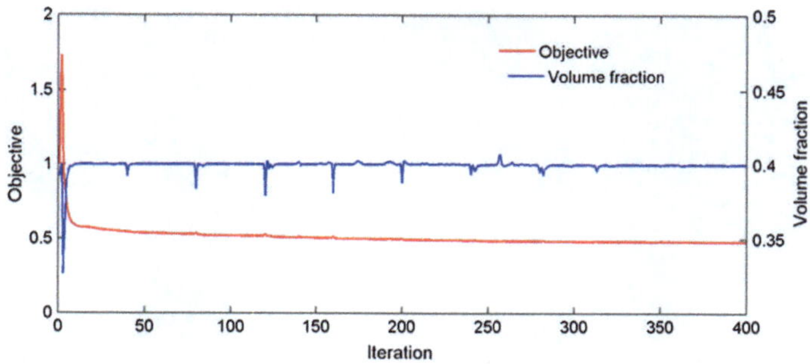

Fig. 5.8 Convergent histories of the objective values and volume constraint, where the objective values are normalized by the objective value corresponding to the initial distribution of γ and Q. The cost of CPU time is 5.01 h

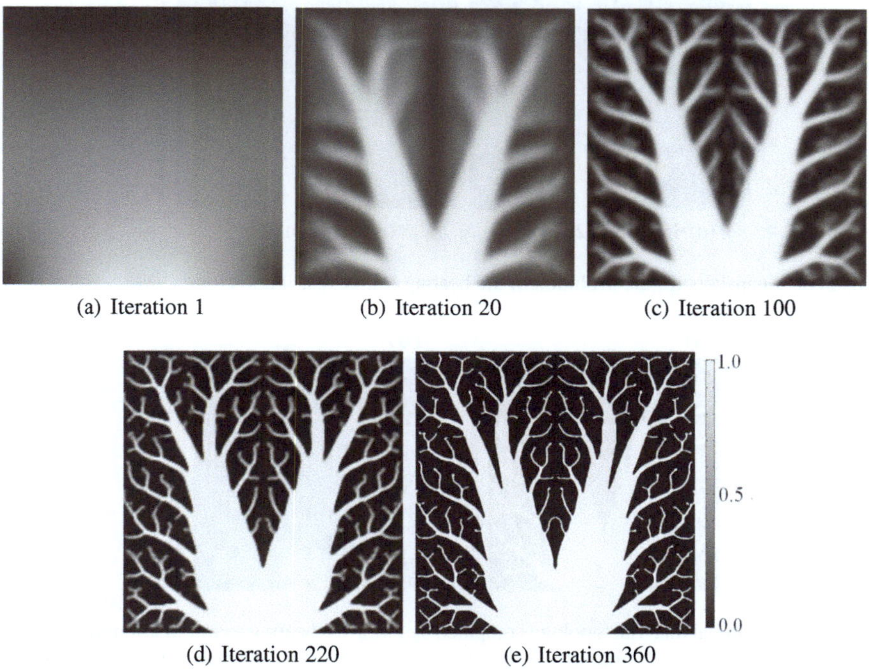

Fig. 5.9 Snapshots for the evolution of the material topology

From the consistency of the values in Table 5.5, the optimality is confirmed for the obtained match between the material topology and heating source distribution. In order to demonstrate the necessity of the combination method furthermore, the material topology is computed with the heating source distribution set as the constant

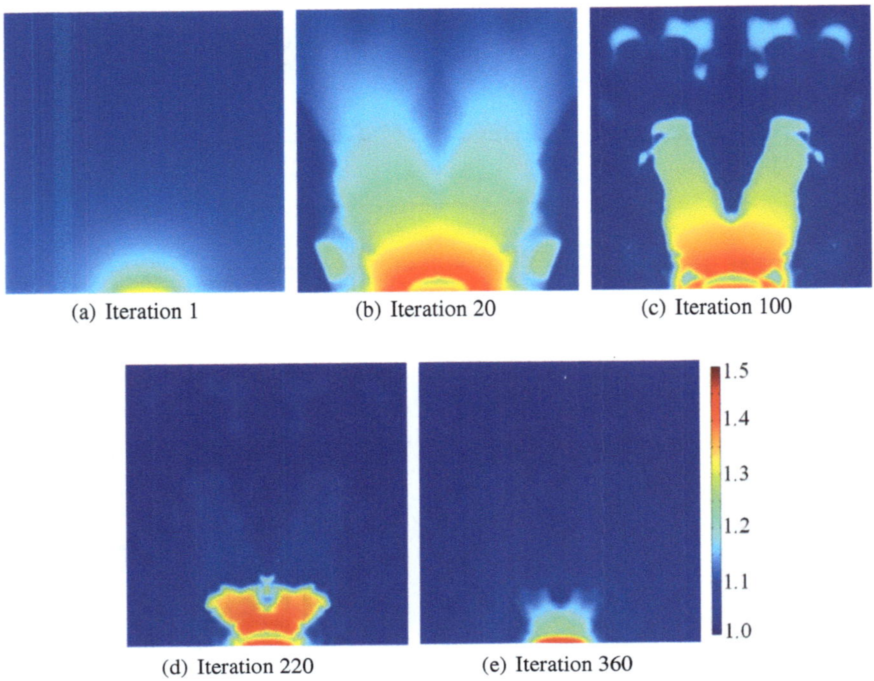

Fig. 5.10 Snapshots for the evolution of the heating source distribution

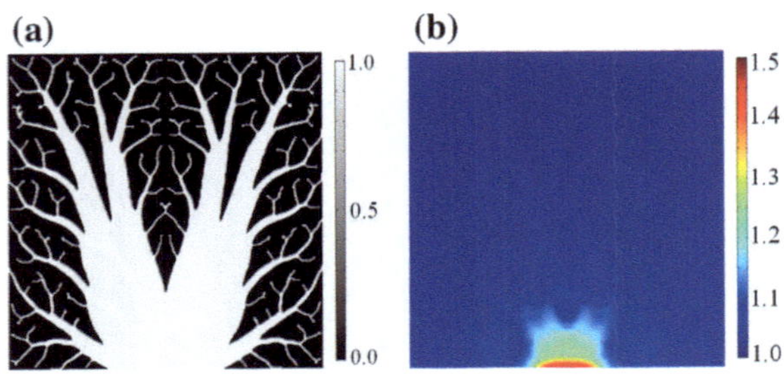

Fig. 5.11 **a** Optimal material topology computed by fixing the heating source distribution as shown in Fig. 5.7c; **b** optimal heating source distribution computed by fixing the material topology as shown in Fig. 5.7b

equal to the lower bond 1 (Fig. 5.12). And the thermal compliance corresponding to the material topology with constant heating source is 1.3890. Compared with the results shown in Fig. 5.7, the thermal compliance is decreased up to 18% by the combination method. Therefore, the necessity of the combination method can be confirmed.

Table 5.5 Objective values corresponding to Figs. 5.7b and 5.11a, b

Figure 5.7b	Figure 5.11a	Figure 5.11b
1.1456	1.1477	1.1409

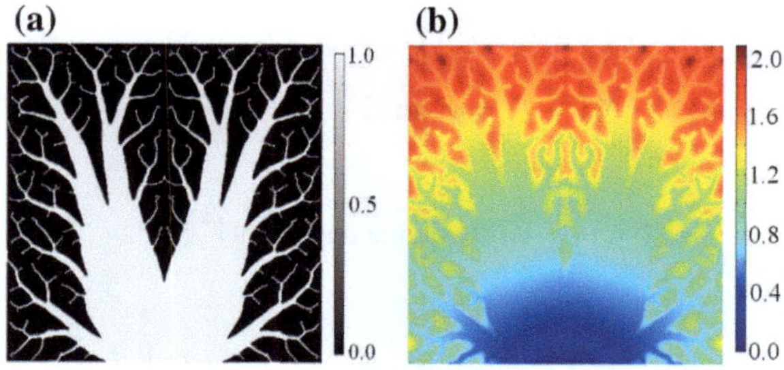

Fig. 5.12 **a** Optimal material topology computed by setting the heating source distribution as the constant equal to the lower bond 1; **b** distribution of the temperature corresponding to the optimal material topology. The corresponding objective value is 1.3890. Compared with the results shown in Fig. 5.7, the combination method decreases the thermal compliance up to 18%

5.4.2 Combination of Topology Optimization and Optimal Boundary Control

5.4.2.1 Fluid Flow

In this section, the combination of topology optimization and optimal boundary control method is tested for the fluid flow problem. At present, the work of the topology optimization in the area of fluid flow was performed in a design domain with fixed boundary condition [7, 17, 22]. Meanwhile, the optimal boundary control of fluid flow was implemented in a fixed fluid domain [44–46]. Therefore, the problem, on how to find the reasonable match between the topology of the fluid domain and boundary control, is encountered. Then the following variational problem is constructed for the fluid flow to find the optimal match between the fluid topology and boundary control:

$$
\begin{aligned}
\text{Find } (\gamma, \mathbf{u}_C) \text{ to minimize } & J(\mathbf{u}, p; \gamma, \mathbf{u}_C) \\
&= \int_\Omega \omega_1 A\left(\mathbf{u}, \nabla\mathbf{u}, p; \overline{\overline{\gamma}}\right) \mathrm{d}\Omega + \int_{\Gamma_D \bigcup \Gamma_N} \omega_2 B\left(\mathbf{u}, p; \overline{\overline{\gamma}}\right) \mathrm{d}\Gamma \\
&+ \int_{\Gamma_C} \omega_3 C\left(p; \overline{\overline{\gamma}}, \mathbf{u}_C\right) \mathrm{d}\Gamma
\end{aligned}
$$

$$\text{subject to} \begin{cases} \begin{cases} \rho \mathbf{u} \cdot \nabla \mathbf{u} - \nabla \cdot \left[\eta \left(\nabla \mathbf{u} + \nabla \mathbf{u}^{\mathrm{T}} \right) \right] + \nabla p = -\alpha \mathbf{u}, \text{ in } \Omega \\ -\nabla \cdot \mathbf{u} = 0, \text{ in } \Omega \\ \mathbf{u} = \mathbf{u}_D, \text{ on } \Gamma_D \\ \mathbf{u} = \mathbf{u}_C, \text{ on } \Gamma_C \\ \left[p\mathbf{I} - \eta \left(\nabla \mathbf{u} + \nabla \mathbf{u}^{\mathrm{T}} \right) \right] \cdot \mathbf{n} = \mathbf{g}, \text{ on } \Gamma_N \end{cases} \\ \begin{cases} -r^2 \nabla \cdot \nabla \tilde{\gamma} + \tilde{\gamma} = \gamma, \text{ in } \Omega \\ \nabla \tilde{\gamma} \cdot \mathbf{n} = 0, \text{ on } \partial \Omega \end{cases} \quad \text{(Filter of } \gamma) \\ \int_\Omega \bar{\tilde{\gamma}} \, \mathrm{d}\Omega \le \theta_{Vol} V_0 \quad \text{(Volume constraint)} \\ 0 \le \gamma \le 1 \quad \text{(Bound of } \gamma) \\ u_{Cl i}(\mathbf{x}) \le u_{Ci} \le u_{Chi}(\mathbf{x}), \ \mathbf{x} \in \Gamma_C \text{ and } i = 1, 2 \text{ or } 3 \quad \text{(Bound of } \mathbf{u}_C) \end{cases} \tag{5.32}$$

where the topology optimization of the flow is also implemented by adding the artificial Darcy force to the Navier-stokes equations; ω_1, ω_2 and ω_3 are the weights of the three integrations of A, B and C in the objective; $\mathbf{u}_C$ is the control variable on the boundary Γ_C; u_{Cli} and u_{Chi}, chosen based on the engineering requirement, are the lower and upper bounds of the i-th component of the control variable $\mathbf{u}_C$, respectively. In Eq. 5.32, the boundary control variable $\mathbf{u}_C$ corresponds to the variable u in Eq. 5.1; the velocity $\mathbf{u}$, the pressure p and filtered design variable $\tilde{\gamma}$ correspond to the components of the state variable y in Eq. 5.1. The design variable γ is filtered by the Helmholtz filter. The filtered design variable is projected by the threshold method (Eq. 5.17). The interpolation of the impermeability α is the same as that in Eq. 5.19. According to Eq. 5.5, the following projection is performed for the boundary control variable:

$$u_{Ci} = u_{Cli} + u_{C01i} \left(u_{Chi} - u_{Cli} \right), \ u_{C01i} \in [0, 1], \ i = 1, 2 \text{ or } 3 \tag{5.33}$$

where u_{C01i} is the i-th component of the projected boundary control variable.

Based on the adjoint method in Sect. 5.2, the adjoint equations of the Navier-Stokes equations are obtained as:

$$\begin{aligned} \rho \left(\nabla \mathbf{u} \cdot \mathbf{u}_a \right) - \rho \mathbf{u} \cdot \nabla \mathbf{u}_a - \nabla \cdot \left[\eta \left(\nabla \mathbf{u}_a + \nabla \mathbf{u}_a^{\mathrm{T}} \right) \right] & \\ + \nabla p_a &= -\omega_1 \left(\frac{\partial A}{\partial \mathbf{u}} - \nabla \cdot \frac{\partial A}{\partial \nabla \mathbf{u}} \right) - \alpha \mathbf{u}_a, \text{ in } \Omega \\ -\nabla \cdot \mathbf{u}_a &= -\omega_1 \frac{\partial A}{\partial p}, \text{ in } \Omega \\ \mathbf{u}_a &= -\omega_2 \frac{\partial B}{\partial p} \mathbf{n}, \text{ on } \Gamma_D \\ \mathbf{u}_a &= -\omega_3 \frac{\partial C}{\partial p} \mathbf{n}, \text{ on } \Gamma_C \end{aligned}$$

$$\left[p_a\mathbf{I} - \eta\left(\nabla\mathbf{u}_a + \nabla\mathbf{u}_a^{\mathrm{T}}\right)\right]\cdot\mathbf{n} = \omega_1\frac{\partial A}{\partial\nabla\mathbf{u}}\cdot\mathbf{n} + \omega_2\frac{\partial B}{\partial\mathbf{u}} + \rho\left(\mathbf{u}\cdot\mathbf{n}\right)\mathbf{u}_a, \text{ on } \Gamma_N \tag{5.34}$$

The adjoint equation of the Helmholtz filter for the design variable is obtained as:

$$\begin{aligned} -r^2\nabla\cdot\nabla\widetilde{\gamma}_a + \widetilde{\gamma}_a &= -\left(\omega_1\frac{\partial A}{\partial\overline{\overline{\gamma}}} + \frac{\partial\alpha}{\partial\overline{\overline{\gamma}}}\mathbf{u}\cdot\mathbf{u}_a\right)\frac{\partial\overline{\overline{\gamma}}}{\partial\widetilde{\gamma}}, \text{ in } \Omega \\ r^2\nabla\widetilde{\gamma}_a\cdot\mathbf{n} &= -\omega_2\frac{\partial B}{\partial\overline{\overline{\gamma}}}\frac{\partial\overline{\overline{\gamma}}}{\partial\widetilde{\gamma}}, \text{ on } \Gamma_D\cup\Gamma_N \\ r^2\nabla\widetilde{\gamma}_a\cdot\mathbf{n} &= -\omega_3\frac{\partial C}{\partial\overline{\overline{\gamma}}}\frac{\partial\overline{\overline{\gamma}}}{\partial\widetilde{\gamma}}, \text{ on } \Gamma_C \end{aligned} \tag{5.35}$$

And the adjoint derivatives are obtained as:

$$\begin{aligned} \left.\frac{DJ}{D\gamma}\right|_{\Omega} &= -\widetilde{\gamma}_a \\ \left.\frac{DJ}{D\mathbf{u}_{C01}}\right|_{\Gamma_C} &= \left[\omega_3\frac{\partial C}{\partial\mathbf{u}_C} + \eta\left(\nabla\mathbf{u}_a + \nabla\mathbf{u}_a^{\mathrm{T}}\right)\cdot\mathbf{n} - p_a\mathbf{n}\right]\cdot\frac{\partial\mathbf{u}_C}{\partial\mathbf{u}_{C01}} \end{aligned} \tag{5.36}$$

To demonstrate the combination of topology optimization and optimal boundary control method for the fluid flow problem, the backstep flow with Reynolds number 100 is optimized. In the optimal boundary control of the backstep flow, the objective is usually chosen to damp out the vorticity development near the corner [36]. At the same time, the dissipation of the backstep flow is usually considered to ensure the smoothness of the fluid channel in topology optimization. Therefore, the objective of the variational problem in Eq. 5.32 is set to be $A = \left[\eta/2\left(\nabla\mathbf{u} + \nabla\mathbf{u}^{\mathrm{T}}\right) : \left(\nabla\mathbf{u} + \nabla\mathbf{u}^{\mathrm{T}}\right) + \alpha\mathbf{u}^2\right] + \left(\nabla\times\mathbf{u}\right)^2/2$, $B = 0$ and $C = \mathbf{u}_C^2/2$ to find the optimal match between the fluid topology of the corner region and the boundary control. The design domain is set as shown in Fig. 5.13a, discretized by 8100 rectangular elements. The parameter choices are listed in Table 5.6. By solving the variational problem, the fluid topology (Fig. 5.13b) and the matched boundary control are obtained (Fig. 5.13c). The convergent histories of the objective values and volume constraint are shown in Fig. 5.14. Snapshots for the evolution of the physical density and boundary control are shown in Figs. 5.15 and 5.16, respectively. From the streamline distribution in Fig. 5.17b, one can conclude that the optimal match between the fluid topology and boundary control has damped out the vorticity development near the corner of the backstep flow effectively.

The optimality is checked for the obtained match between the fluid topology and boundary control. When the boundary control on Γ_C is fixed (Fig. 5.13c), the topology optimization is performed and the topology is obtained as shown in Fig. 5.18a; when the topology is fixed (Fig. 5.13b), the optimal boundary control method is implemented and the boundary control is derived as shown in Fig. 5.18b. The objective

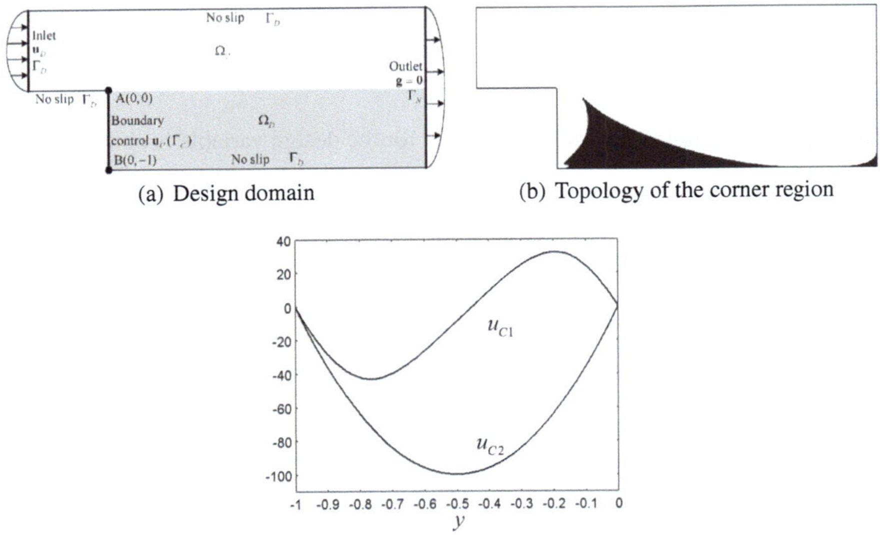

(a) Design domain (b) Topology of the corner region

(c) Boundary control $\mathbf{u}_C = (u_{C1}, u_{C2})$

Fig. 5.13 **a** Design domain of the backstep flow, where Ω_D defined with the design variable γ is the design domain and $\Omega = \Omega_C \cup \Omega_D$ is the computational domain; **b** fluid topology of the corner region in the optimal match; **c** boundary control of the optimal match

Table 5.6 Parameter setting for the optimization of the backstep flow. s is the reference coordinate on the corresponding boundary

ρ	η	$\mathbf{u}_D$	$\mathbf{g}$	r	θ_{Vol}	α_S	q
1	1	$(400s(1-s), 0)$	$\mathbf{0}$	1×10^{-2}	0.8	1×10^4	1
ξ	ω_1	ω_2	ω_3	u_{Cl1}	u_{Ch1}	u_{Cl2}	u_{Ch2}
0.5	0.4	0	0.2	$-400s(1-s)$	$400s(1-s)$	$-400s(1-s)$	$400s(1-s)$

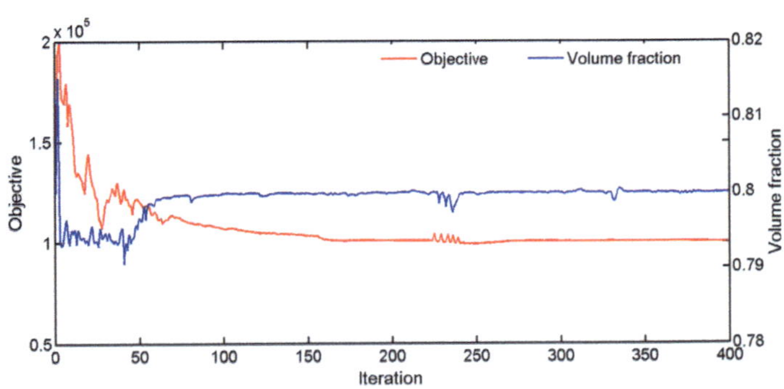

Fig. 5.14 Convergent histories of the objective values and volume constraint. The objective value corresponding to the optimality is 1.0016×10^5. The cost of CPU time is 5.08 h

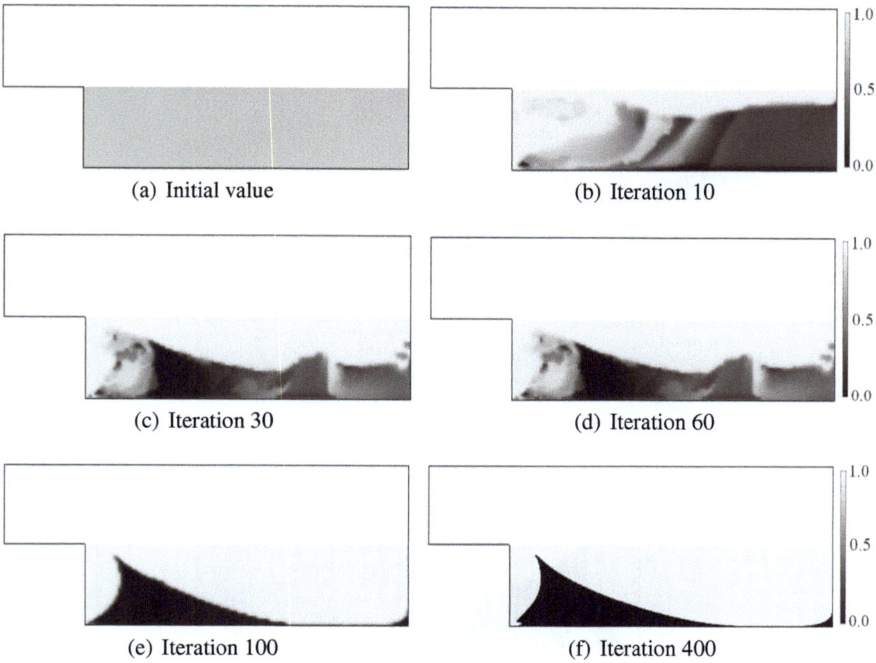

Fig. 5.15 Snapshots for the evolution of the physical density

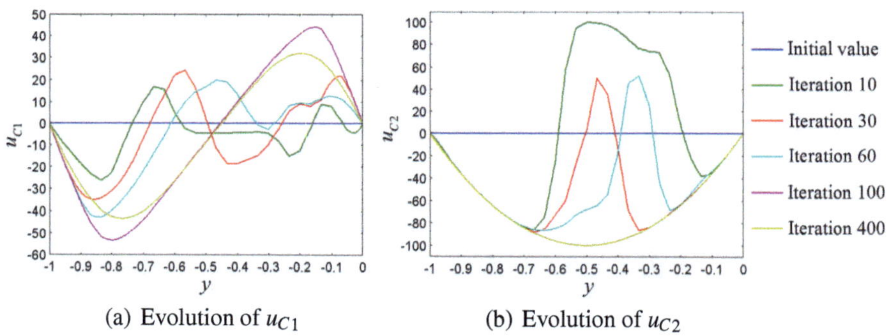

Fig. 5.16 Snapshots for the evolution of $\mathbf{u}_C$ on Γ_C

values corresponding to the results in Figs. 5.13 and 5.18a, b are listed in Table 5.7. Based on the results in Figs. 5.13 and 5.18 and the consistency of the objective values in Table 5.7, the optimality is confirmed for the obtained match between the fluid topology and boundary control.

For the backstep flow, the material topology and boundary control are also computed separately using the topology optimization and optimal control method. The material topology and boundary control are obtained as shown in Figs. 5.19 and 5.20. When only the topology optimization method is used, the material topology

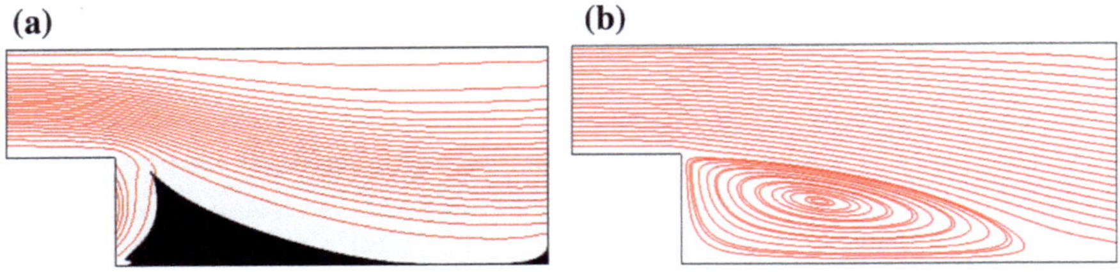

Fig. 5.17 **a** Streamline distribution in the backstep flow optimized by combining the topology optimization and optimal boundary control method; **b** streamline distribution in the backstep flow without optimization

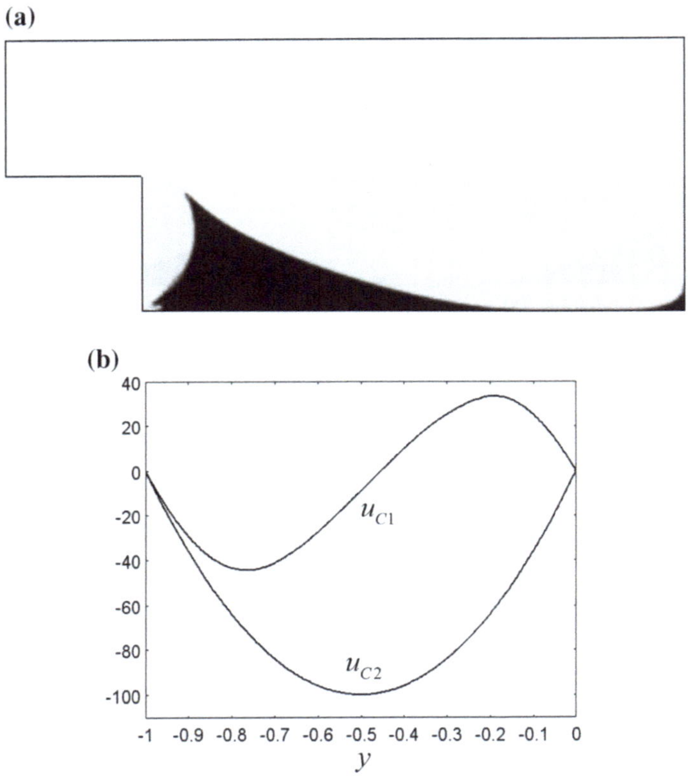

Fig. 5.18 **a** Optimal topology of the backstep flow obtained by the topology optimization method, where the boundary control on Γ_C is fixed as shown in Fig. 5.13c; **b** optimal boundary control obtained by the optimal boundary control method, where the topology of the backstep flow is fixed as shown in Fig. 5.13b

Table 5.7 Objective values corresponding to Figs. 5.13b, and 5.18a, b

Figure 5.13b	Figure 5.18a	Figure 5.18b
1.0008×10^5	1.0011×10^5	1.0009×10^5

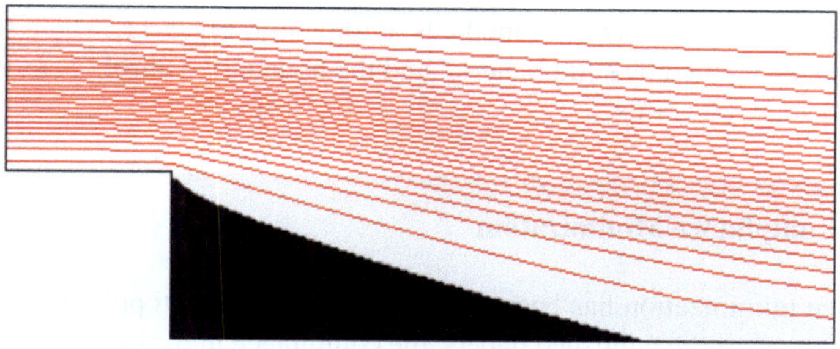

Fig. 5.19 Topology and streamline distribution of the backstep flow obtained by the topology optimization method

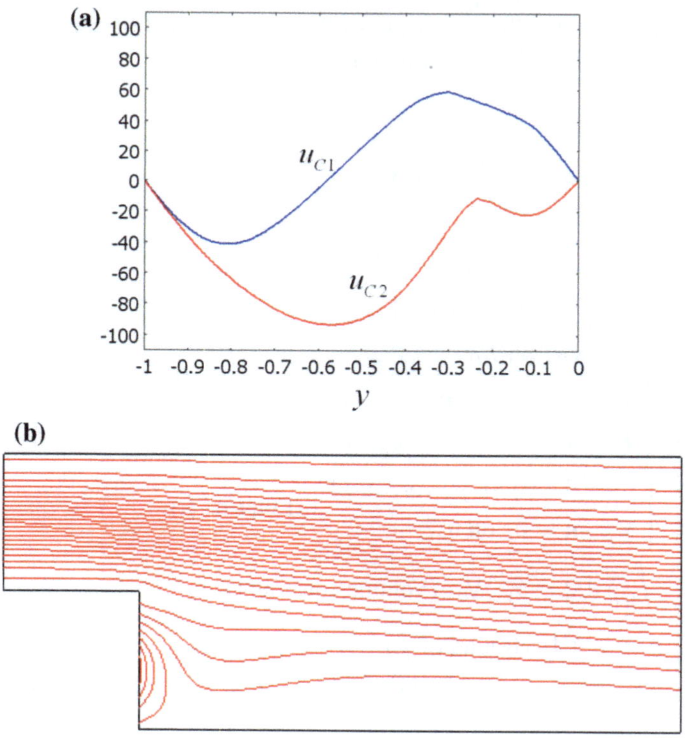

Fig. 5.20 **a** Optimal control of the backstep flow obtained by the optimal boundary control method; **b** streamline distribution of the backstep flow with the optimal control

is obtained as the corner of the backstep cut off directly to remove the vortex and decrease the dissipation of the flow; when only the optimal control method is used, the boundary control helps to avoid the closed streamlines and remove the vortex. Being different, the combination method combines the effects of both the topology optimization and optimal control methods as shown in Fig. 5.13, where the corner of the backstep is filled by solid material and the boundary control opens the closed streamlines to avoid vortex.

5.4.2.2 Compliance Minimization

Compliance minimization has been the most mature area in topology optimization [47–52]. In most of the published papers, the compliance minimization is performed under the specified load, i.e. the optimal design is found to fit the given load. In this section, the combination of topology optimization and optimal boundary control method is used to find the optimal match between the material topology of compliant structure and distribution of the surface force loaded on the boundary. The corresponding variational problem is constructed as:

$$
\begin{aligned}
&\text{Find}: \ (\gamma, \boldsymbol{\sigma}_C) \\
&\text{Min}: \ J(\mathbf{v}; \gamma, \boldsymbol{\sigma}_C) = \int_\Omega \omega_1 A\left(\mathbf{v}, \nabla\mathbf{v}; \bar{\tilde{\gamma}}\right) \mathrm{d}\Omega + \int_{\Gamma_D \bigcup \Gamma_N} \omega_2 B\left(\mathbf{v}; \bar{\tilde{\gamma}}\right) \mathrm{d}\Gamma \\
&\qquad\qquad + \int_{\Gamma_C} \omega_3 C\left(\mathbf{v}; \bar{\tilde{\gamma}}, \boldsymbol{\sigma}_C\right) \mathrm{d}\Gamma \\
&\text{S.t.} \ \begin{cases}
\begin{cases}
\nabla \cdot \left[-\mathbf{K}\left(\nabla\mathbf{v} + \nabla\mathbf{v}^\mathrm{T}\right)\right] = \mathbf{0}, \ \text{in } \Omega \\
\mathbf{v} = \mathbf{v}_D, \ \text{on } \Gamma_D \\
\mathbf{K}\left(\nabla\mathbf{v} + \nabla\mathbf{v}^\mathrm{T}\right) \cdot \mathbf{n} = \boldsymbol{\sigma}_C, \ \text{on } \Gamma_C \\
\mathbf{K}\left(\nabla\mathbf{v} + \nabla\mathbf{v}^\mathrm{T}\right) \cdot \mathbf{n} = \boldsymbol{\sigma}, \ \text{on } \Gamma_N
\end{cases} \quad \text{(Elastics)} \\
\begin{cases}
-r^2 \nabla \cdot \nabla\tilde{\gamma} + \tilde{\gamma} = \gamma, \ \text{in } \Omega \\
\nabla\tilde{\gamma} \cdot \mathbf{n} = 0, \ \text{on } \partial\Omega
\end{cases} \quad \text{(Filter of } \gamma) \\
\int_\Omega \bar{\tilde{\gamma}} \, \mathrm{d}\Omega \le \theta_{Vol} V_0 \quad \text{(Volume constraint)} \\
0 \le \gamma \le 1 \quad \text{(Bound of } \gamma) \\
\sigma_{Cli}(\mathbf{x}) \le \sigma_{Ci} \le \sigma_{Chi}(\mathbf{x}), \ \mathbf{x} \in \Gamma_C, \ i = 1, 2 \text{ or } 3 \quad \text{(Bound of } \boldsymbol{\sigma}_C)
\end{cases}
\end{aligned}
\tag{5.37}
$$

where the topology optimization of the compliance is implemented using the SIMP method [47, 51]; $\mathbf{K}$ is the stiffness tensor of the isotropic linear elastic material; $\mathbf{v}$ is the displacement; $\mathbf{v}_D$ is the known displacement on Γ_D; $\boldsymbol{\sigma}_C$, representing stress distribution, is the control variable on the boundary Γ_C; $\boldsymbol{\sigma}$ is the known stress distribution on the boundary Γ_N; σ_{Cli} and σ_{Chi}, chosen based on the engineering requirement, are the lower and upper bounds of the i-th component of the boundary control variable $\boldsymbol{\sigma}_C$, respectively. In Eq. 5.37, the boundary control variable $\boldsymbol{\sigma}_C$ corresponds to the control variable u in Eq. 5.1; the displacement $\mathbf{v}$ and filtered design variable $\tilde{\gamma}$ correspond to the components of the state variable y in Eq. 5.1. The design variable

γ is filtered and projected with the same method as that in the former sections. The interpolation of Young's module E in the stiffness tensor $\mathbf{K}$ is:

$$E = E_0 \overline{\overline{\gamma}}^m \tag{5.38}$$

where E_0 is the Young's module of the material; m is the penalty chosen to be 3 [51]. For the boundary control variable, the following projection is performed according to Eq. 5.5:

$$\sigma_{Ci} = \sigma_{Cli} + \sigma_{C01i}\left(\sigma_{Chi} - \sigma_{Cli}\right), \ \sigma_{C01i} \in [0, 1], \ i = 1, 2 \text{ or } 3 \tag{5.39}$$

where σ_{C01i} is the i-th component of the projected boundary control variable.

Based on the adjoint method in Sect. 5.2, the adjoint equation of the elastics equation is obtained as:

$$\begin{aligned}
\nabla \cdot \left[-\mathbf{K}\left(\nabla \mathbf{v}_a + \nabla \mathbf{v}_a^{\mathrm{T}}\right)\right] &= -\omega_1 \left(\frac{\partial A}{\partial \mathbf{v}} - \nabla \cdot \frac{\partial A}{\partial \nabla \mathbf{v}}\right), \text{ in } \Omega \\
\mathbf{v}_a &= \mathbf{0}, \text{ on } \Gamma_D \\
\mathbf{K}\left(\nabla \mathbf{v}_a + \nabla \mathbf{v}_a^{\mathrm{T}}\right) \cdot \mathbf{n} &= -\left(\omega_1 \frac{\partial A}{\partial \nabla \mathbf{v}} \cdot \mathbf{n} + \omega_2 \frac{\partial B}{\partial \mathbf{v}}\right), \text{ on } \Gamma_N \\
\mathbf{K}\left(\nabla \mathbf{v}_a + \nabla \mathbf{v}_a^{\mathrm{T}}\right) \cdot \mathbf{n} &= -\left(\omega_1 \frac{\partial A}{\partial \nabla \mathbf{v}} \cdot \mathbf{n} + \omega_3 \frac{\partial C}{\partial \mathbf{v}}\right), \text{ on } \Gamma_C
\end{aligned} \tag{5.40}$$

where $\mathbf{v}_a$ is the adjoint variable of the displacement $\mathbf{v}$. The adjoint equation of the Helmholtz filter for the design variable is:

$$\begin{aligned}
-r^2 \nabla \cdot \nabla \tilde{\gamma}_a + \tilde{\gamma}_a &= -\left[\omega_1 \frac{\partial A}{\partial \overline{\overline{\gamma}}} + \frac{\partial \mathbf{K}}{\partial \overline{\overline{\gamma}}}\left(\nabla \mathbf{v} + \nabla \mathbf{v}^{\mathrm{T}}\right) : \nabla \mathbf{v}_a\right] \frac{\partial \overline{\overline{\gamma}}}{\partial \tilde{\gamma}}, \text{ in } \Omega \\
r^2 \nabla \tilde{\gamma}_a \cdot \mathbf{n} &= -\omega_2 \frac{\partial B}{\partial \overline{\overline{\gamma}}} \frac{\partial \overline{\overline{\gamma}}}{\partial \tilde{\gamma}}, \text{ on } \Gamma_D \cup \Gamma_N \\
r^2 \nabla \tilde{\gamma}_a \cdot \mathbf{n} &= -\omega_3 \frac{\partial C}{\partial \overline{\overline{\gamma}}} \frac{\partial \overline{\overline{\gamma}}}{\partial \tilde{\gamma}}, \text{ on } \Gamma_C
\end{aligned} \tag{5.41}$$

And the adjoint derivatives are:

$$\begin{aligned}
\left.\frac{DJ}{D\gamma}\right|_{\Omega} &= -\tilde{\gamma}_a \\
\left.\frac{DJ}{D\boldsymbol{\sigma}_{C01}}\right|_{\Gamma_C} &= \left(\omega_3 \frac{\partial C}{\partial \boldsymbol{\sigma}_C} - \mathbf{v}_a\right) \cdot \frac{\partial \boldsymbol{\sigma}_C}{\partial \boldsymbol{\sigma}_{C01}}
\end{aligned} \tag{5.42}$$

To demonstrate the combination of topology optimization and optimal boundary control method for compliant mechanism design, a cantilever is optimized to find

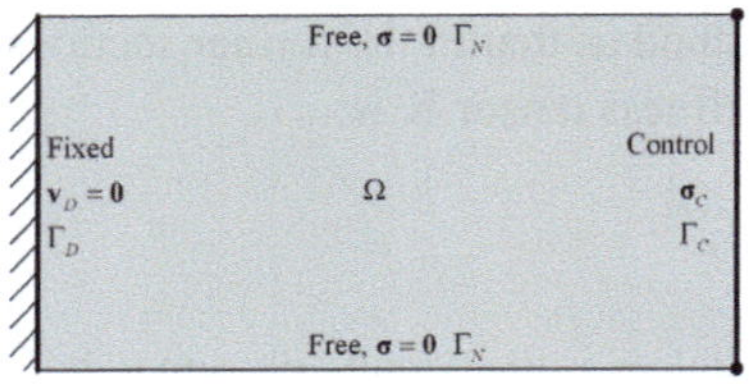

(a) Design domain

(b) Material topology

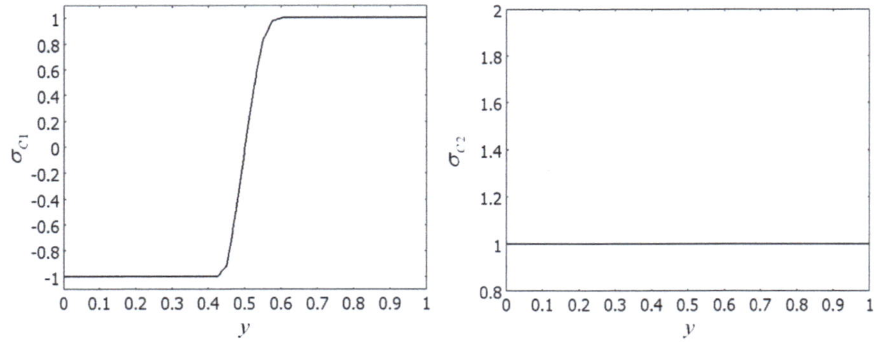

(c) Distribution of the surface force $\sigma_C = (\sigma_{C1}, \sigma_{C2})$

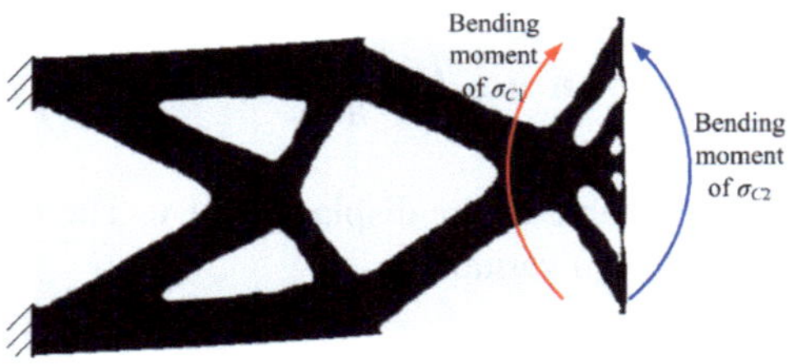

(d) Bending moment of the obtained distribution of the surface force

Fig. 5.21 **a** Design domain of the cantilever, where the size is 2×1; **b** material topology of the cantilever; **c** distribution of the surface force $\boldsymbol{\sigma}_C = (\sigma_{C1}, \sigma_{C2})$; **d** bending moments of σ_{C1} and σ_{C2}. The objective value corresponding to the optimality is 1.2685×10^{-10}. The cost of CPU time is 0.2141 h

the optimal match between the material topology and distribution of the surface force. The design domain of the cantilever is shown in Fig. 5.21a discretized by 80×40 rectangular elements, and the parameter choices are listed in Table 5.8. The objective is chosen to minimize the strain energy in the cantilever, by setting $A = 1/2\mathbf{K}\left(\nabla\mathbf{v} + \nabla\mathbf{v}^{\mathrm{T}}\right) : \left(\nabla\mathbf{v} + \nabla\mathbf{v}^{\mathrm{T}}\right)$, $B = 0$ and $C = 0$. After solving the variational problem, the optimal match of the material topology and distribution of the surface force are obtained as shown in Fig. 5.21b, c. The convergent histories of the objective values and volume constraint are shown in Fig. 5.22. Snapshots for the evolution of the physical density is shown in Fig. 5.23.

Table 5.8 Parameter setting for solving the variational problem corresponding to the compliance minimization of the cantilever

$\mathbf{v}_D$	$\boldsymbol{\sigma}$	E_0	ν	m	r	θ_{Vol}	ξ	ω_1	ω_2	ω_3	σ_{l1}	σ_{h1}	σ_{l2}	σ_{h2}
0	**0**	2.0×10^{11}	0.33	3	1×10^{-2}	0.5	0.5	1	0	0	−1	1	1	2

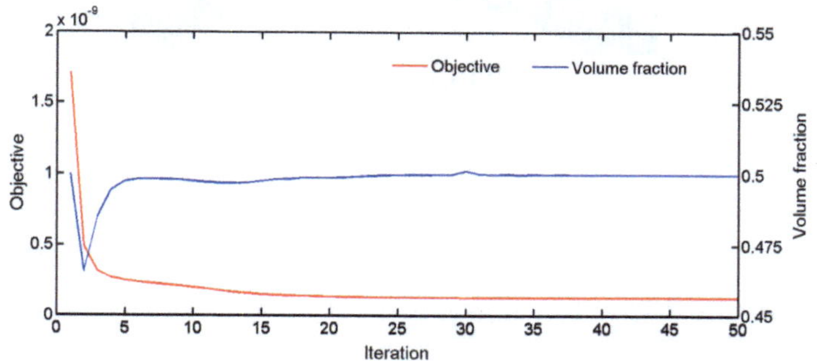

Fig. 5.22 Convergent histories of the objective values and volume constraint

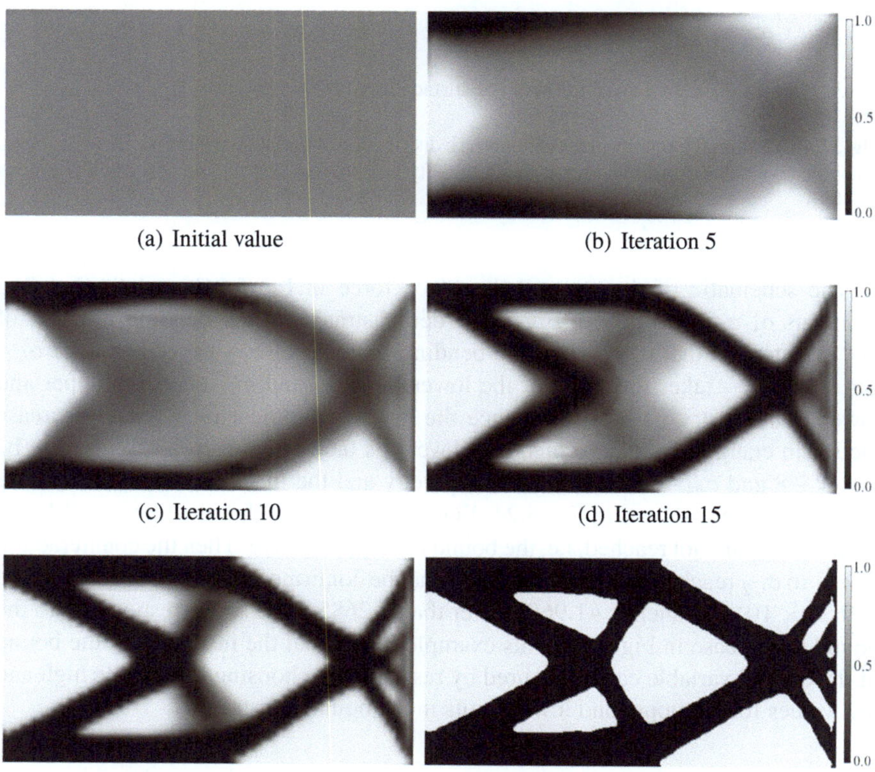

(a) Initial value (b) Iteration 5

(c) Iteration 10 (d) Iteration 15

(e) Iteration 20 (f) Iteration 50

Fig. 5.23 Snapshots for the evolution of the material topology

(a) Material topology

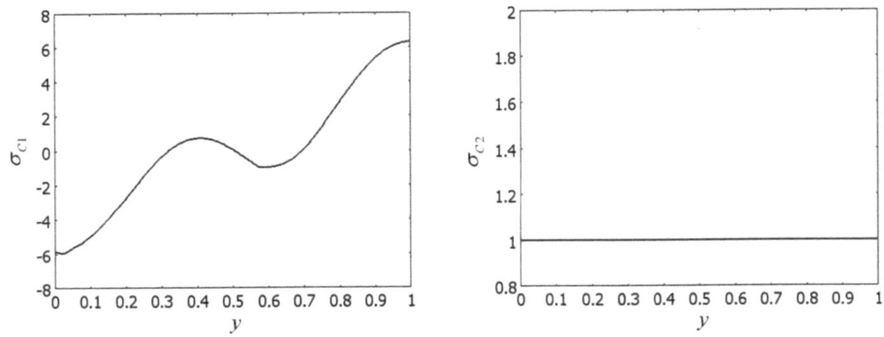

(b) Distribution of the surface force $\boldsymbol{\sigma}_C = (\sigma_{C1}, \sigma_{C2})$

Fig. 5.24 **a** Material topology of the cantilever; **b** distribution of the surface force $\boldsymbol{\sigma}_C = (\sigma_{C1}, \sigma_{C2})$. The objective value corresponding to the optimality is 7.3620×10^{-11}. The cost of CPU time is 0.6109 h

The schematic of the controlled surface force in Fig. 5.21b and the bending moments of σ_{C1} and σ_{C2} in Fig. 5.21d demonstrate that the bending moment of σ_{C1} counteracts that of σ_{C2} and the bending moment of σ_{C2} exceeds that of σ_{C1}. Therefore, σ_{C2} takes the value of the lower limit σ_{l2} and σ_{C1} takes the upper and lower limits to try its best to balance the bending moment of σ_{C2} and decrease the strain energy of the deformed cantilever. As the bound of σ_{C1} is relaxed to be $\sigma_{l1} = -8$ and $\sigma_{h1} = 8$, the material topology and the matched distribution of the surface force are shown in Fig. 5.24. Figure 5.24b shows that the upper and lower limits of σ_{C1} are not reached, i.e. the bound of σ_{C1} is inactive. Then the counteraction of σ_{C1} to σ_{C2} reaches the maximum. This can be confirmed from the objective value 7.3620×10^{-11}, which is 41.96% lower than 1.2685×10^{-10} corresponding to the bound active case in Fig. 5.21. This example shows that the inactivity of the bound of the control variable can be ensured by respectively choosing reasonable high and low values for the upper and lower limits in the bound constraint.

5.4.3 Test for Transient Problem

Transient problems are ubiquitous in engineering. This section is focused on the combination of topology optimization and optimal control method for the transient problem. The topology optimization and optimal control method have been developed for the transient problem, respectively. For the transient problems, the topology optimization is performed in the interpolated design domain with fixed distribution and boundary inputs; and optimal control is implemented in the fixed domain defined with physical field. To find the optimal match of the material topology and control variable for the transient problem, the topology optimization and optimal control method are combined and tested for the unsteady fluid flow problem. Topology optimization method for the unsteady flow has been developed in [8, 53], recently. Then based on the topology optimization and optimal control method for the unsteady flow, the following variational problem is constructed to find the optimal match:

$$
\begin{aligned}
&\text{Find}: \ (\gamma, \mathbf{f}, \mathbf{u}_C) \\
&\text{Min}: \ J(\mathbf{u}, p; \gamma, \mathbf{f}, \mathbf{u}_C) = \int_0^{\Theta} \int_{\Omega} \omega_1 A\left(\mathbf{u}, \nabla \mathbf{u}, p; \overline{\widetilde{\gamma}}, \widetilde{\mathbf{f}}\right) \mathrm{d}\Omega \mathrm{d}t \\
&\qquad + \int_0^{\Theta} \int_{\Gamma_D \bigcup \Gamma_N} \omega_2 B\left(\mathbf{u}, p; \overline{\widetilde{\gamma}}, \widetilde{\mathbf{f}}\right) \mathrm{d}\Gamma \mathrm{d}t \\
&\qquad + \int_0^{\Theta} \int_{\Gamma_C} \omega_3 C\left(p; \overline{\widetilde{\gamma}}, \widetilde{\mathbf{f}}, \mathbf{u}_C\right) \mathrm{d}\Gamma \mathrm{d}t \\
&\text{S.t.} \left\{
\begin{array}{l}
\left\{
\begin{array}{l}
\rho \dfrac{\partial \mathbf{u}}{\partial t} + \rho \mathbf{u} \cdot \nabla \mathbf{u} - \nabla \cdot \left[\eta \left(\nabla \mathbf{u} + \nabla \mathbf{u}^{\mathrm{T}}\right)\right] + \nabla p = -\alpha \mathbf{u} + \widetilde{\mathbf{f}} \overline{\widetilde{\gamma}}^m, \ \text{in } \Omega \\
-\nabla \cdot \mathbf{u} = 0, \ \text{in } \Omega \\
\mathbf{u}|_{t=0} = \mathbf{u}_0, \ \text{in } \Omega \\
\mathbf{u} = \mathbf{u}_D, \ \text{on } \Gamma_D \\
\mathbf{u} = \mathbf{u}_C, \ \text{on } \Gamma_C \\
\left[p\mathbf{I} - \eta \left(\nabla \mathbf{u} + \nabla \mathbf{u}^{\mathrm{T}}\right)\right] \cdot \mathbf{n} = \mathbf{g}, \ \text{on } \Gamma_N
\end{array}
\right. \\
\left\{
\begin{array}{l}
-r^2 \nabla \cdot \nabla \widetilde{\gamma} + \widetilde{\gamma} = \gamma, \ \text{in } \Omega \\
\nabla \widetilde{\gamma} \cdot \mathbf{n} = 0, \ \text{on } \partial \Omega
\end{array}
\right. \quad \text{(Filter of } \gamma) \\
\left\{
\begin{array}{l}
-r^2 \nabla \cdot \nabla \widetilde{\mathbf{f}} + \widetilde{\mathbf{f}} = \mathbf{f}, \ \text{in } \Omega \\
\nabla \widetilde{\mathbf{f}} \cdot \mathbf{n} = \mathbf{0}, \ \text{on } \partial \Omega
\end{array}
\right. \quad \text{(Filter of } \mathbf{f}) \\
\int_{\Omega} \overline{\widetilde{\gamma}} \, \mathrm{d}\Omega \le \theta_{Vol} V_0 \quad \text{(Volume constraint)} \\
0 \le \gamma \le 1 \quad \text{(Bound of } \gamma) \\
f_{li}(\mathbf{x}) \le f_i \le f_{hi}(\mathbf{x}), \ \mathbf{x} \in \Omega, \ i = 1, 2 \text{ or } 3 \quad \text{(Bound of } \mathbf{f}) \\
u_{Cli}(\mathbf{x}) \le u_{Ci} \le u_{Chi}(\mathbf{x}), \ \mathbf{x} \in \Gamma_C, \ i = 1, 2 \text{ or } 3 \quad \text{(Bound of } \mathbf{u}_C)
\end{array}
\right.
\end{aligned}
\tag{5.43}
$$

where t is the time; $[0, \Theta]$ is the considered time interval; $\mathbf{u}_0$ is the initial distribution of the fluid velocity. When solving the unsteady incompressible Navier-Stokes equations, the design variable γ is time independent because the layout of the fluid domain is kept unchanged; and $\mathbf{f}$ and $\mathbf{u}_C$ are time dependent, because different velocity distributions at different time needs different distributions of the controls. The design variable γ and the control variable $\mathbf{f}$ are filtered by Helmholtz filters. The projection of the filtered design variable $\tilde{\gamma}$ is performed by the threshold method as in Eq. 5.17. The physical distribution control is penalized by $\overline{\overline{\gamma}}^m$, where the penalty m is chosen to be 3. And the bound constraints of the distribution control $\mathbf{f}$ and boundary control $\mathbf{u}_c$ are projected using Eqs. 5.20 and 5.33, respectively.

Based on the adjoint method in Sect. 5.2, the adjoint equations of the Navier-Stokes equations are obtained as:

$$
\begin{aligned}
-\rho\frac{\partial \mathbf{u}_a}{\partial t} + \rho\left(\nabla\mathbf{u}\cdot\mathbf{u}_a\right) - \rho\mathbf{u}\cdot\nabla\mathbf{u}_a - \nabla\cdot\left[\eta\left(\nabla\mathbf{u}_a + \nabla\mathbf{u}_a^{\mathrm{T}}\right)\right] & \\
+\nabla p_a &= -\omega_1\left(\frac{\partial A}{\partial \mathbf{u}} - \nabla\cdot\frac{\partial A}{\partial \nabla\mathbf{u}}\right) - \alpha\mathbf{u}_a, \text{ in } \Omega \\
-\nabla\cdot\mathbf{u}_a &= -\omega_1\frac{\partial A}{\partial p}, \text{ in } \Omega \\
\mathbf{u}_a|_{t=\Theta} &= \mathbf{0}, \text{ in } \Omega \\
\mathbf{u}_a &= -\omega_2\frac{\partial B}{\partial p}\mathbf{n}, \text{ on } \Gamma_D \\
\mathbf{u}_a &= -\omega_3\frac{\partial C}{\partial p}\mathbf{n}, \text{ on } \Gamma_C \\
\left[p_a\mathbf{I} - \eta\left(\nabla\mathbf{u}_a + \nabla\mathbf{u}_a^{\mathrm{T}}\right)\right]\cdot\mathbf{n} &= \omega_1\frac{\partial A}{\partial \nabla\mathbf{u}}\cdot\mathbf{n} + \omega_2\frac{\partial B}{\partial \mathbf{u}} + \rho\left(\mathbf{u}\cdot\mathbf{n}\right)\mathbf{u}_a, \text{ on } \Gamma_N
\end{aligned}
\tag{5.44}
$$

The transient adjoint Eq. 5.44 are terminal value problems, where the value of $\mathbf{u}_a$ at the terminal time Θ is specified and the transient solver is implemented from the time Θ to 0. The adjoint equations of the Helmholtz filters can be obtained as:

$$
\begin{aligned}
-r^2\nabla\cdot\nabla\tilde{\gamma}_a + \tilde{\gamma}_a &= -\frac{1}{\Theta}\int_0^{\Theta}\left(\omega_1\frac{\partial A}{\partial \overline{\overline{\gamma}}} + \frac{\partial \alpha}{\partial \overline{\overline{\gamma}}}\mathbf{u}\cdot\mathbf{u}_a - m\overline{\overline{\gamma}}^{m-1}\tilde{\mathbf{f}}\cdot\mathbf{u}_a\right)\frac{\partial \overline{\overline{\gamma}}}{\partial \tilde{\gamma}}\,\mathrm{d}t, \text{ in } \Omega \\
r^2\nabla\tilde{\gamma}_a\cdot\mathbf{n} &= -\frac{1}{\Theta}\int_0^{\Theta}\omega_2\frac{\partial B}{\partial \overline{\overline{\gamma}}}\frac{\partial \overline{\overline{\gamma}}}{\partial \tilde{\gamma}}\,\mathrm{d}t, \text{ on } \Gamma_D\cup\Gamma_N \\
r^2\nabla\tilde{\gamma}_a\cdot\mathbf{n} &= -\frac{1}{\Theta}\int_0^{\Theta}\omega_3\frac{\partial C}{\partial \overline{\overline{\gamma}}}\frac{\partial \overline{\overline{\gamma}}}{\partial \tilde{\gamma}}\,\mathrm{d}t, \text{ on } \Gamma_C
\end{aligned}
\tag{5.45}
$$

$$
\begin{aligned}
-r^2\nabla\cdot\nabla\widetilde{\mathbf{f}}_a+\widetilde{\mathbf{f}}_a &= -\omega_1\frac{\partial A}{\partial\widetilde{\mathbf{f}}}+\overline{\overline{\gamma}}^m\mathbf{u}_a, \text{ in } \Omega \\
r^2\nabla\widetilde{\mathbf{f}}_a\cdot\mathbf{n} &= -\omega_2\frac{\partial B}{\partial\widetilde{\mathbf{f}}}, \text{ on } \Gamma_D\cup\Gamma_N \\
r^2\nabla\widetilde{\mathbf{f}}_a\cdot\mathbf{n} &= -\omega_3\frac{\partial C}{\partial\widetilde{\mathbf{f}}}, \text{ on } \Gamma_C
\end{aligned}
\tag{5.46}
$$

From Eqs. 5.45 and 5.46, one can see that the adjoint variable $\widetilde{\gamma}_a$ is time independent and $\widetilde{\mathbf{f}}_a$ is time dependent. The adjoint derivatives of the objective functional in Eq. 5.43 are:

$$
\begin{aligned}
\left.\frac{DJ}{D\gamma}\right|_{\Omega} &= -\Theta\widetilde{\gamma}_a \\
\left.\frac{DJ}{D\mathbf{f}_{01}}\right|_{\Omega} &= -\widetilde{\mathbf{f}}_a\cdot\frac{\partial\mathbf{f}}{\partial\mathbf{f}_{01}} \\
\left.\frac{DJ}{D\mathbf{u}_{C01}}\right|_{\Gamma_C} &= \left[\omega_3\frac{\partial C}{\partial\mathbf{u}_C}+\eta\left(\nabla\mathbf{u}_a+\nabla\mathbf{u}_a^{\mathrm{T}}\right)\cdot\mathbf{n}-p_a\mathbf{n}\right]\cdot\frac{\partial\mathbf{u}_C}{\partial\mathbf{u}_{C01}}
\end{aligned}
\tag{5.47}
$$

In the following, a flow rectifier is optimized to demonstrate the combination of topology optimization and optimal control method for the transient problem. The design domain is shown in Fig. 5.26a discretized by 100 × 100 rectangular elements. The parameter choices are listed in Table 5.9. The integrated functionals in the objective are set to be $A = 1/2\eta\left(\nabla\mathbf{u}+\nabla\mathbf{u}^{\mathrm{T}}\right):\left(\nabla\mathbf{u}+\nabla\mathbf{u}^{\mathrm{T}}\right)+\alpha\mathbf{u}^2+\left(\widetilde{\mathbf{f}}\overline{\gamma}\right)^2$ in Ω, $B = (\mathbf{u}\cdot\mathbf{n}-u_{rn})^2$ on Γ_N and $C = \mathbf{u}_C^2$ on Γ_C, where u_{rn} is the desired velocity distribution in the outward normal direction of the outlet Γ_N. This setting of the multi objective is to rectify the outward normal velocity distribution on Γ_N approaching the desired distribution u_{rn} and minimize the dissipation power of the flow and the scope of the physical control $\widetilde{\mathbf{f}}$ and $\mathbf{u}_C$, simultaneously. u_{rn} is set as $4s(1-s)V(t)$, where s is the reference coordinate on Γ_N and $V(t)$ represents the time dependent relation shown in Fig. 5.25. The known velocity imposed on the inlet is $\mathbf{u}_D = (4s(1-s)U(t), 0)$, where $U(t)$ represents the time dependent relation shown in Fig. 5.25.

By solving the variational problem in the time interval [0, 1], the optimal match of the fluid topology (Fig. 5.26b), physical distribution control (Figs. 5.27 and 5.28) and boundary control (Fig. 5.29) are obtained. From the obtained fluid topology (Fig. 5.26b) and boundary control (Fig. 5.29), one can see that a ramose channel connected to the boundary Γ_C is produced to compensate the difference of the flux at the inlet Γ_D and outlet Γ_N and enforce the flux rate at the outlet Γ_N to be consistent with that of the desired velocity distribution. From the physical distribution control in Figs. 5.27 and 5.28, one can see that the physical distribution control mainly distributes near the outlet Γ_N to enforce the outward normal velocity distribution to be consistent with the desired case.

Table 5.9 Parameter setting for solving the variational problem of the unsteady flow rectifier. s is the reference coordinate on the corresponding boundary; $U(t)$ and $h(t)$ are functions of the time t as shown in Fig. 5.25

ρ	η	$\mathbf{u}_D$	$\mathbf{g}$	Θ	m	r	θ_{Vol}	α_S	q	ξ
1	1	$(4s(1-s)U(t), 0)$	$\mathbf{0}$	1	3	1×10^{-2}	0.4	1×10^{4}	1	0.5
ω_1	ω_2	ω_3	f_{l1}	f_{h1}	f_{l2}	f_{h2}	u_{Cl1}	u_{Ch1}	u_{Cl2}	u_{Ch2}
1×10^{-5}	$1 - 2\omega_1$	ω_1	-1	1	-1	1	$-4s(1-s)h(t)$	$4s(1-s)h(t)$	$-4s(1-s)h(t)$	$4s(1-s)h(t)$

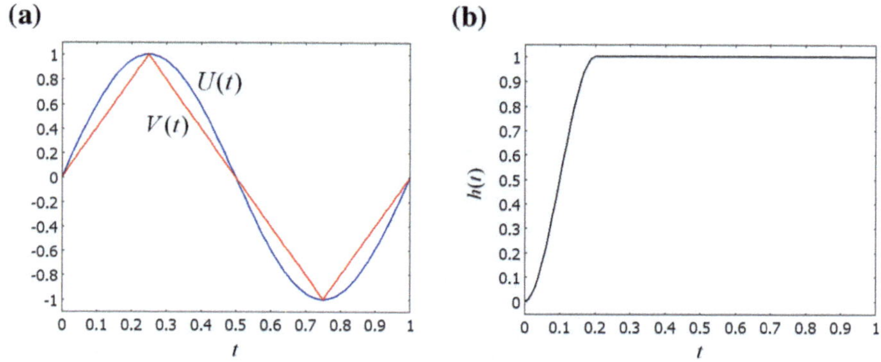

Fig. 5.25 Plotting of the functions $U(t)$, $V(t)$ and $h(t)$

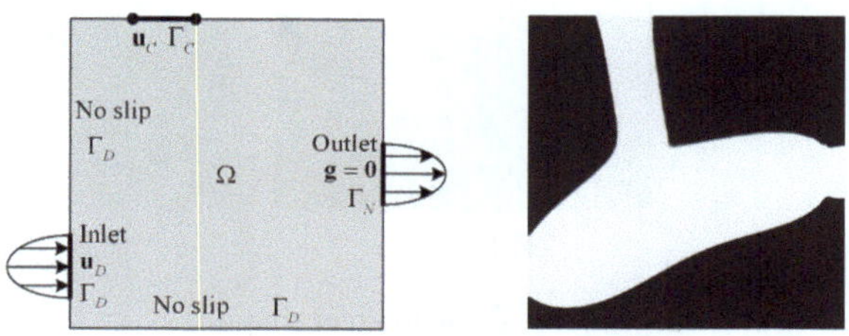

(a) Design domain

(b) Topology of the flow rectifier

Fig. 5.26 **a** Design domain of the flow rectifier; **b** topology of the flow rectifier in the optimal match. The objective value corresponding to the optimality is 2.9663×10^{-4}. The CPU time cost is 49.9901 h

5.5 Summary

The topology optimization and optimal control method has been combined. By the presented combination, the optimal match between the material topology and control has been achieved. In this combination, the control variable is constrained by a bound constraint and the bound constraint is projected onto [0, 1], which is the same as the bound constraint for the design variable in topology optimization, and this is convenient to use the method of moving asymptotes to evolve the design and control variables simultaneously. The corresponding variational problems are analyzed and solved using the adjoint method and iterative approach, respectively. In the numerical examples, the combination of topology optimization and optimal distribution control has been tested for the steady flow and heat transfer problems; the combination of topology optimization and optimal boundary control has been demonstrated by the steady flow and compliance minimization problems; and the combination method for the transient problem has been also discussed and demonstrated by the unsteady flow problem.

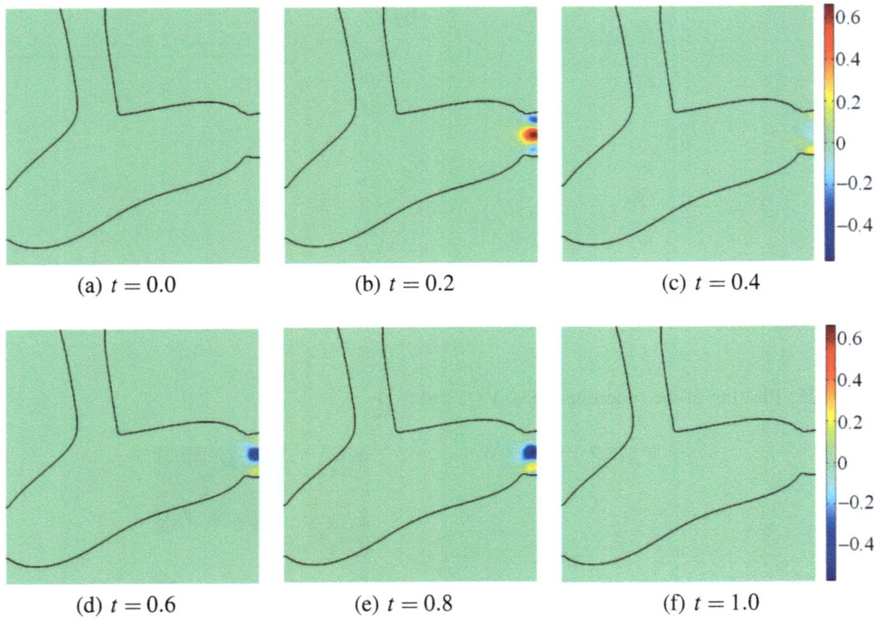

Fig. 5.27 Physical distribution control $\widetilde{f_1}\overline{\overline{\gamma}}^m$ at different time

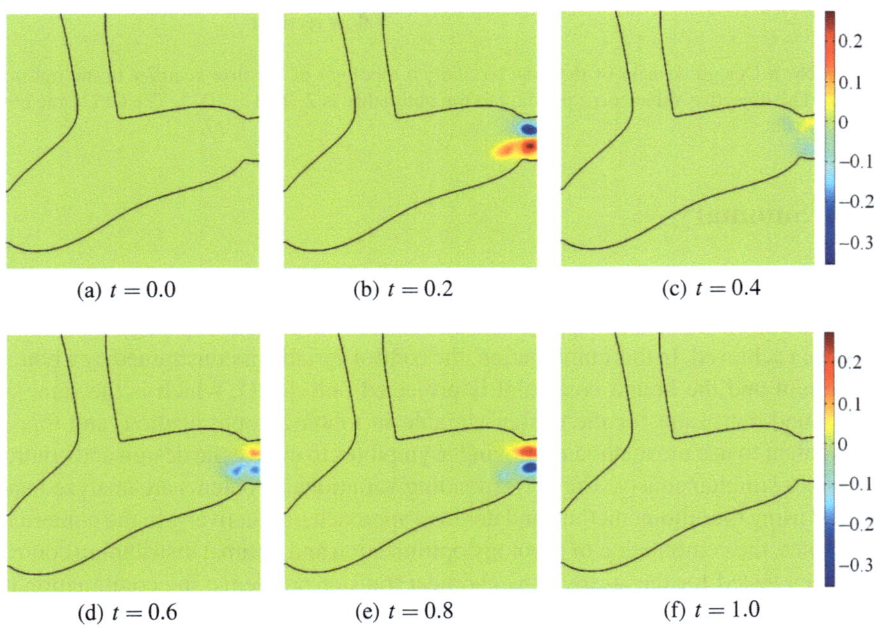

Fig. 5.28 Physical distribution control $\widetilde{f_2}\overline{\overline{\gamma}}^m$ at different time

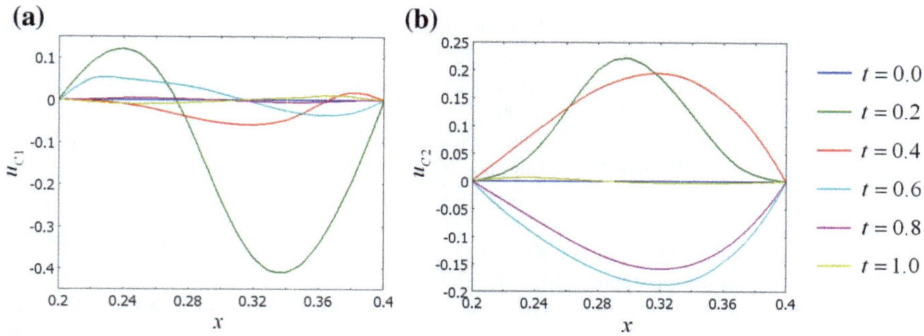

(a) u_{C1} (b) u_{C2}

Fig. 5.29 Boundary control $\mathbf{u}_C$ at different time

References

1. E. Zeidler, *Nonlinear Functional Analysis and Its Applications. I, Fixed-Point Theorems* (Springer, Berlin, 1986)
2. K. Svanberg, The method of moving asymptotes: a new method for structural optimization. Int. J. Numer. Meth. Eng. **24**, 359–373 (1987)
3. M.B. Giles, N.A. Pierce, An introduction to the adjoint approach to design. Flow Turbul. Combust. **65**, 393–415 (2000)
4. B. Mohammadi, O. Pironneau, *Applied Shape Optimization for Fluids* (OXFORD, 2010)
5. M. Hinze, R. Pinnau, M. Ulbrich, S. Ulbrich, *Optimization with PDE Constraints* (Springer, Berlin, 2009)
6. http://www.comsol.com
7. L.H. Olesen, F. Okkels, H. Bruus, A high-level programming-language implementation of topology optimization applied to steady-state Navier-Stokes flow. Int. J. Numer. Meth. Eng. **65**, 975–1001 (2006)
8. Y. Deng, Z. Liu, P. Zhang, Y. Liu, Y. Wu, Topology optimization of unsteady incompressible Navier-Stokes flow. J. Comput. Phys. **230**, 6688–6708 (2011)
9. B. Lazarov, O. Sigmund, Filters in topology optimization based on Helmholtz type differential equations. Int. J. Numer. Methods Eng. (2010). doi:10.1002/nme.3072
10. F. Wang, B.S. Lazarov, O. Sigmund, On projection methods, convergence and robust formulations in topology optimization. Struct. Multidiscip. Optim. **43**, 767–784 (2011)
11. S. Xu, Y. Cai, G. Cheng, Volume preserving nonlinear density filter based on heaviside funtions. Struct. Multidiscip. Optim. **41**, 495–505 (2010)
12. J. Guest, J. Prevost, T. Belytschko, Achieving minimum length scale in topology optimization using nodal design variables and projection functions. Int. J. Numer. Methods Eng. **61**, 238–254 (2004)
13. O. Sigmund, Morphology-based black and white filters for topology optimization. Struct. Multidiscip. Optim. **33**, 401–424 (2007)
14. A. Kawamoto, T. Matsumori, S. Yamasaki, T. Nomura, T. Kondoh, S. Nishiwaki, Heaviside projection based topology optimization by a PDE-filtered scalar function. Struct. Multidiscip. Optim. (2010). doi:10.1007/s00158-010-0562-2
15. H.C. Elman, D.J. Silvester, A.J. Wathen, *Finite Elements and Fast Iterative Solvers: with Applications in Incompressible Fluid Dynamics* (OXFORD, 2006)
16. R.L. Panton, *Incompressible Flow* (Wiley, 1984)
17. T. Borrvall, J. Petersson, Topology optimization of fluid in Stokes flow. Int. J. Numer. Meth. Fluids **41**, 77–107 (2003)

18. N. Aage, T.H. Poulsen, A. Gersborg-Hansen, O. Sigmund, Topology optimization of large scale stokes flow problems. Struct. Multidiscip. Optim. **35**, 175–180 (2008)
19. P.H. Guillaume, K.S. Idris, Topological sensitivity and shape optimization for the Stokes equations. SIAM J. Cont. Opti. **43**, 1–31 (2004)
20. J.K. Guest, J.H. Proevost, Topology optimization of creeping fluid flows using a Darcy-Stokes finite element. Int. J. Numer. Meth. Eng. **66**, 461–484 (2006)
21. N. Wiker, A. Klarbring, T. Borrvall, Topology optimization of regions of Darcy and Stokes flow. Int. J. Numer. Meth. Eng. **69**, 1374–1404 (2007)
22. A. Gersborg-Hansen, O. Sigmund, R.B. Haber, Topology optimization of channel flow problems. Struct. Multidiscip. Optim. **29**, 1–12 (2005)
23. A. Evgrafov, Topology optimization of slightly compressible fluids. ZAMM **86**, 46–62 (2006)
24. S. Zhou, Q. Li, A variational level set method for the topology optimization of steady-state Navier-Stokes flow. J. Comput. Phys. **227**, 10178–10195 (2008)
25. X. Duan, Y. Ma, R. Zhang, Shape-topology optimization for Navier-Stokes problem using variational level set method. J. Comput. Appl. Math. **222**, 487–499 (2008)
26. G. Pingen, K. Maute, Optimal design for non-Newtonian flows using a topology optimization approach. Comput. Math. Appl. **59**, 2340–2350 (2010)
27. F. Okkels, H. Bruus, Scaling behavior of optimally structured catalytic microfluidic reactors. Phys. Rev. E **75**, 1–4 (2007)
28. C.S. Andreasen, A.R. Gersborg, O. Sigmund, Topology optimization of microfluidic mixers. Int. J. Numer. Meth. Fluids **61**, 498–513 (2008)
29. Y. Deng, Z. Liu, P. Zhang, Y. Wu, J.G. Korvink, Optimization of no-moving part fluidic resistance microvalves with low Reynolds number, in *Proceedings IEEE International Conference on Micro Electro Mechanical Systems (MEMS)* (2010), pp. 67–70
30. Y. Deng, Y. Wu, M. Xuan, J.G. Korvink, Z. Liu, Dynamic optimization of valveless micropump, in *Proceedings IEEE International Conference on Solid-State Sensors, Actuators, and Microsystems (Transducers)* (2011), pp. 442–445
31. Z. Liu, Y. Deng, S. Lin, M. Xuan, Optimization of micro Venturi diode in steady flow at low Reynolds number. Eng. Optim. **44**, 1389–1404 (2012)
32. Y. Deng, Z. Liu, P. Zhang, Y. Liu, Q. Gao, Y. Wu, A flexible layout design method for passive micromixers. Biomed. Microdevices **14**, 929–945 (2012)
33. M. Sandro, Optimal boundary and distribution controls for the velocity tracking problem for Navier-Stokes flows. Ph.D. Thesis, Blacksburg, Virginia, May, 1997
34. T. Slawig, PDE-constrained control using FEMLAB—Control of the Navier-Stokes equations. Numer. Algorithms **42**, 107–126 (2006)
35. Y. Deng, Z. Liu, P. Zhang, Y. Wu, Y. Liu, Topology optimization of steady and unsteady incompressible Navier-Stokes flows driven by body forces. Struct. Multidiscip. Optim. **47**, 555–570 (2013)
36. S.S. Sritharan, *Optimal Control of Viscous Flow* (SIAM, Philadelphia, 1998)
37. D.M. Nosenchuck, G.L. Brown, Discrete spatial control of wall shear stress in a turbulent boundary layer, in *Proceedings of the International Conference on Near Wall Turbulence* (1993)
38. C. Foias, R. Temam, Remarques sur les equations de Navier-Stokes stationnaireset les phenomenes successifs de bifurcation. Annali Scuola Norm. Sup. Di Pisa **V**(1), 29–63 (1978)
39. Q. Li, G.P. Steven, Y.M. Xie, O.M. Querin, Evolutionary topology optimization for temperature reduction of heat conducting fields. Int. J. Heat Mass Transfer **47**(23), 5071–5083 (2004)
40. A. Gersborg-Hansen, M.P. Bendsoe, O. Sigmund, Topology optimization of heat conduction problems using the finite volume method. Struct. Multidiscip. Optim. **31**, 251–259 (2006)
41. Q. Li, G.P. Steven, O.M. Querin, Y.M. Xie, Shape and topology design for heat conduction by evolutionary structural optimization. Int. J. Heat Mass Transfer **42**, 3361–3371 (1999)
42. C. Zhuang, Z. Xiong, H. Ding, A level set method for topology optimization of heat conduction problem under multiple load cases. Comput. Methods Appl. Mech. Eng. **196**, 1074–1084 (2007)
43. F. Troltzsch, An SQP method for the optimal control of a nonlinear heat equation. Control Cybern. **23**, 268–288 (1994)

44. M.D. Gunzburger, L. Hou, T.P. Svobodny, Boundary velocity control of incompressible flow with an application to viscous drag reduction. SIAM J. Control Optim. **30**, 167–181 (1992)
45. S.S. Biringen, Active control of transition by periodic suction-blowing. Phys. Fluids **27**, 1345–1347 (1984)
46. H.O. Fattorini, S.S. Biringen, Existence of optimal controls for viscous flow problems. Proc. Roy. Soc. London Series A **439**, 81–102 (1992)
47. M.P. Bendsoe, N. Kikuchi, Generating optimal topologies in optimal design using a homogenization method. Comput. Methods Appl. Mech. Eng. **71**, 197–224 (1988)
48. G.I.N. Rozvany, *Shape and Layout Optimization of Structural Systems and Optimality Criteria Methods* (Springer, 1994)
49. G.I.N. Rozvany, *Topology Optimization in Structural Mechanics* (Springer, 1997)
50. O. Sigmund, On the design of compliant mechanisms using topology optimization. Mech. Struct. Mach. **25**, 495–526 (1997)
51. O. Sigmund, A 99-line topology optimization code written in Matlab. Struct. Multidiscip. Optim. **21**, 120–127 (2000)
52. A. Saxena, Topology design of large displacement compliant mechanisms with multiple materials and multiple output ports. Struct. Multidiscip. Optim. **30**, 477–490 (2005)
53. S. Kreissl, G. Pingen, K. Maute, Topology optimization for unsteady flow. Int. J. Numer. Meth. Eng. **87**, 1229–1253 (2011)

Chapter 6
Inverse Design of Microfluidics Using Topology Optimization

Abstract In this chapter, several typical microfluidic components are inversely designed, to demonstrate the power of the topology optimization for microfluidics. And the typical microfluidic components include the passive micromixer, electroosmotic micromixer, no-moving-part microvalve, and valveless micropump.

6.1 Topology Optimization-Based Inverse Design of Passive Micromixer

This section discusses the flexible layout design method of passive micromixers based on topology optimization of fluidic flows. Being different from the trial and error method, this method obtains the detailed layout of a passive micromixer according to the desired mixing performance by solving a topology optimization problem. Therefore, the dependence on the experience of the designer is weaken, when this method is used to design a passive micromixer with acceptable mixing performance. Several design disciplines for the passive micromixers are considered to demonstrate the flexibility of the layout design method for passive micromixers. These design disciplines include the approximation of the real 3D micromixer, the manufacturing feasibility, the spacial periodic design, and effects of the Péclet number and Reynolds number on the designs obtained by this layout design method. The capability of this design method is validated by several comparisons performed between the obtained layouts and the optimized designs in the published literatures, where the values of the mixing measurement is improved up to 40.4% for one cycle of the micromixer.

6.1.1 Modeling

When a micromixer is used to mix two fluidic flows with different solutes, the desired effect is the mixing of the two flows with anticipated distribution of the concentration at the outlet of the micromixer. The anticipated distribution of the concentration at the outlet can be specified by the designer based on the required performance

Y. Deng et al., *Topology Optimization Theory for Laminar Flow*,
https://doi.org/10.1007/978-981-10-4687-2_6

(e.g. hydrodynamic focusing and sufficient mixing) of the micromixer. The mixing performance of the micromixer can be measured by the least square variance between the obtained concentration c and the anticipated concentration c_a at the outlet [1, 2]:

$$\Psi(c) = \int_{\Gamma_o} (c - \overline{c})^2 \, \mathrm{d}\Gamma \bigg/ \int_{\Gamma_i} (c_0 - \overline{c})^2 \, \mathrm{d}\Gamma \tag{6.1}$$

where c_0 is the reference distribution of the solute concentration at the inlet Γ_i; Γ_o is the outlet of the micromixer. If the required performance of the micromixer is hydrodynamic focusing, the anticipated concentration $\overline{c}$ is specified to be a function of the coordinates; if the required performance is to achieve the sufficient mixing of the two solutes, the anticipated concentration $\overline{c}$ is specified to be the average concentration. Here, it is focused on the performance of sufficient mixing. Therefore, $\overline{c}$ is the ideal concentration distribution of the solute at the outlet after sufficient mixing. Assuming that the concentration of the solute at the inlet of the two flows are c_i and 0 respectively, the ideal concentration distribution of the solute at the outlet, which is also the average concentration, is

$$\overline{c} = c_i Q_{f1} / \left(Q_{f1} + Q_{f2}\right) \tag{6.2}$$

where Q_{f1} and Q_{f2} are the fluxes of the two flows at the inlets of the micromixer, respectively. The definition of the measurement of the mixing performance in Eq. 6.1 means that a lower value of $\Psi(c)$ corresponds to better mixing performance. Therefore, the design task is to find the reasonable layout of a micromixer corresponding to the lowest value of the mixing measurement in a designer specified design domain. And this layout can be obtained using the topology optimization method. Under the hypothesis of continuum, microfluidic flow is described by the Navier-Stokes equations, and mixing of flows is described by the convection-diffusion equation. In the topology optimization method, which was proposed by Borvall and Petersson for the Stokes flow [3] and extended to the Navier-Stokes flow by Gersborg et al. [4] and Olesen et al. [5], an artificial friction force is added to the Navier-Stokes equations. This method begins with the homogeneous distribution of an artificial porous material; then, the artificial porous material is evolved to solid and liquid phases; at last, the solid and liquid phases are approximated by the artificial materials with sufficient high and zero impermeability, respectively. The artificial friction force is expressed as $\mathbf{f} = -\alpha(\gamma)\mathbf{u}$, where α is the impermeability of the artificial porous medium, γ is the design variable and $\mathbf{u}$ is the velocity of the fluidic flow. The design variable γ is bounded in the interval [0, 1], where 0 and 1 corresponds to the solid and fluid, respectively. The impermeability of the porous medium α is interpolated by the design variable γ as [3–5]

$$\alpha(\gamma) = \alpha_{\min} + (\alpha_{\max} - \alpha_{\min})\, q\,(1 - \gamma) / (q + \gamma) \tag{6.3}$$

where $\alpha_{\min}$ and $\alpha_{\max}$ are the artificial impermeability of the fluid and solid, respectively; q is a positive parameter used to tune the convex property of the interpolating equation. In the topology optimization of fluidic flows, $\alpha_{\min}$ is chosen to be 0, and $\alpha_{\max}$ should be chosen to be a high but finite number to well approximate the solid and ensure the numerical stability simultaneously.

Based on the above description, the layout of a passive micromixer can be obtained by finding the distribution of the design variable γ for the topology optimization problem expressed as

$$\text{Find } \gamma \in [0, 1] \text{ to minimize } \Psi(c), \text{ subject to}$$
$$\begin{cases} -\nabla \cdot \left[-p\mathbf{I} + \eta\left(\nabla\mathbf{u} + \nabla\mathbf{u}^{\mathrm{T}}\right)\right] + \rho(\mathbf{u}\cdot\nabla)\mathbf{u} = \mathbf{f}, \text{ in } \Omega \\ -\nabla\cdot\mathbf{u} = 0, \text{ in } \Omega \\ \nabla\cdot\left(-D_f\nabla c\right) + \mathbf{u}\cdot\nabla c = 0, \text{ in } \Omega \end{cases} \tag{6.4}$$

where ρ, η and D are the density, viscosity and diffusion constant of the fluid, respectively; $\mathbf{u}$ is the fluidic velocity; p is the fluidic pressure; $\Omega = \Omega_D \cup \Omega_C$ is the computational domain; Ω_D is the design domain; Ω_C is the channels connected to the inlet and outlet of the design domain; and the body forces in the design domain and channels are $-\alpha(\gamma)\,\mathbf{u}$ and $\mathbf{0}$, respectively. The boundary conditions for the Navier-Stokes equations are the parabolic velocity distribution at the inlet, the zero pressure condition at the outlet, the no slip boundary condition at the wall, and the continuity condition at the interior boundaries. The boundary conditions for the convection-diffusion equation are the concentration distribution at the inlet, zero normal diffusion at the wall and outlet boundaries, and the continuity condition at the interior boundaries.

6.1.2 Analyzing

The problem (Eq. 6.4) used to find the reasonable layout of the micromixer is an optimization problem with partial differential equation constraints, and it can be analyzed using the adjoint method [6]. For the detailed stabilization and adjoint analysis of the problem, one can refer to Andreasen's work in [7]. The problem in Eq. 6.4 is solved by the finite element method. To use the linear elements for solving both the Navier-Stokes equations and the convection-diffusion equation, these equations are stabilized using the generalized least squares (GLS) and the streamline upwind Petrov-Galerkin (SUPG) technologies, respectively [8]. The weak form of the adjoint equations of the convection-diffusion equation and the Navier-Stokes equations in the problem is: find $\mathbf{u}_a \in (\mathscr{H}(\Omega))^3$, $p_a \in \mathscr{L}_2(\Omega)$ and $c_a \in \mathscr{H}(\Omega)$ satisfying

$$\int_{\Omega}\left[c_a\mathbf{u}\cdot\nabla\hat{c}_a+D_f\nabla c_a\cdot\nabla\hat{c}_a+\mathbf{u}\cdot\nabla c_a\tau_{SUPG}\left(\mathbf{u}\cdot\nabla\hat{c}_a-D_f\Delta\hat{c}_a\right)\right]\mathrm{d}\Omega$$
$$+\int_{\Gamma_o}2\left(c-\overline{c}\right)\hat{c}_a\,\mathrm{d}\Gamma\Big/\int_{\Gamma_o}\left(c_0-\overline{c}\right)^2\mathrm{d}\Gamma=0,\ \text{with } c_a=0 \text{ on } \Gamma_i,\ \forall\hat{c}_a\in\mathscr{H}\left(\Omega\right) \tag{6.5}$$

and

$$\int_{\Omega}\nabla\mathbf{u}_a\cdot\eta\left(\nabla\hat{\mathbf{u}}_a+\nabla\hat{\mathbf{u}}_a^{\mathrm{T}}\right)-p_a\nabla\cdot\hat{\mathbf{u}}_a+\mathbf{u}_a\cdot\left(\hat{\mathbf{u}}_a\cdot\rho\nabla\mathbf{u}+\mathbf{u}\cdot\rho\nabla\hat{\mathbf{u}}_a\right)-\frac{\partial\mathbf{f}}{\partial\mathbf{u}}\mathbf{u}_a\cdot\hat{\mathbf{u}}_a$$
$$-\hat{p}_a\nabla\cdot\mathbf{u}_a\,\mathrm{d}\Omega+\sum_{i=1}^{N_e}\int_{\Omega_i}\tau_{GLS}\nabla p_a\cdot\nabla\hat{p}_a\,\mathrm{d}\Omega=-\int_{\Omega}c_a\hat{\mathbf{u}}_a\cdot\nabla c$$
$$+\left(\frac{\partial\tau_{SUPG}}{\partial\mathbf{u}}\cdot\hat{\mathbf{u}}_a\right)\mathbf{u}\cdot\nabla c_a\left(\mathbf{u}\cdot\nabla c-D_f\Delta c\right)+\tau_{SUPG}\hat{\mathbf{u}}_a\cdot\nabla c_a\left(\mathbf{u}\cdot\nabla c-D_f\Delta c\right)$$
$$+\tau_{SUPG}\mathbf{u}\cdot\nabla c_a\left(\hat{\mathbf{u}}_a\cdot\nabla c\right)\mathrm{d}\Omega,\ \text{with } \mathbf{u}_a=0 \text{ on } \Gamma_i\cup\Gamma_w,$$
$$\forall\hat{\mathbf{u}}_a\in\left(\mathscr{H}\left(\Omega\right)\right)^3 \text{ and } \forall\hat{p}_a\in\mathscr{L}_2\left(\Omega\right) \tag{6.6}$$

where $\mathbf{u}_a$, p_a and c_a, are the adjoint variables of $\mathbf{u}$, p, and c, respectively; Γ_w is the wall boundary; $\mathscr{H}(\Omega)$ and $\mathscr{L}_2(\Omega)$ are the first-order Hilbert space and the second-order Lebesgue integrable functional space, respectively; τ_{GLS} and τ_{SUPG} are the stabilization parameters choosing as in [8]; N_e is the number of elements used to discretize the design domain; and Ω_i is the domain of the i-th element. The first-order adjoint derivative of the optimization problem is

$$\frac{d\Psi}{d\gamma}=\frac{d\alpha}{d\gamma}\mathbf{u}\cdot\mathbf{u}_a,\ \mathbf{x}\in\Omega_D \tag{6.7}$$

Because the admissible value of the design variable γ is in [0, 1], the adjoint derivative at the solution of Eq. 6.4 satisfies [6]

$$\gamma=P_{[0,1]}\left(\gamma-\lambda\frac{d\Psi}{d\gamma}\right),\ \text{for } \forall\lambda>0 \tag{6.8}$$

where $P_{[0,1]}(\cdot)=\min\left(\max\left(0,\cdot\right),1\right)$ is a projection operator. According to [6], Eq. 6.8 is equivalent to

$$\begin{cases}\dfrac{d\Psi}{d\gamma}=0, & \mathbf{x}\in\{\mathbf{x}\in\Omega_D|0<\gamma(\mathbf{x})<1\}\\ \dfrac{d\Psi}{d\gamma}\geq 0, & \mathbf{x}\in\{\mathbf{x}\in\Omega_D|\gamma(\mathbf{x})=0\}\\ \dfrac{d\Psi}{d\gamma}\leq 0, & \mathbf{x}\in\{\mathbf{x}\in\Omega_D|\gamma(\mathbf{x})=1\}\end{cases} \tag{6.9}$$

6.1.3 Solving

Solving the problem for the layout design of micromixers is implemented using the gradient-based iterative approach. In the iterative procedure, the Navier-Stokes equations, the convection-diffusion equation, and the corresponding adjoint equations in the weak form are solved by the finite element method using the commercial software *COMSOL Multiphysics* with linear elements [9]. Then, the adjoint derivative can be obtained according to Eq. 6.7. As noted by Jameson [10], a smoothed gradient often can be employed to speed up the convergence of an iterative procedure. The discretized adjoint derivative is smoothed by Laplacian smoothing [11]. The discretized design variable is updated by the superlinear convergent quasi-Newtonian method, in which the Hessian approximation is used to mimic the property of the true Hessian matrix. Therefore, the search direction is obtained by using the Hessian approximation in place of the exact Hessian matrix. According to Eq. 6.8, the discretized design variable can be updated by

$$\boldsymbol{\gamma}_{k+1} = P_{[0,1]}\left(\boldsymbol{\gamma}_k + \lambda_k \mathbf{p}_k\right) \tag{6.10}$$

where λ_k, a positive number, is the step length; $\mathbf{p}_k = -\mathbf{H}_k \mathbf{g}_k$ is the descent direction; $\mathbf{H}_k$ and $\mathbf{g}_k$ are the inverse Hessian approximation and discretized adjoint derivative corresponding to the discretized design variable $\boldsymbol{\gamma}_k$ in the k-th iteration, respectively. The inverse Hessian approximation $\mathbf{H}_k$ is updated using the LBFGS formula [12–14]. The step length λ_k is chosen by the backtracking approach, where the initial value of λ_k is chosen to be the natural value 1 of the quasi-Newtonian method [12]. The Wolfe condition is considered to terminate the search procedure of the step length λ_k [12]. According to Eqs. 6.9 and 6.10 can be implemented as

$$\gamma_{k+1}^i = \begin{cases} P_{[0,1]}\left(\gamma_k^i\right), & g_k^i \geq 0 \text{ and } \gamma_k^i \leq \varepsilon \\ P_{[0,1]}\left(\gamma_k^i\right), & g_k^i \leq 0 \text{ and } \gamma_k^i \geq 1-\varepsilon \\ P_{[0,1]}\left(\gamma_k^i + \lambda_k p_k^i\right), & \text{others} \end{cases} \tag{6.11}$$

for $i = 1, 2, \ldots, N$, where γ_{k+1}^i is the i-th component of the discretized design variable $\boldsymbol{\gamma}_{k+1}$ in the $(k+1)$-th iteration; γ_k^i, p_k^i and g_k^i are the i-th component of $\boldsymbol{\gamma}_k$, $\mathbf{p}_k$ and $\mathbf{g}_k$, respectively; ε is the admissible tolerance, which is a low positive number; N is the element nodes number. The stopping conditions for the iterative procedure are

$$\|\lambda_k \mathbf{p}_k\|_1 \leq 1 \times 10^{-6} \tag{6.12}$$

and the maximal iteration number 3×10^3. If one of these two conditions is satisfied, the iterative procedure will be stopped. The procedure of solving the problem for the layout design of micromixers is listed in Table 6.1.

Table 6.1 Procedure of the iterative approach for solving the topology optimization problem

1.	Give the initial value of the design variable γ
2.	Solve the Navier-Stokes equations and convection-diffusion equation by the finite element method
3.	Solve the adjoint equations; compute the adjoint derivative and the corresponding objective value
4.	Smooth the adjoint derivative
5.	Update the design variable by the quasi-Newtonian method
6.	Check for convergence; if the stopping conditions are not satisfied, go to 2; and
7.	Post processing

6.1.4 Results and Discussion

In this section, the layouts of several typically used passive micromixers are investigated using the layout design method. In the passive micromixers, the inconsistent cross-sections along the flow direction give rise to chaotic advection and the distortion of the streamlines, which is an interplay between initial, centrifugal, and viscous effects of the fluidic flow [15]. Chaotic advection results in the manifold of the streamline and increases the mixing length and the area of the interface between the two mixing flows. Considering manufacturing feasibility, micromixers usually are designed with a limited number of layers with user specified layout of blocks, i.e., the staggered herringbone micromixer (SHM) [16], or they accomplish the mixing based on the multivortex in the cross section of a Dean flow, which is induced by the centrifugal force loaded on the fluid in the bend microchannel [15]. Based on the above, one can see that the inconsistent cross section of the micromixer is the main factor to be considered for improving the transversal mixing process. Therefore, the layout design of a micromixer should be performed in three dimensions, and an inconsistent cross section should be used to enhance the mixing performance.

Consideration of the manufacturing feasibility can be performed by adding a corresponding constraint into the problem in Eq. 6.4. The values of the corresponding parameters for solving Eq. 6.4 are listed in Table 6.2, if there is no special specification. The values of $\alpha_{\max}$ and q are chosen based on numerical experiments [5], and L is the size of the cross section of the design domain for the micromixers. In the following, the design domains are discretized by $20 \times 20 \times 20$ hexahedral elements per cubic space with an edge size equals to L; the concentration of the solutes in the two inflows with equal flux are set to be dimensionless value 2 and 0, respectively. Based on the specification of the concentration of the solutes at the inlet, the ideal concentration distribution at the outlet can be computed to be $\overline{c} = 1$

Table 6.2 Parameter choice for the layout design of micromixers

$\alpha_{\max}$	q	ρ	η	D	L	Re	Pe
1×10^7	1	1×10^3 kg/m^3	1×10^{-3} Pa s	1×10^{-9}	400 μm	2.5	2500

according to Eq. 6.2. All the computation are performed on a DELL workstation (DELL PRECISION T5500, two Intel Xeon Quad X5550 CPUs, 48 gigabyte memory).

6.1.4.1 Free 3D Design

First, a micromixer in a straight channel is designed to demonstrate the mixing mechanism and the capability of the layout design method. The design domain is shown in Fig. 6.1a. By solving the problem in Eq. 6.4, the layout of the micromixer is obtained, as shown in Fig. 6.1b, and the corresponding value of the mixing measurement is 0.573. The history of the values of the mixing measurement in the iterative procedure is plotted in Fig. 6.1c. From the plotting of the concentration and projected

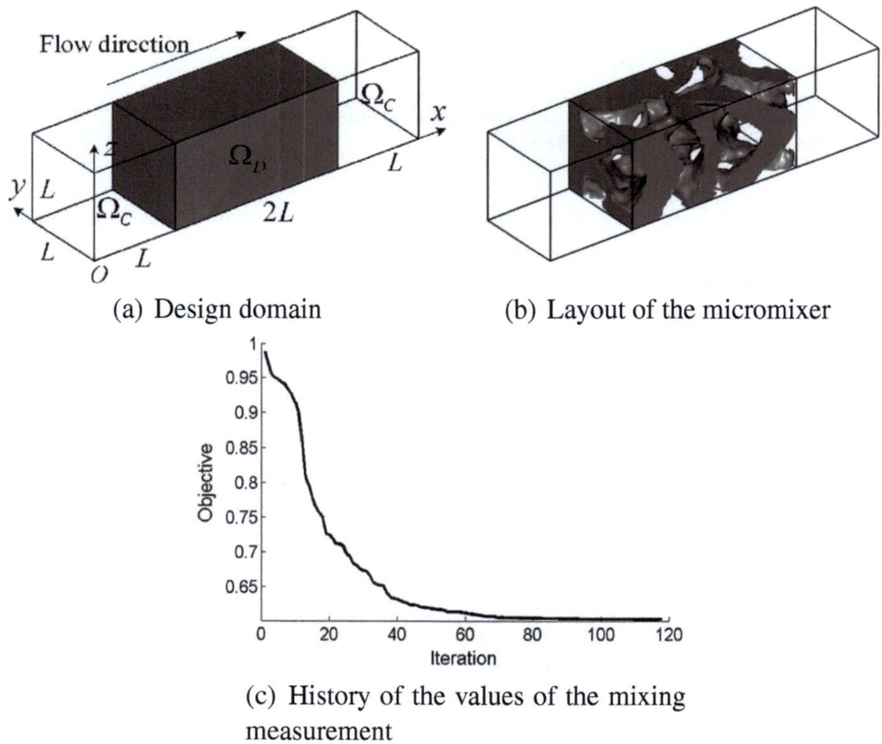

Fig. 6.1 **a** Design domain, where Ω_C is the channels connected to the design domain Ω_D and the length of Ω_D is twice of the width L; **b** obtained layout of the micromixer; **c** history of the values of the mixing measurement in the iterative procedure. The value of the mixing measurement corresponding to the obtained layout is 0.573. The number of iteration steps is 118, and the cost of the CPU time is 18.4 h. The fluid velocity at the inlet is $(40 s_1 (1 - s_1) s_2 (1 - s_2) U_{max}, 0, 0)$ m/s, where s_1 and s_2 are the arc length parameters at the inlet; the viscosity is 1×10^{-3} Pa s; and the minimum feature size is 100 μm

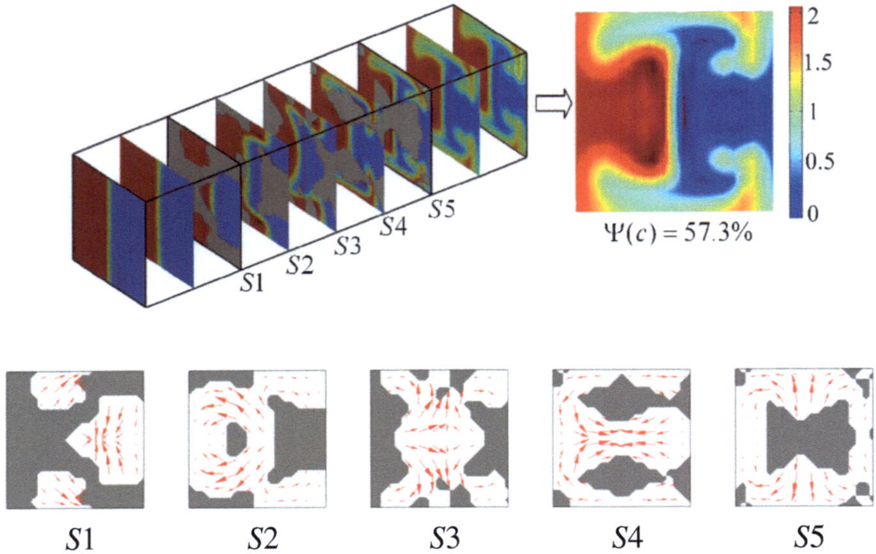

Fig. 6.2 Concentration and projected velocity vector distribution in the cross sections corresponding to the obtained micromixer in Fig. 6.1b

velocity vectors in several cross sections (Fig. 6.2), one can confirm that the chaotic advection caused by the inconsistent cross section of the micromixer can improve the mixing performance effectively. From Fig. 6.1b, one can see that the obtained layout has a complicated, three-dimensional layout, which is unfeasible for the commonly available manufacturing processing in the area of micro-electro-mechanical systems (MEMS).

6.1.4.2 Manufacturability

When the manufacturability is considered, one needs an additional constraint for the problem in Eq. 6.4. Microfluidic designs should be compatible with the planar, layer-by-layer geometries that are imposed by current, lithography-based techniques of microfabrication [16]. Therefore, the manufacturing constraint used is that the design variable has a consistent value in the depth direction of the design domain. For the design domain in Fig. 6.1a, the z axis is the depth direction, and the nodes with the same x and y coordinates have the same design variable value; then, the design variable is transferred to be a function defined in the xOy plane. Therefore, the number of the design variable is reduced after the finite element discretization. By the manufacturing constraint, the adjoint derivative in Eq. 6.7 is transformed into the following form:

$$\frac{d\Psi}{d\gamma} = -\frac{1}{z_t - z_b} \int_{z_b}^{z_t} \frac{d\alpha}{d\gamma} \mathbf{u} \cdot \mathbf{u}_a \, dz, \ \ \mathbf{x} \in \Omega_{Dxy} \tag{6.13}$$

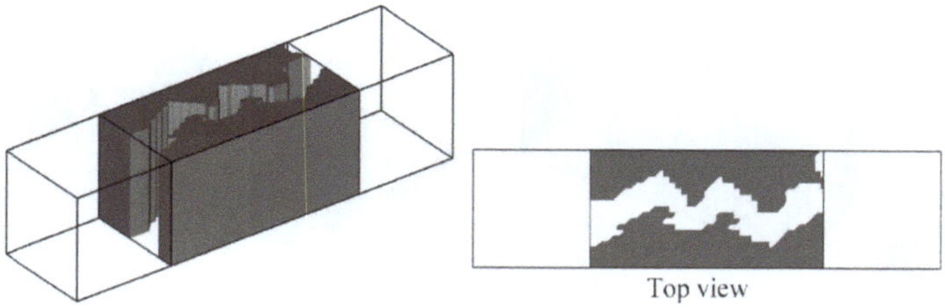

Fig. 6.3 Layout of the micromixer considering manufacturability. The value of the mixing measurement corresponding to the obtained design is 0.879. The number of iteration steps is 25, and CPU time is 3.2 h. The fluid velocity at the inlet is $(40s_1(1-s_1)s_2(1-s_2), 0, 0)$ m/s, where s_1 and s_2 are the arc length parameters at the inlet; the viscosity is 1×10^{-3} Pa s; and the minimum feature size is 100 μm

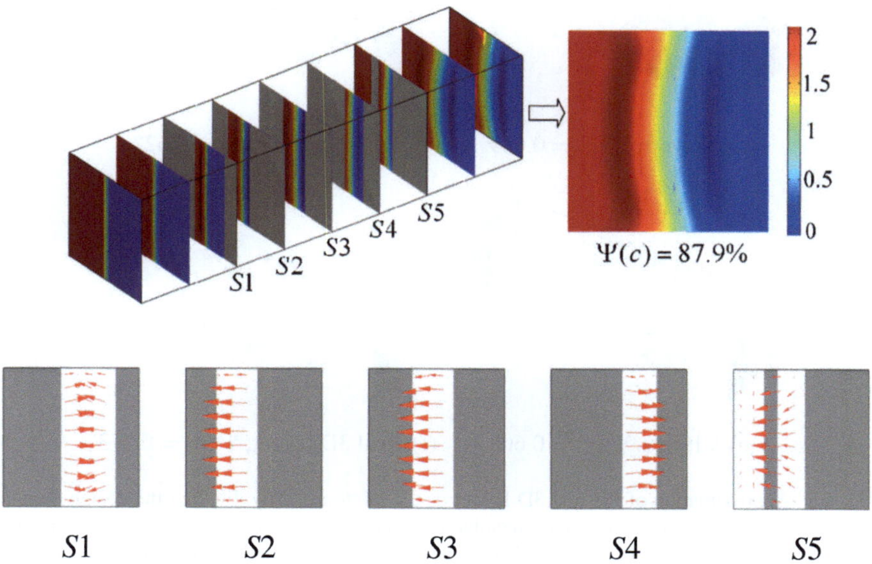

Fig. 6.4 Concentration and projected velocity vector distribution in the cross sections corresponding to the obtained layout with manufacturing constraint in Fig. 6.3

where z values of the bottom and top surfaces of the design domain are z_b and z_t ($z_t > z_b$), respectively; Ω_{Dxy} is the projection of the design domain Ω_D in the xOy plane. Based on the above manufacturing constraint, the micromixer in the design domain shown in Fig. 6.1a is redesigned. Figures 6.3 and 6.4 show the obtained layout of the micromixer and the distribution of the concentration, and the projected velocity vectors in the cross sections, respectively. The corresponding value of the mixing measurement is 0.879. Compared with the case without manufacturing constraint in Fig. 6.1, the mixing performance is weakened significantly by the manufacturing constraint. Therefore, a key design task is to balance the mixing performance and the manufacturability.

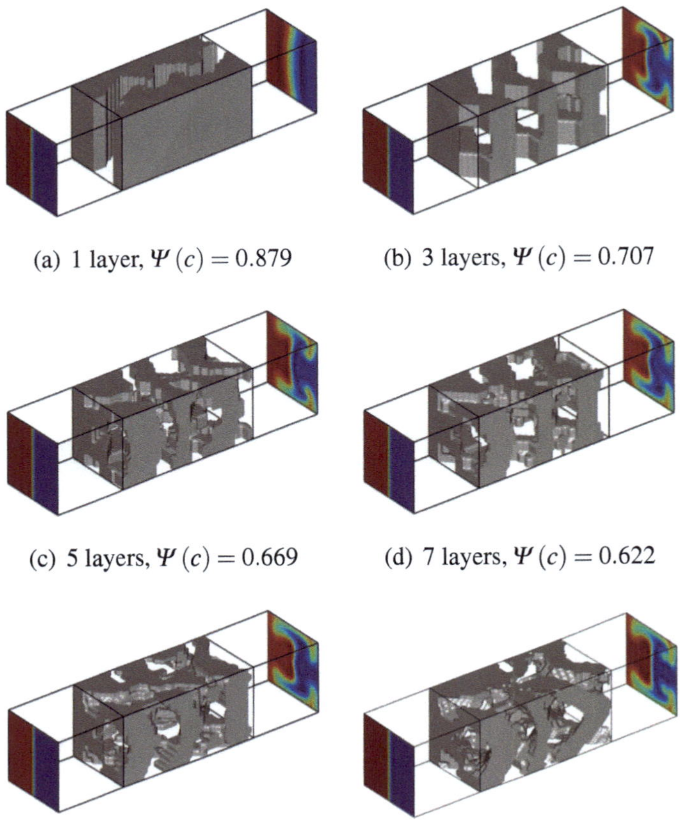

(a) 1 layer, $\Psi(c) = 0.879$ (b) 3 layers, $\Psi(c) = 0.707$

(c) 5 layers, $\Psi(c) = 0.669$ (d) 7 layers, $\Psi(c) = 0.622$

(e) 9 layers, $\Psi(c) = 0.605$ (f) Real 3D layout, $\Psi(c) = 0.573$

Fig. 6.5 Approximation of the real 3D layout of the micromixer by dividing the design domain into different numbers of layers with manufacturing constraint. The fluid velocity at the inlet is $(40s_1(1-s_1)s_2(1-s_2), 0, 0)$ m/s, where s_1 and s_2 are the arc length parameters at the inlet; the viscosity is 1×10^{-3} Pa s; and the minimum feature size is 100 μm

It is necessary to research a method to approximate the obtained real 3D layout so that designers can balance the manufacturing constraint and the mixing performance for a design of the micromixer obtained by the layout design method. One approach is to divide the design domain into several layers along the z-axis. At the same time, the design variable in each layer is constrained by the manufacturing constraint. To validate this method, the design domain corresponding to the obtained design in Fig. 6.1b is divided into different numbers of layers. After solving the problems for the design domain divided into different numbers of layers, the corresponding layouts of the micromixers are obtained as shown in Fig. 6.5, where a larger number of divided layers produces better mixing performance. However, more layers means more challenges in the manufacturing process. It is valuable to determine an

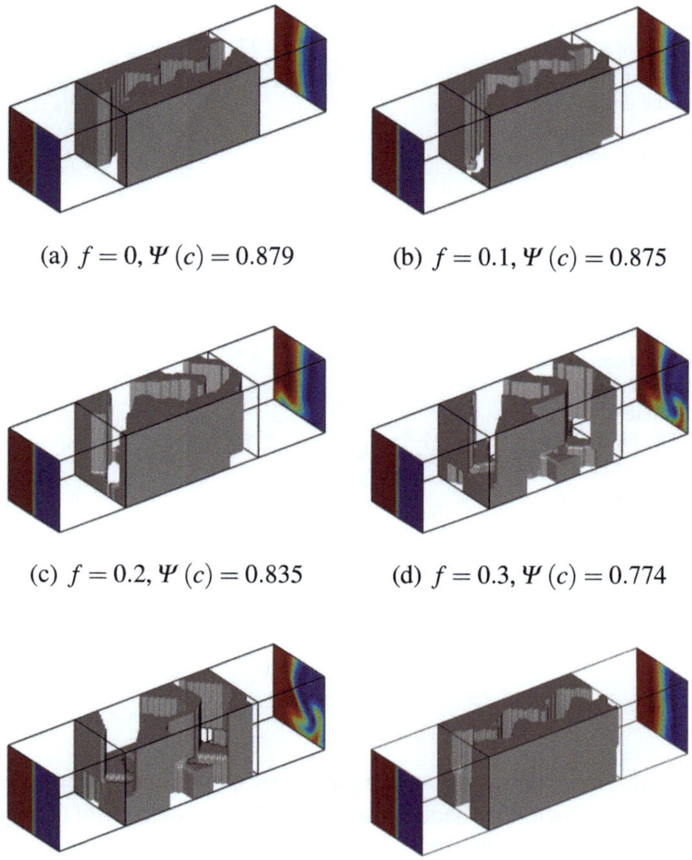

(a) $f = 0, \Psi(c) = 0.879$ (b) $f = 0.1, \Psi(c) = 0.875$

(c) $f = 0.2, \Psi(c) = 0.835$ (d) $f = 0.3, \Psi(c) = 0.774$

(e) $f = 0.4, \Psi(c) = 0.764$ (f) $f = 0.5, \Psi(c) = 0.879$

Fig. 6.6 Layouts of micromixers obtained by dividing the design domain into two layers. The fluid velocity at the inlet is $(40s_1(1-s_1)s_2(1-s_2), 0, 0)$ m/s, where s_1 and s_2 are the arc length parameters at the inlet; the viscosity is 1×10^{-3} Pa s; and the minimum feature size is 100 μm

appropriate balance between mixing performance and the complexity of manufacturing. The design domain divided into two layers is feasible for manufacture in most cases. By dividing the design domain into two layers, the layouts of micromixers are obtained as shown in Fig. 6.6, where f is the fraction of the bottom layer height. Figure 6.6 shows that a more reasonable choice for the value of f is 0.4.

In [16], the design domain is set to be the bottom layer of a straight channel as shown in Fig. 6.7, where nL and f are the length and fraction of the height of the design domain, respectively. The problems in Eq. 6.4 with manufacturing constraint are solved for different values of n and f; the values of the mixing measurement are plotted in Fig. 6.8, which shows that choosing a larger n or a larger f helps to improve the mixing performance of the micromixers with layouts obtained by the layout

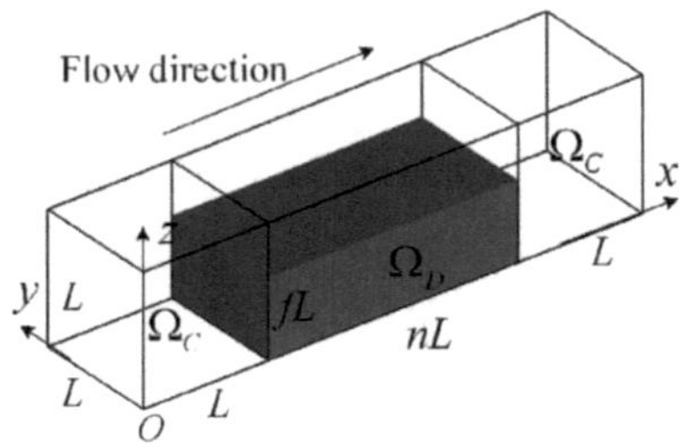

Fig. 6.7 Design domain at the *bottom layer* of a straight channel. Ω_C is the channels connected to the design domain Ω_D; the length and height of Ω_D are n and f times of L, respectively

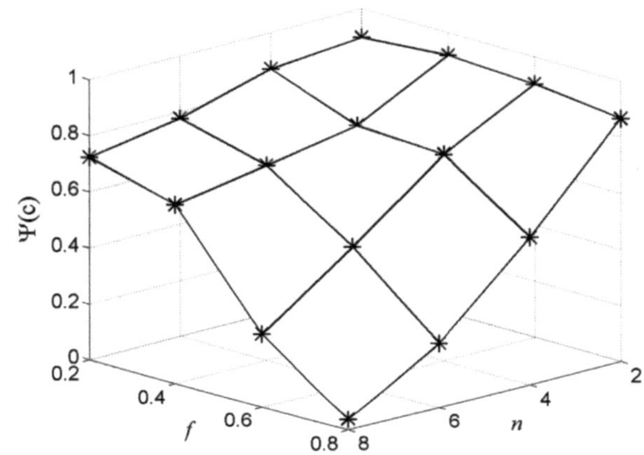

Fig. 6.8 Plotting of the mixing measurement of the obtained micromixers in the design domains as shown in Fig. 6.7 for different values of n and f. The fluid velocity at the inlet is $(40s_1(1-s_1)s_2(1-s_2), 0, 0)$m/s, where s_1 and s_2 are the arc length parameters at the inlet; the viscosity is 1×10^{-3} Pa s; and the minimum feature size is 100 μm

design method. This is because that a larger n corresponds to a longer mixing length and a larger f strengthens the chaotic advection in the cross section of the micromixer. The obtained layouts of the micromixer corresponding to $n = 8$, $f = 0.6$, and $n = 8$, $f = 0.8$ are shown in Fig. 6.9, which also shows the concentration distributions in the cross sections. The layout corresponding to $f = 0.8$ (Fig. 6.9b) shows that the structure in the bottom layer of the micromixer has a large aspect ratio, which increases the difficulty of manufacturing. On the other hand, the thick bottom layer with $f = 0.8$ corresponds to a thin top layer, for which the mesh must be refined to ensure the accuracy of the solution; thereby, computational requirement is increased. Therefore, the fraction of the height of the design domain for the micromixer is chosen to be $f = 0.6$.

Inspired by the obtained micromixers shown in Fig. 6.6, the mixing performance of the micromixer in Fig. 6.9a can be improved additionally if the top layer is included in the design domain. Then the micromixer shown in Fig. 6.10 is obtained, and the corresponding value of the mixing measurement is 0.031. The comparison of the results in Figs. 6.9 and 6.10 demonstrates the flexibility on the choice of the design domain in the design of the micromixer by the layout design method, and a reasonable choice of the design domain is helpful for improving the mixing performance.

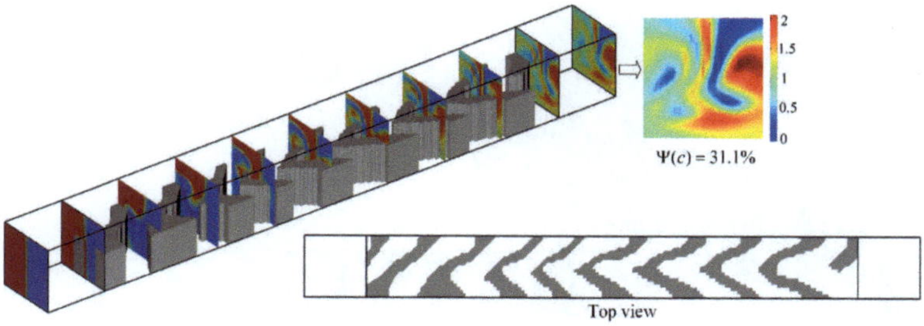

(a) $n = 8, f = 0.6$

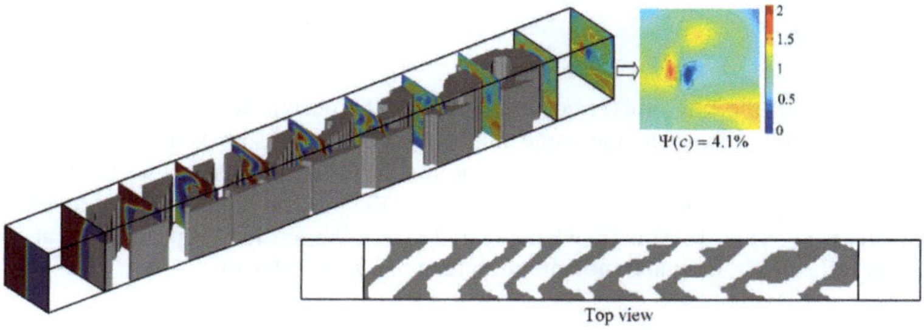

(b) $n = 8, f = 0.8$

Fig. 6.9 **a** Layout of the micromixer for the design domain as shown in Fig. 6.7, where $n = 8$, $f = 0.6$, and the mixing measurement corresponding to the obtained design is $\Psi(c) = 0.267$; the number of iteration steps is 56; and the CPU time is 9.3 h; **b** layout of the micromixer for the design domain as shown in Fig. 6.7, where $n = 8$, $f = 0.8$, and the mixing measurement corresponding to the obtained design is $\Psi(c) = 0.041$; the number of iteration steps is 105; and the CPU time is 20.4 h. The fluid velocity at the inlet is $(40 s_1 (1 - s_1) s_2 (1 - s_2), 0, 0)$ m/s, where s_1 and s_2 are the arc length parameters at the inlet; the viscosity is 1×10^{-3} Pa s; and the minimum feature size is 100 μm

6.1.4.3 Spatial Periodic Layout

Spatial periodic layout is a usual choice for the design of micromixers, because one needs merely to design one period of the layout for a micromixer and then use the series-wound method to obtain the micromixer. Currently, most of the micromixers in the literatures are based on the spatial periodic design, e.g., the design of the SHM [16–32], the 3D serpentine micromixer [21, 33–38], and the rhombic micromixer [39, 40]. To obtain a micromixer with a spatial periodic layout, the periodic constraint needs to be added to the problem in Eq. 6.4, and the periodic constraint is expressed as

$$\gamma(\mathbf{x}_0) = \gamma(\mathbf{x}_0 + i\mathbf{r}), \ \forall \mathbf{x}_0 \in \Omega_0, \ i = 1, 2 \ldots, m - 1 \tag{6.14}$$

where $\mathbf{r}$ is the known vector fixed by two corresponding nodes in two neighboring periods; Ω_0 is the first period; and m is the number of periods. The adjoint derivative for the problem in Eq. 6.4 with periodic constraint is calculated as

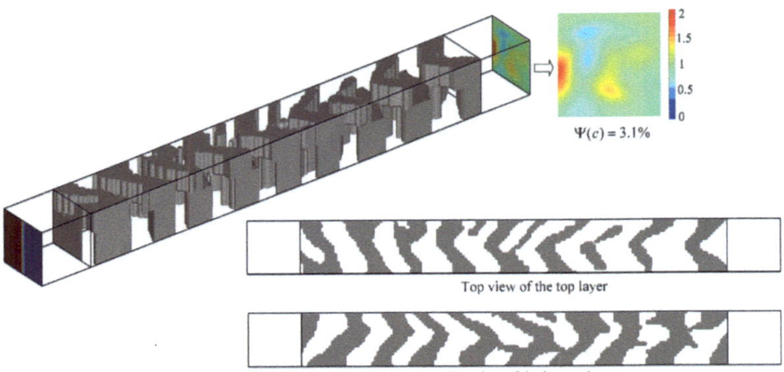

Fig. 6.10 Layout of the micromixer corresponding to the design domain divided into two layers, where the *bottom layer* has the same thickness as the one shown in Fig. 6.9a. The value of the mixing measurement corresponding to the obtained design is $\Psi(c) = 0.031$. The number of iteration steps is 193, and the CPU time is 33.2 h. The fluid velocity at the inlet is $(40s_1(1-s_1)s_2(1-s_2), 0, 0)$ m/s, where s_1 and s_2 are the arc length parameters at the inlet; the viscosity is 1×10^{-3} Pa s; and the minimum feature size is 100 μm

$$\left.\frac{d\Psi}{d\gamma}\right|_{\Omega_0} = \frac{1}{m}\sum_{i=0}^{m-1}\left.\left(-\frac{d\alpha}{d\gamma}\mathbf{u}\cdot\mathbf{u}_a\right)\right|_{\mathbf{x}=\mathbf{x}_0+i\mathbf{r},\mathbf{x}_0\in\Omega_0} \tag{6.15}$$

Based on Eqs. 6.14 and 6.15, the micromixers are designed with different numbers of periods for the design domain in Fig. 6.7, where $n = 8$ and $f = 0.6$. Figure 6.11 shows that different numbers of periods correspond to different layouts of micromixers, and the mixing performance become worse when compared to the no periodic constraint case in Fig. 6.9. For different choices of the value of n in the design domain shown in Fig. 6.7, a series of layouts of micromixers with the same periodic length $2L$ are computed as shown in Fig. 6.12. This example demonstrates that the obtained layouts are different from each other. After sequentially connecting the obtained layouts in Fig. 6.12a–c to have the same length as the one in Fig. 6.12d, the value of the mixing measurement is computed as shown in Fig. 6.13, which shows that the layout in Fig. 6.13d has the best mixing performance. Therefore, the periodic design of micromixers based on the series-wound method is the suboptimal choice. This confirms the conclusion in [7, 32].

For the convenience of the periodic design of a micromixer in a long design domain, the design of the layout of a micromixer by the series-wound method is still a widely used strategy, although this is a suboptimal choice. One can use the layout design method to design one cycle of the periodic micromixer and enhance the mixing performance by series-wound extending the obtained cycle. Two different mixing cycles are supplied as shown in Fig. 6.14, i.e., one bend cycle and one split cycle. Based on the obtained cycle, one can connect several cycles sequentially by the series-wound method to achieve better mixing performance. Figure 6.15 shows the sequential connection of the cycles in Fig. 6.14, where the values of the mixing

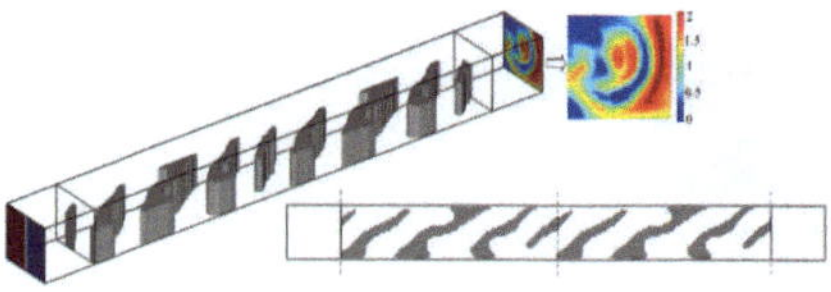

(a) 2 periods, $\Psi(c) = 0.300$, iteration steps 121, CPU time 11.4 hours

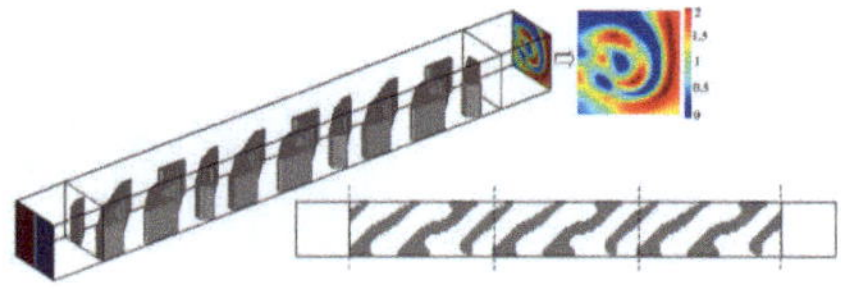

(b) 3 periods, $\Psi(c) = 0.294$, iteration steps 173, CPU time 21.1 hours

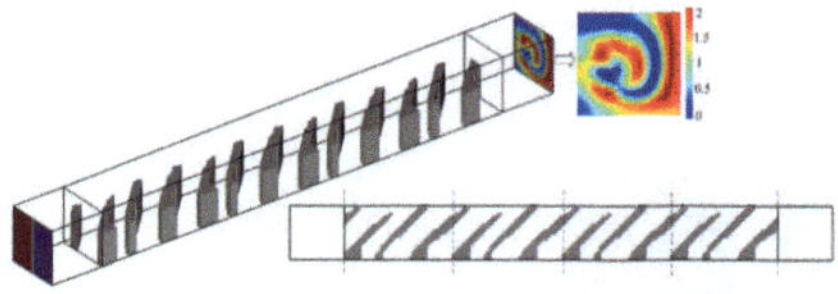

(c) 4 periods, $\Psi(c) = 0.388$, iteration steps 391, CPU time 28.3 hours

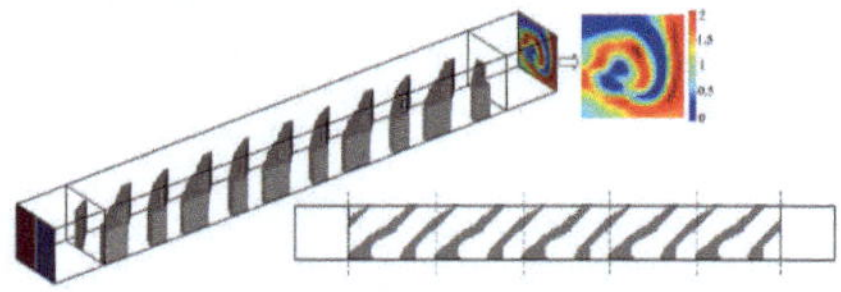

(d) 5 periods, $\Psi(c) = 0.354$, iteration steps 88, CPU time 7.5 hours

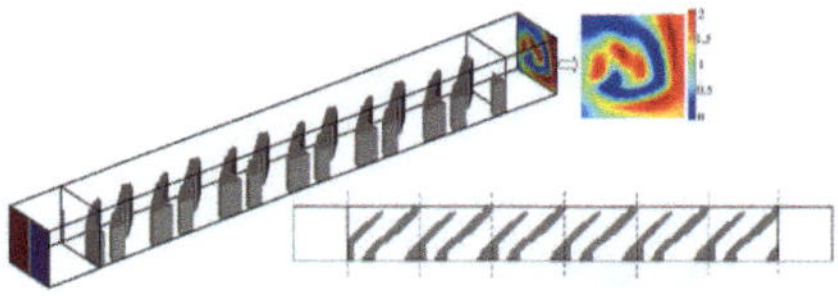

(e) 6 periods, $\Psi(c) = 0.387$, iteration steps 70, CPU time 7.0 hours

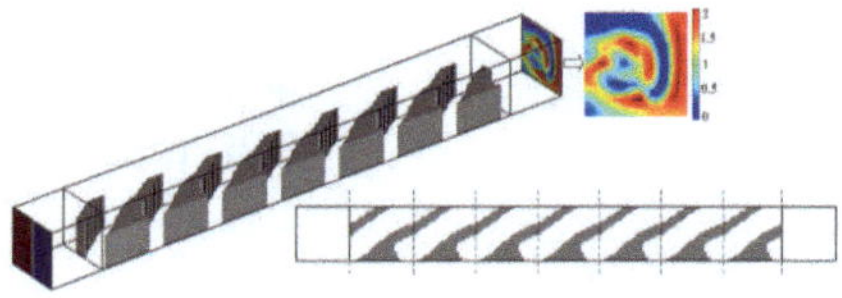

(f) 7 periods, $\Psi(c) = 0.334$, iteration steps 65, CPU time 6.2 hours

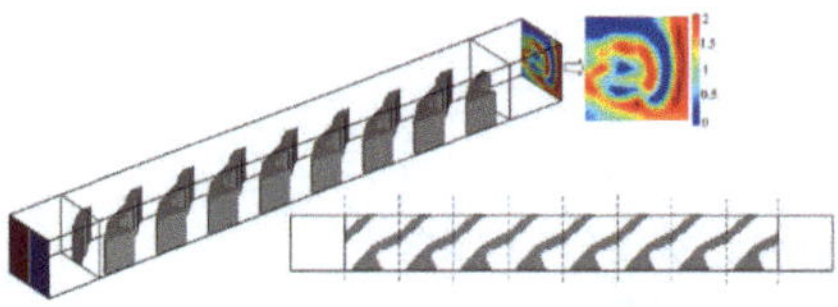

(g) 8 periods, $\Psi(c) = 0.320$, iteration steps 96, CPU time 8.4 hours

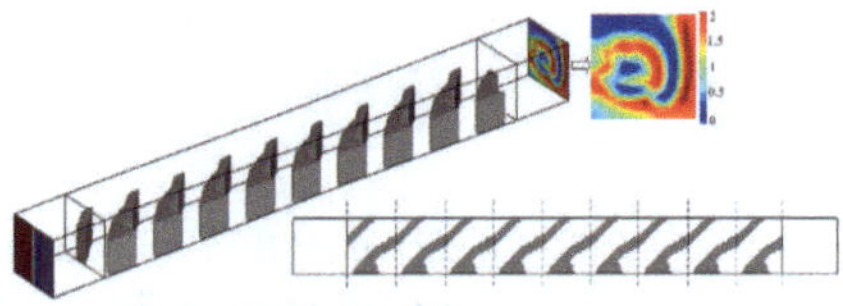

(h) 9 periods, $\Psi(c) = 0.330$, iteration steps 92, CPU time 8.8 hours

Fig. 6.11 Layouts of the periodic micromixers with different numbers of periods in the design domain in Fig. 6.7, where $n = 8$ and $f = 0.6$. The fluid velocity at the inlet is $(40s_1(1-s_1)s_2(1-s_2), 0, 0)$ m/s, where s_1 and s_2 are the arc length parameters at the inlet; the viscosity is 1×10^{-3} Pa s; and the minimum feature size is 100 μm

measurement are both less than 0.050 after sequentially connecting 9 and 3 cycles respectively. Although the mixing length of the micromixer with series-wound bend cycles is longer than the mixing length with series-wound split cycles, the micromixer with series-wound bend cycles takes up the same spatial distance in the x-axis direction (0.76 cm) as the one with series-wound split cycles. In [1], the threshold level of mixing is defined as 0.050, which is assumed to be complete mixing when the value of the mixing measurement is less than 0.050. Therefore, these two micromixers with series-wound cycles reach the threshold defined in [1]. The comparison of the values of the mixing measurement corresponding to the results in Fig. 6.15 shows that the split cycle is a better choice than the bend cycle.

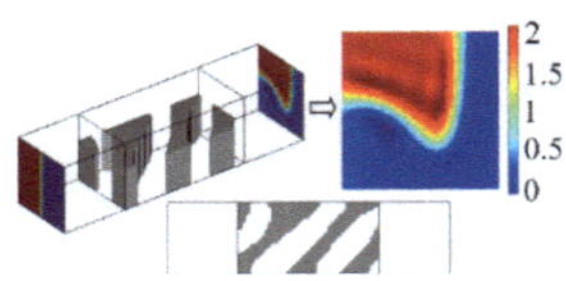

(a) $n = 2$, $f = 0.6$, 1 period, $\Psi(c) = 0.903$, iteration steps 174, CPU time 5.8 hours

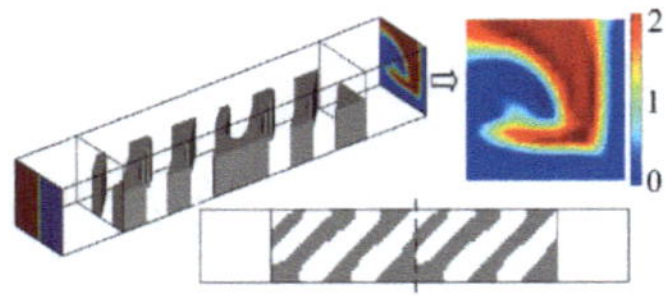

(b) $n = 4$, $f = 0.6$, 2 periods, $\Psi(c) = 0.774$, iteration steps 69, CPU time 5.3 hours

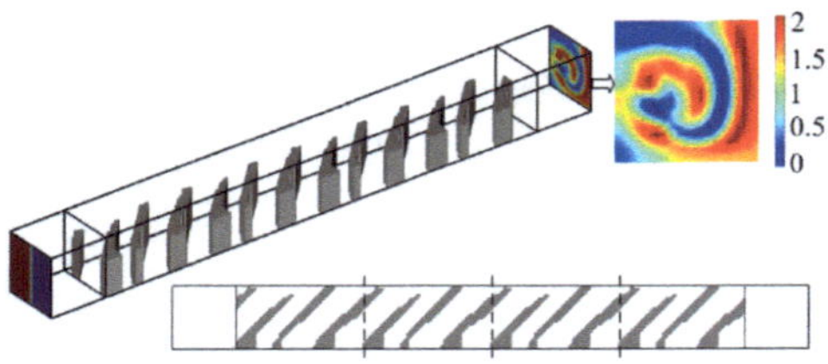

(c) $n = 8$, $f = 0.6$, 4 periods, $\Psi(c) = 0.388$, iteration steps 391, CPU time 27.5 hours

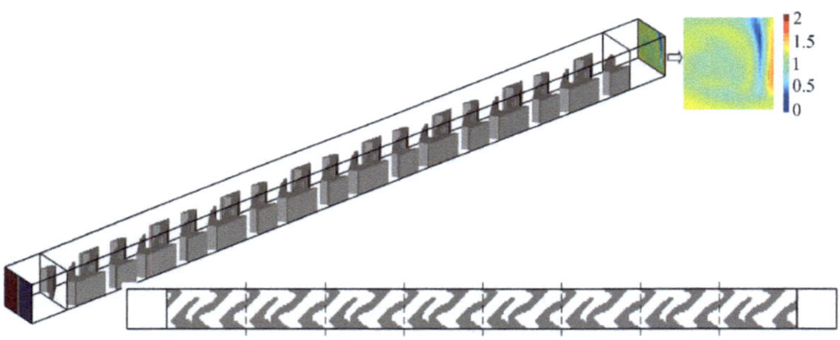

(d) $n = 16$, $f = 0.6$, 8 periods, $\Psi(c) = 0.030$, iteration steps 187, CPU time 30.6 hours

Fig. 6.12 Layouts of the periodic micromixers with the same period length in the design domain in Fig. 6.7. The fluid velocity at the inlet is $(40s_1(1-s_1)s_2(1-s_2), 0, 0)$ m/s, where s_1 and s_2 are the arc length parameters at the inlet; the viscosity is 1×10^{-3} Pa s; and the minimum feature size is 100 μm

(a) Series-wound layout in Fig. 6.12a; $\Psi(c) = 0.063$

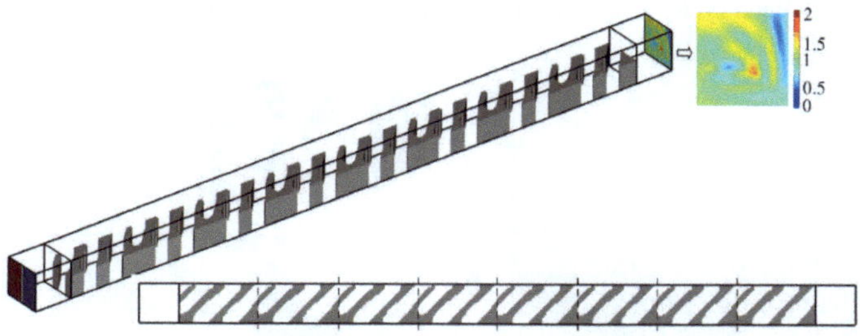

(b) Series-wound layout in Fig. 6.12b; $\Psi(c) = 0.041$

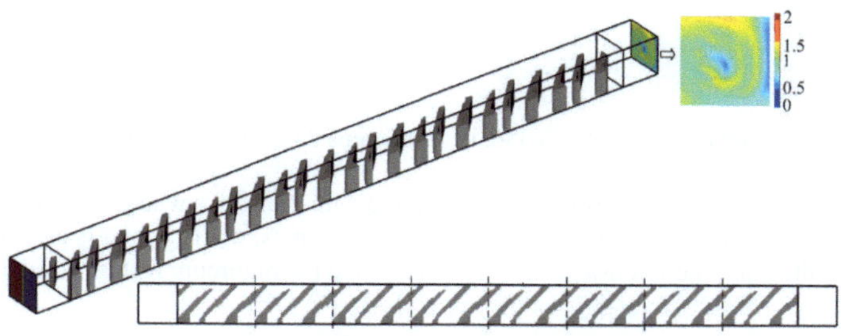

(c) Series-wound layout in Fig. 6.12c; $\Psi(c) = 0.036$

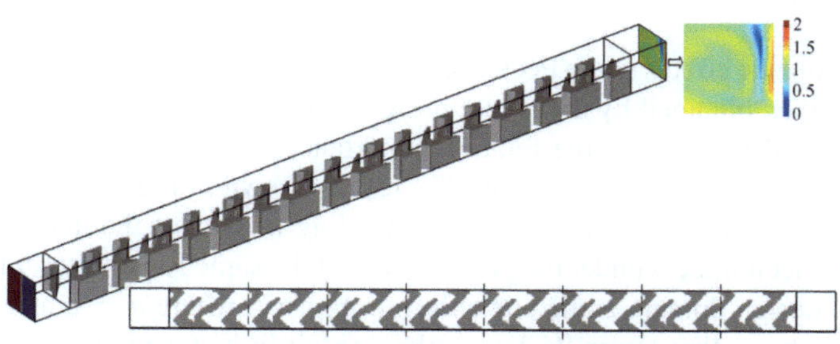

(d) Layout with 8 periods in Fig. 6.12d; $\Psi(c) = 0.030$

Fig. 6.13 Comparison of the series-wound layouts of the periodic micromixers in Fig. 6.12

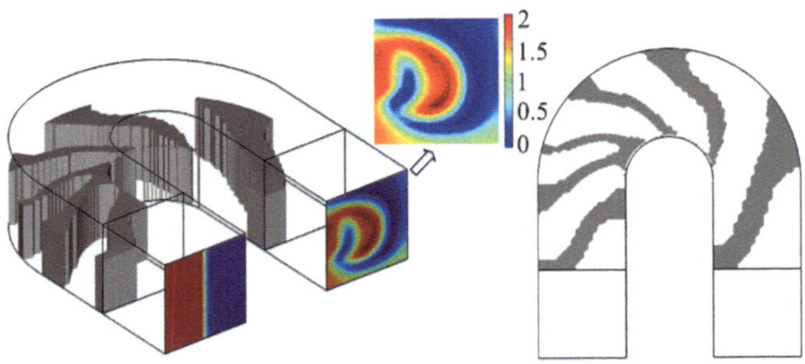

(a) Bend cycle, $\Psi(c) = 0.541$, iteration steps 66, CPU time 5.3 hours

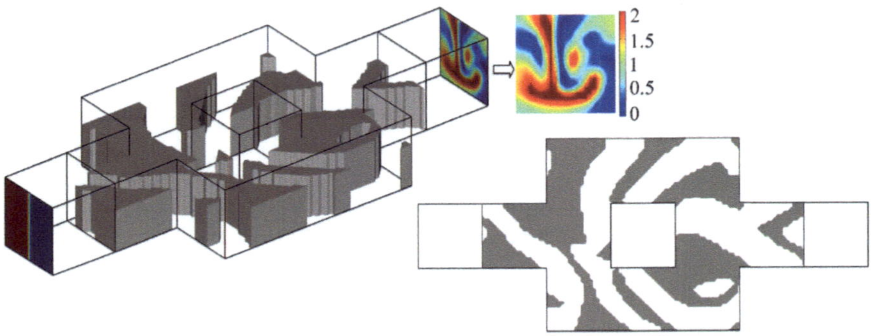

(b) Split cycle, $\Psi(c) = 0.386$, iteration steps 239, CPU time 27.8 hours

Fig. 6.14 Layouts of the periodic cycles for the periodic micromixers, where f is equal to 0.6. The fluid velocity at the inlet is $(40s_1(1-s_1)s_2(1-s_2), 0, 0)$ m/s, where s_1 and s_2 are the arc length parameters at the inlet; the viscosity is 1×10^{-3} Pa s; and the minimum feature size is 100 μm

6.1.4.4 Effects of Péclet Number and Reynolds Number

In this section, the effects of the Péclet number and Reynolds number on the layout of the micromixer, obtained by the layout design method, are investigated. The Péclet number is used to measure the relative importance of the convection of the flow and diffusion for the mixing progress. The results shown in Fig. 6.16 are the values of the mixing measurement and the corresponding layouts of the micromixers for different Péclet numbers under the precondition of the same Reynolds number of the flows. The design domain is the one shown in Fig. 6.7, where $n = 8$ and $f = 0.6$. Figure 6.16 shows that lower Péclet numbers result in better mixing performance. This is because lower Péclet numbers correspond to higher values of the diffusive constant under the precondition of consistent Reynolds number. Therefore, the value of the mixing measurement decreases, i.e. the mixing performance becomes better as the Péclet number decreases.

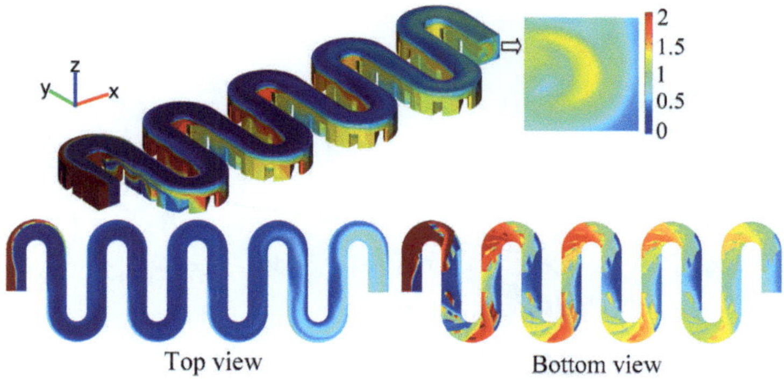

(a) Series-wound bend cycles, $\Psi(c) = 0.045$

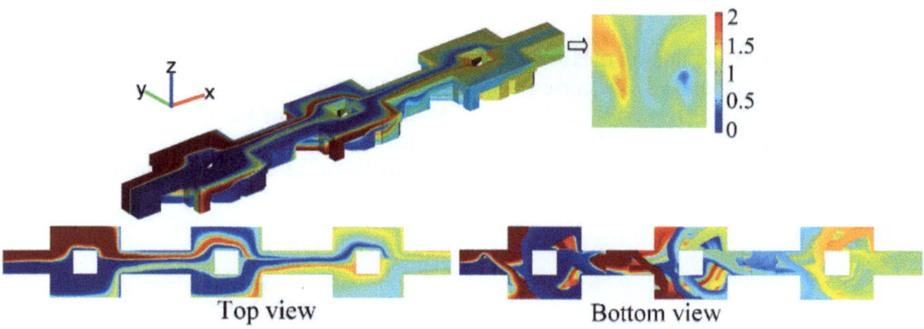

(b) Series-wound split cycles, $\Psi(c) = 0.040$

Fig. 6.15 Concentration distribution in the periodic layouts of the micromixers with series-wound cycles in Fig. 6.12

The Reynolds number is used to present the relative importance of the convection and viscosity of the fluidic flow. Because convection is an important aspect for the micromixing, the Reynolds number influences the layout of the micromixer obtained by the layout design method. As shown in Fig. 6.17, the layouts of the micromixers with different Reynolds numbers are computed in the design domain shown in Fig. 6.7, where $n = 8$, $f = 0.6$, and the Péclet number is 2500. Figure 6.17 shows that different Reynolds numbers correspond to different layouts of the micromixers for the same Péclet number. In the low Reynolds number flows, the mixing progress is influenced mainly by the diffusion, and the convection of the flow can be strengthened by the obtained layout of the micromixer. Figure 6.17 shows that the obtained layout corresponding to low Reynolds number has relatively better performance than the case corresponding to relatively high Reynolds number. One reason for this effect is that the high Reynolds number represents high fluidic velocity, which means short mixing time in a design domain with a fixed length, even when the micromixer has reasonable blocks at the bottom layer of the micromixer (Tables 6.3 and 6.4).

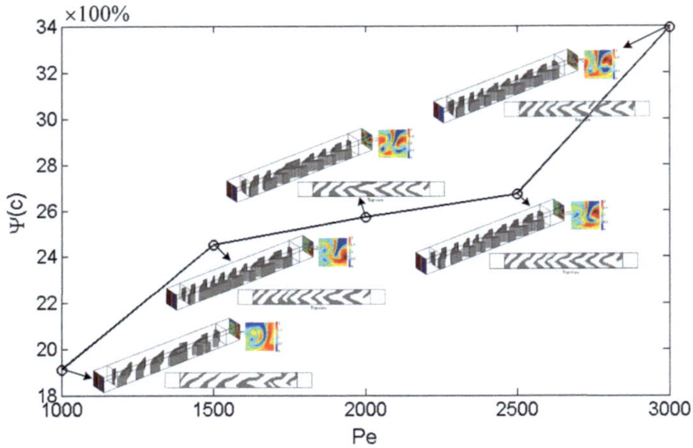

Fig. 6.16 Values of the mixing measurement and the corresponding layouts of the micromixers with different Péclet numbers. The corresponding Reynolds number is 2.5; the fluid velocity at the inlet is $(16s_1(1-s_1)s_2(1-s_2)U_{max}, 0, 0)$ m/s, where s_1 and s_2 are the arc length parameters at the inlet, U_{max} is the maximal value of the velocity and its values correspond to different Péclet numbers are listed in Table 6.3; the viscosity is 1×10^{-3} Pa s; and the minimum feature size is 100 μm

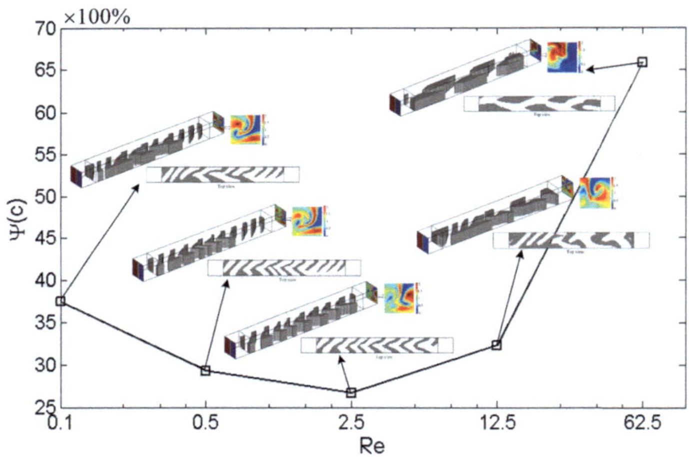

Fig. 6.17 Values of the mixing measurement and the corresponding layouts of the micromixers with different Reynolds numbers. The corresponding Péclet number is 2500; the fluid velocity at the inlet is $(40s_1(1-s_1)s_2(1-s_2), 0, 0)$ m/s, where s_1 and s_2 are the arc length parameters at the inlet; the values of the viscosity correspond to different Reynolds numbers are listed in Table 6.4; and the minimum feature size is 100 μm

6.1.4.5 Comparison

To validate the layout design method of the discussed micromixer, comparisons are performed between the obtained layouts of the passive micromixers and the results

Table 6.3 List of the maximal values of the velocity at the inlet of the micromixers with different Péclet numbers

Pe	1000	1500	2000	2500	3000
U_{max} (m/s)	1.0×10^{-2}	1.5×10^{-2}	2.0×10^{-2}	2.5×10^{-2}	3.0×10^{-2}

Table 6.4 List of the fluid viscosity corresponding to different Reynolds numbers

Re	0.1	0.5	2.5	12.5	62.5
η (Pa s)	2.5×10^{-2}	5×10^{-3}	1×10^{-3}	2×10^{-4}	4×10^{-5}

in the literatures. The comparison can be performed from two points of view, i.e. (1) comparing the values of the mixing measurement at the outlets of the micromixers with the same mixing length, and (2) comparing the necessary mixing length (or mixing time) for the same value of the mixing measurement at the outlets [1]. In the following, the first point of view is used.

During the comparison, the compared designs are drawn and computed with the sizes and parameters according to the corresponding literature. One of the well-known passive micromixers is the staggered herringbone mixer (SHM), which was proposed by Stroock et al. [16]. In the past, the SHM was investigated in [17–21, 23, 32] and optimized in [22, 24–31]. Cortes-Quiroza et al. conducted one of the latest shape optimization works for the SHM [22, 26]. Based on the optimized shape they obtained, the value of the mixing measurement of one cycle of the SHM is computed as 0.258 for the flow with the Reynolds number 1 (Fig. 6.18a), and the hydrodynamic pressure drop across the micromixer is 56 Pa. The value of the mixing measurement is 0.209 for the micromixer obtained by the layout design method in the design domain with the same cross section size as the one shown in Fig. 6.18b, and the corresponding hydrodynamic pressure drop across the micromixer is 60 Pa. Therefore, the value of the mixing measurement is improved up to 19.0% by paying the cost of increasing 4 Pa of the hydrodynamic pressure.

Recently, Hossain et al. [29] and Choudhary et al. [23] proposed to use a bilayer SHM to enhance the mixing performance. In [29], the top-bottom symmetrical bilayer SHM is optimized by the shape optimization method. However, the conclusion in [23] shows that the top-bottom asymmetrical bilayer SHM is more efficient, and an overall enhancement of the mixing performance is observed as the degree of asymmetry of the layout for the bilayer SHM is increased. Therefore, the top-bottom asymmetrical bilayer SHM is a more reasonable choice. Based on the simulation parameters in [23], the mixing performance of one cycle of the top-bottom asymmetrical bilayer SHM is computed, as shown in Fig. 6.19a; the corresponding value of the mixing measurement is 0.235, and the hydrodynamic pressure drop across the micromixer is 28 Pa. The top-bottom asymmetrical bilayer SHM is computed by the layout design method in the design domain with the same parameters as in [23]. The obtained layout of the top-bottom asymmetrical bilayer SHM is shown in Fig. 6.19b; the corresponding value of the mixing measurement is 0.140; and the hydrodynamic

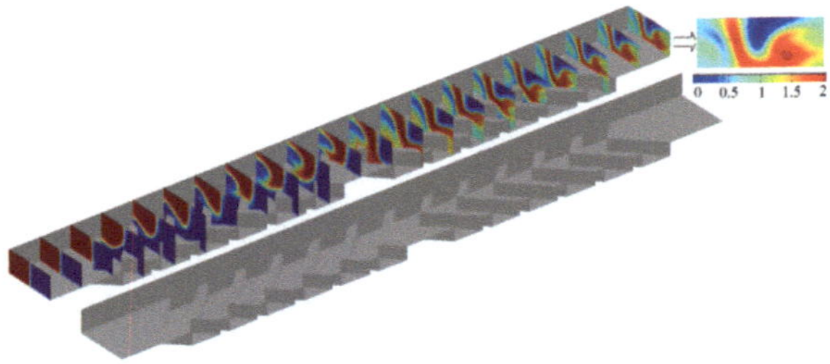

(a) Optimized SHM in [32]

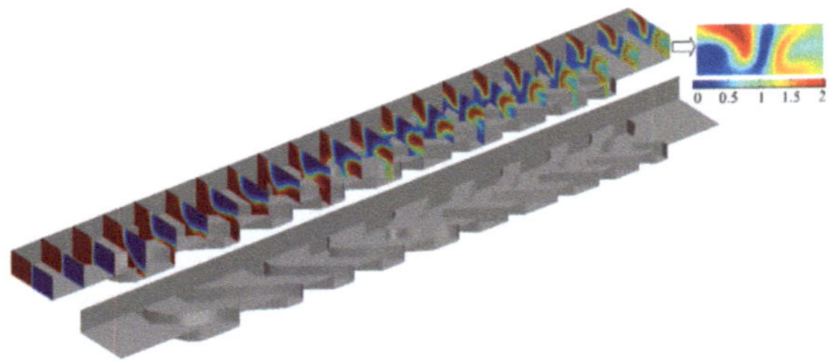

(b) SHM obtained by the layout design method

Fig. 6.18 **a** Concentration distribution in the SHM optimized by Cortes-Quiroza and Zangeneh in [26], where the value of the mixing measurement is 0.258, and the hydrodynamic pressure drop across the micromixer is 56 Pa; **b** concentration distribution in the SHM obtained by the layout design method, where the value of the mixing measurement is 0.209, and the hydrodynamic pressure drop across the micromixer is 60 Pa. By the layout design method, the value of the mixing measurement is improved up to 19.0% by paying the cost of increasing 4 Pa of the hydrodynamic pressure

pressure drop across the micromixer is 53 Pa. Therefore, the value of the mixing measurement is improved up to 40.4% by paying the cost of increasing 25 Pa of the hydrodynamic pressure.

6.1.5 *Summary*

The layout design method of passive micromixers based on the topology optimization of fluidic flows has been discussed,m with considering several design disciplines. The manufacturability of the obtained layouts is ensured by the manufacturing constraint. The mixing performance depends significantly on manufacturability. For the design of a micromixer with two-layer manufacturing constraint, a thicker bottom layer helps enhance the mixing performance when the design domain is chosen to be the bottom layer of a two-layer microchannel; and choosing both layers of the two-layer microchannel to be the design domain is a more reasonable choice. This demonstrates the flexibility of the layout design method on the choice of the design domain.

In order to balance the computational cost of numerical design procedure and the mixing performance of the obtained micromixer, one can choose a reasonable long design domain with a user-specified number of spacial periods or use the

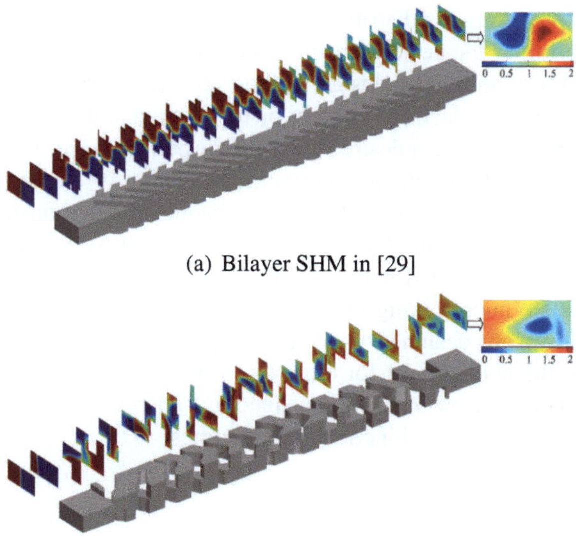

(a) Bilayer SHM in [29]

(b) Bilayer SHM obtained by the layout design method

Fig. 6.19 **a** Concentration distribution in the bilayer SHM in [23], where the value of the mixing measurement is 0.235, and the hydrodynamic pressure drop across the micromixer is 28 Pa; **b** concentration distribution in the bilayer SHM obtained by the layout design method, where the value of the mixing measurement is 0.140, and the hydrodynamic pressure drop across the micromixer is 53 Pa. By the layout design method, the value of the mixing measurement is improved improved up to 40.4% by paying the cost of increasing 25 Pa of the hydrodynamic pressure

series-wound method to extend one cycle of the layout obtained by the layout design method, although the periodic design of a micromixer is the suboptimal choice.

The effects of Péclet number and Reynolds number on the layout of micromixer obtained by the layout design method have been investigated. The capability of this design method is validated by the comparisons performed between the designed layouts and the optimized designs in the recently published literatures, where the values of the mixing measurement is improved up to 40.4% for one cycle of the micromixer. The comparisons show that the layout design method is effective and flexible for designing passive micromixers. In addition, this method can be extended to design more general micromixers, such as the passive micromixers in the centrifugal chips and the active micromixers. These will be investigated in the future.

6.2 Inverse Determination of Electrode Distribution for Electroosmotic Micromixer

Electroosmosis is one of the most used actuation mechanism for the microfluidics in the current active lab-on-chip devices. It is generated on the induced charged microchannel walls in contact with an electrolyte solution. Electrode patterns play the key role on providing the distribution of the external electric field for electroosmosis, and determines the performance of electroosmotic microfluidics. Therefore, this section proposes an inverse design method for the electrodes of electroosmotic microfluidics, where the electrode patterns on the microchannel wall can be inversely determined to achieve designer desired microfluidic performance. This inverse design method is carried out by implementing the interpolation of electric insulation and electric potential on the specified walls of microchannels. To demonstrate the capability of this inverse design method, the typical electroosmotic microdevice, electroosmotic micromixer, is modeled with several electrode patterns derived. The results show that complete micromixing of microflows can be achieved by the electrodes with patterns inversely designed using the proposed method. And this proposed method permits the applications for other types of electroosmotic microdevices in chemicals and biochemistry due to its outstanding capability on inversely determining the performance of electrokinetical microfluidics.

6.2.1 Modeling

The investigated electroosmotic micromixer is modeled as sketched in Fig. 6.20. A fluid flux is imposed on the inlet of the sketched micromixer with known concentration distribution; the anode and cathode for the electroosmosis is localized on the top and bottom surfaces, respectively. Under the hypothesis of continuum, the Navier-Stokes equations have been widespread used to describe the motion of fluid in microfluidics as

$$\begin{aligned} &\rho \mathbf{u} \cdot \nabla \mathbf{u} + \nabla \cdot \left[-\eta \left(\nabla \mathbf{u} + \nabla \mathbf{u}^{\mathrm{T}} \right) + p\mathbf{I} \right] = \mathbf{0}, \text{ in } \Omega \\ &- \nabla \cdot \mathbf{u} = 0, \text{ in } \Omega \end{aligned} \tag{6.16}$$

where $\mathbf{u}$, p are the fluid velocity and pressure, respectively; ρ and η are the density and dynamic viscosity of the electrolyte solution, respectively; Ω is the computational domain sketched in Fig. 6.20, with the inlet boundary Γ_i, wall boundary Γ_w and outlet boundary Γ_o satisfying $\Gamma_i \cup \Gamma_o \cup \Gamma_w = \partial \Omega$.

Under the assumption that the thickness of the Debye layer is much smaller than the feature size of microfluidics, the Helmholtz-Smoluchowski relation is satisfied at the walls of electroosmotic microchannel, where the slip velocity is proportional to the tangential component of the electric field intensity imposed by electrodes [41]

$$\mathbf{u} = -\mu_{eo} \left[\nabla V - (\mathbf{n} \cdot \nabla V) \, \mathbf{n} \right], \text{ on } \Gamma_w \tag{6.17}$$

where V is the distribution of the external electrical potential imposed by the electrodes; $\mu_{eo} = -\varepsilon_r \varepsilon_0 \frac{\zeta_0}{\eta}$ is the electroosmotic mobility, with ε_r, ε_0 and ζ_0 respectively representing the relative permittivity, permittivity of free space and Zeta potential; $\mathbf{n}$ is the outward unitary normal at $\partial \Omega$.

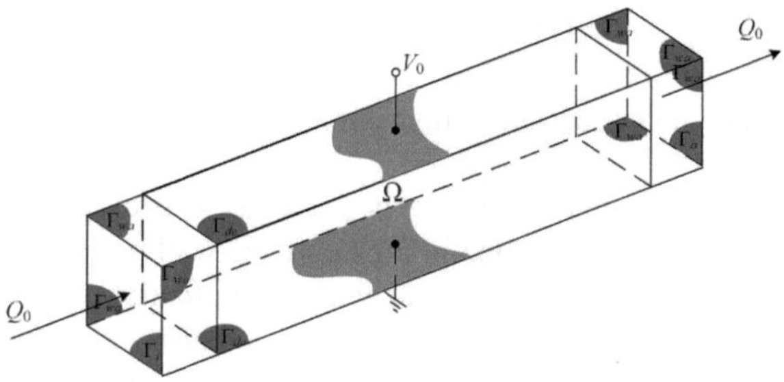

Fig. 6.20 Sketch of the computational domain Ω for electroosmotic micromixer, where Γ_i is the inlet, Γ_o is the outlet, Γ_w composed of Γ_{wa} and Γ_{de} is the of microchannel wall, Q_0 is the fluid flux imposed on the inlet of the microchannel, V_0 is the electrical potential imposed on the anode and the cathode is connected to ground. The electrode patterns are determined on Γ_{de}

For computational domain of the electroosmotic microflow, the inlet and outlet boundaries is used to truncate Ω, which is a part of a system or a network. For electroosmotics, slip velocity can exist at the sides of the inlet, because of the presence of the tangential component of the electric field intensity. And this can cause the unsuitability of the usually used parabolic velocity distribution with no-slip velocity at the sides of the inlet. Comparatively, the defective boundary condition has non-essentiality to enforce the velocity distribution on the corresponding boundary [42, 43]. Therefore, the defective boundary condition is imposed on the inlet of the electroosmotic microflow with specified flux

$$\int_{\Gamma_i} -\mathbf{u} \cdot \mathbf{n} \, \mathrm{d}\Gamma = Q_0, \text{ on } \Gamma_i \tag{6.18}$$

where $Q_0 = U \int_{\Gamma_i} 1 \, \mathrm{d}s$ is the know flux at the inlet Γ_i; U is the average velocity. At the outlet, the open boundary condition can be imposed as

$$\left[-\eta \left(\nabla \mathbf{u} + \nabla \mathbf{u}^{\mathrm{T}}\right) + p\mathbf{I}\right] \cdot \mathbf{n} = \mathbf{0}, \text{ on } \Gamma_o \tag{6.19}$$

For electroosmotic micromixers, the two factors that influence the mixing performance of a micromixer are diffusion and chaotic advection. The mixing of two species diluted in electroosmotic microflows can be described by the convection-diffusion equation

$$\mathbf{u} \cdot \nabla c - \nabla \cdot (D \nabla c) = 0, \text{ in } \Omega \tag{6.20}$$

where c is the concentration; D is the diffusion constant. The imposed boundary conditions for the convection-diffusion equation are the known concentration distribution at the inlet

$$c = c_i(\mathbf{x}), \text{ on } \Gamma_i \tag{6.21}$$

and the diffusion insulation at the wall and outlet

$$-(D\nabla c)\cdot\mathbf{n}=0,\ \text{on}\ \Gamma_w\cup\Gamma_o \tag{6.22}$$

where c_i is the known concentration distribution at the inlet of the electroosmotic micromixer.

For electroosmotics, the distribution of the external electrical potential is described by the Laplace equation. As sketched in Fig. 6.20, the wall boundary Γ_w is split into two parts, i.e. Γ_{wa} and Γ_{de}. For the external electrical potential, Γ_{wa} is electric insulation boundary with $-(\sigma\nabla V)\cdot\mathbf{n}=0$; and Γ_{de}, the design domain for the electrodes, is the union of electric insulation and electric potential boundary parts. To distinguish those two types of boundary parts, the variable, nominated physical density, is utilized, and it is valued in [0, 1] with value 0 and 1 respectively representing electric potential and electric insulation boundary types. Then the boundary condition on Γ_{de} can be expressed to be the interpolation of electric insulation and electric potential

$$-(\sigma\nabla V)\cdot\mathbf{n}=\alpha\left(V-V_0\right),\ \text{on}\ \Gamma_{de} \tag{6.23}$$

where σ is electric conductivity; V_0 is the specified electric potential on the electrodes; α is the penalization expressed to be [3]

$$\alpha=\alpha_{max}\frac{q\left(1-\gamma_{fp}\right)}{q+\gamma_{fp}} \tag{6.24}$$

with γ_{fp}, α_{max} and q respectively representing the physical density variable, the penalization parameter and the parameter used to tune the convexity of the penalization. The value of α_{max} should be chosen to be large enough to ensure the domination of the term $(V-V_0)$ in Eq. 6.23, when the physical density is valued 0. Meanwhile, Eq. 6.23 degenerates into the electric insulation boundary condition, when the physical density is valued 1. Based on numerical tests, α_{max} and q are chosen to be 1×10^5 and 1×10^{-3}, respectively. In sum, the external electrical potential can be described to be

$$\begin{aligned}&-\nabla\cdot(\sigma\nabla V)=0,\ \text{in}\ \Omega\\&-(\sigma\nabla V)\cdot\mathbf{n}=0,\ \text{on}\ \Gamma_i\cup\Gamma_o\cup\Gamma_{wa}\\&-(\sigma\nabla V)\cdot\mathbf{n}=\alpha\left(V-V_0\right),\ \text{on}\ \Gamma_{de}\end{aligned} \tag{6.25}$$

where the electric insulation boundary condition is imposed on the inlet and outlet of the electroosmotic microflows.

On the physical density variable, it is derived from the design variable defined on Γ_{de}, using the procedure introduced in [44]: the design variable is filtered using the Helmholtz filter to ensure the minimum scale of the implicitly expressed structural pattern

$$\begin{aligned}&-r^2\nabla_s\cdot\nabla_s\gamma_f+\gamma_f=\gamma,\ \text{in}\ \Gamma_{de}\\&-r^2\nabla_s\gamma_f\cdot\mathbf{n}_s=0,\ \text{on}\ \partial\Gamma_{de}\end{aligned} \tag{6.26}$$

where γ is the design variable, γ_f is the filtered design variable, ∇_s is the gradient operator defined for the local coordinate system on Γ_{de}, r is the filter radius, and $\mathbf{n}_s$ is the outward unitary normal on $\partial \Gamma_{de}$; then the filtered design variable is projected using the threshold method to remove the intermediary values between 0 and 1 and derive the physical density variable

$$\gamma_{fp} = \frac{\tanh(\beta\xi) + \tanh\left(\beta\left(\gamma_f - \xi\right)\right)}{\tanh(\beta\xi) + \tanh(\beta(1-\xi))} \tag{6.27}$$

where β and ξ are projection parameters. For the value choice of the projection parameters, one can refer to [45].

Based on the above description, the desired performance of the electroosmotic micromixers can be derived by inversely designing the electrode patterns, where the 0–1 distribution of the physical density representing the electrode patterns can be determined on Γ_{de} by solving a constructed variational problem.

When an electroosmotic micromixer is used to mix two fluids with different solutes, the desired effect is the mixing with average concentration at the outlet of the micromixer. The mixing performance of the electroosmotic micromixer can be evaluated by the normalized least square variance between the obtained concentration and the anticipated concentration at the outlet, where the normalizing factor is the least square difference between the concentration distribution at the inlet and anticipated concentration at the outlet [1, 46, 47]. Then, the design target can be to find the reasonable electrode patterns corresponding to the smallest value of the mixing measurement, where the convection-diffusion equation is used to describe the electroosmotic micromixing. Therefore, the variational problem is constructed as

$$\begin{aligned}
&\text{Find } \gamma \in [0, 1] \text{ to minimize } \Psi = \frac{1}{\Psi_0} \int_{\Gamma_o} (c - \bar{c})^2 \, \mathrm{d}s \\
&\text{Subject to } \left\{ \begin{array}{l}
\left\{ \begin{array}{l}
-\nabla \cdot (\sigma \nabla V) = 0, \text{ in } \Omega \\
-(\sigma \nabla V) \cdot \mathbf{n} = 0, \text{ on } \Gamma_i \cup \Gamma_o \cup \Gamma_{wa} \\
-(\sigma \nabla V) \cdot \mathbf{n} = \alpha (V - V_0), \text{ on } \Gamma_{de}
\end{array} \right. \\
\left\{ \begin{array}{l}
\rho \mathbf{u} \cdot \nabla \mathbf{u} + \nabla \cdot \left[-\eta \left(\nabla \mathbf{u} + \nabla \mathbf{u}^{\mathrm{T}}\right) + p\mathbf{I}\right] = \mathbf{0}, \text{ in } \Omega \\
-\nabla \cdot \mathbf{u} = 0, \text{ in } \Omega \\
\int_{\Gamma_i} -\mathbf{u} \cdot \mathbf{n} \, \mathrm{d}\Gamma = Q_0, \text{ on } \Gamma_i \\
\left[-\eta \left(\nabla \mathbf{u} + \nabla \mathbf{u}^{\mathrm{T}}\right) + p\mathbf{I}\right] \mathbf{n} = \mathbf{0}, \text{ on } \Gamma_o \\
\mathbf{u} = -\mu_{eo} \left[\nabla V - (\mathbf{n} \cdot \nabla V) \mathbf{n}\right], \text{ on } \Gamma_w
\end{array} \right. \\
\left\{ \begin{array}{l}
\mathbf{u} \cdot \nabla c - \nabla \cdot (D \nabla c) = 0, \text{ in } \Omega \\
c = c_i(\mathbf{x}), \text{ on } \Gamma_i \\
-(D \nabla c) \cdot \mathbf{n} = 0, \text{ on } \Gamma_w \cup \Gamma_o
\end{array} \right. \\
\left\{ \begin{array}{l}
-r^2 \nabla_s \cdot \nabla_s \gamma_f + \gamma_f = \gamma, \text{ in } \Gamma_{de} \\
-r^2 \nabla_s \gamma_f \cdot \mathbf{n}_s = 0, \text{ on } \partial \Gamma_{de}
\end{array} \right.
\end{array} \right.
\end{aligned} \tag{6.28}$$

where $\Psi_0 = \int_{\Gamma_i} (c_i - \bar{c})^2 \, \mathrm{d}s$ is the least square difference between the concentration distribution c_i at the inlet and anticipated concentration $\bar{c}$ at the outlet.

6.2.2 Analyzing and Solving

The variational problem for determination of the electrode patterns of electroosmotic micromixers is solved using the iterative approach, where the design variable is evolved based on the gradient information of the variational problem in Eq. 6.28. The gradient information of the variational problem can be derived from the adjoint analysis of the variational problem.

To solve the partial differential equations (PDEs) in the constraints of Eq. 6.28, the linear element-based finite element method is utilized for the weak forms of these PDEs [8]

$$\text{find } V \in \mathscr{H}(\Omega) \text{ satisfying } \int_\Omega \sigma \nabla V \cdot \nabla \hat{V} \, \mathrm{d}v + \int_{\Gamma_{de}} \alpha V \hat{V} \, \mathrm{d}s = 0, \; \forall \hat{V} \in \mathscr{H}(\Omega) \tag{6.29}$$

$$\begin{aligned}
&\text{find } \mathbf{u} \in (\mathscr{H}(\Omega))^3, \; p \in \mathscr{L}^2(\Omega), \; \boldsymbol{\lambda}_f \in \left(\mathscr{H}^{\frac{1}{2}}(\Gamma_w)\right)^3 \text{ and } \lambda_Q \in \mathbb{R} \text{ satisfying} \\
&\int_\Omega \rho (\mathbf{u} \cdot \nabla \mathbf{u}) \cdot \hat{\mathbf{u}} + \left[\eta \left(\nabla \mathbf{u} + \nabla \mathbf{u}^{\mathrm{T}}\right) - p\mathbf{I}\right] : \nabla \hat{\mathbf{u}} - \hat{p} \nabla \cdot \mathbf{u} \, \mathrm{d}v \\
&- \sum_{i=1}^{N_e} \int_{\Omega_i} \tau_{gls} \nabla p \cdot \nabla \hat{p} \, \mathrm{d}v + \lambda_Q \int_{\Gamma_i} \hat{\mathbf{u}} \cdot \mathbf{n} \, \mathrm{d}s + \hat{\lambda}_Q \left(\int_{\Gamma_i} \mathbf{u} \cdot \mathbf{n} \, \mathrm{d}s - Q_0\right) \\
&+ \int_{\Gamma_w} \{\mathbf{u} + \mu_{eo} [\nabla V - (\mathbf{n} \cdot \nabla V) \mathbf{n}]\} \cdot \hat{\boldsymbol{\lambda}}_f + \left(\boldsymbol{\lambda}_f - p\mathbf{n}\right) \cdot \hat{\mathbf{u}} \, \mathrm{d}s = 0, \\
&\forall \hat{\mathbf{u}} \in (\mathscr{H}(\Omega))^3, \; \forall \hat{p} \in \mathscr{L}^2(\Omega), \; \forall \hat{\boldsymbol{\lambda}}_f \in \left(\mathscr{H}^{-\frac{1}{2}}(\Gamma_w)\right)^3 \text{ and } \forall \hat{\lambda}_Q \in \mathbb{R}
\end{aligned} \tag{6.30}$$

$$\begin{aligned}
&\text{find } c \in \mathscr{H}(\Omega) \text{ and } \lambda_c \in \mathscr{H}^{\frac{1}{2}}(\Gamma_i) \text{ satisfying} \\
&\int_\Omega \mathbf{u} \cdot \nabla c \hat{c} + D \nabla c \cdot \nabla \hat{c} \, \mathrm{d}v + \sum_{i=1}^{N_e} \int_{\Omega_i} \tau_{supg} \left(\mathbf{u} \cdot \nabla c - D \nabla^2 c\right) \left(\mathbf{u} \cdot \nabla \hat{c}\right) \mathrm{d}v \\
&+ \int_{\Gamma_i} (c - c_i) \hat{\lambda}_c + \lambda_c \hat{c} \, \mathrm{d}s = 0, \; \forall \hat{c} \in \mathscr{H}(\Omega) \text{ and } \forall \hat{\lambda}_c \in \mathscr{H}^{-\frac{1}{2}}(\Gamma_i)
\end{aligned} \tag{6.31}$$

$$\text{find } \gamma_f \in \mathscr{H}(\Gamma_{de}) \text{ satisfying } \int_{\Gamma_d} r^2 \nabla_s \gamma_f \cdot \nabla_s \hat{\gamma}_f + \gamma_f \hat{\gamma}_f \, \mathrm{d}s = 0, \; \forall \hat{\gamma}_f \in \mathscr{H}(\Gamma_{de}) \tag{6.32}$$

where $\boldsymbol{\lambda}_f$ and λ_Q are the Lagrangian multipliers used to impose the velocity boundary condition and defective boundary condition respectively on Γ_w and Γ_i, for the Navier-Stokes equations; λ_c is the Lagrangian multiplier used to impose the known concentration boundary condition on Γ_i for the convection-diffusion equation; $\hat{c}$, $\hat{\mathbf{u}}$, $\hat{p}$, $\hat{V}$, $\hat{\gamma}_f$,

$\hat{\boldsymbol{\lambda}}_f$, $\hat{\lambda}_Q$ and $\hat{\lambda}_c$ are the test functions of the corresponding state variables, respectively; $\mathscr{H}(\Omega)$ and $\mathscr{L}^2(\Omega)$ are the first-order Hilbert space and the second-order Lebesgue integrable functional space defined on Ω, respectively; $\mathscr{H}^{-\frac{1}{2}}(\Gamma_w)$ is the dual space of the trace space $\mathscr{H}^{\frac{1}{2}}(\Gamma_w)$; $\mathscr{H}(\Gamma_{de})$ is the first-order Hilbert space defined on Γ_{de}; $\mathbb{R}$ is the real number field the Navier-Stokes equations and convection-diffusion equation are stabilized using the generalized least squares (GLS) and the streamline upwind Petrov-Galerkin (SUPG) technologies, respectively [8]; $\tau_{gls} = h_i^2/(12\eta)$ and $\tau_{supg} = \left[4/\left(h_i^2 D\right) + 2|\mathbf{u}|/h_i\right]^{-1}$ are the stabilization parameters chosen as in [8], with h_i representing the size of the i-th element Ω_i; and N_e is the number of elements used to discretize Ω.

Based on the adjoint method for partial differential equation constrained optimization problem [6], the variational problem in Eq. 6.28 can be analyzed for deriving the adjoint derivative as the gradient information to evolve the design variable; and the weak forms of the adjoint equations of the convection-diffusion equation, the Navier-Stokes equations, the electrical potential equation and the Helmholtz filter are derived to be

$$
\begin{aligned}
&\text{find } c_a \in \mathscr{H}(\Omega) \text{ and } \lambda_{ca} \in \mathscr{H}^{-\frac{1}{2}}(\Gamma_i) \text{ satisfying} \\
&\int_\Omega \mathbf{u}\cdot\nabla\hat{c}_a c_a + D\nabla c_a\cdot\nabla\hat{c}_a \,\mathrm{d}v + \sum_{i=1}^{N_e}\int_{\Omega_i}\tau_{supg}\left(\mathbf{u}\cdot\nabla c_a\right)\left(\mathbf{u}\cdot\nabla\hat{c}_a - D\nabla^2\hat{c}_a\right)\mathrm{d}v \\
&+ \int_{\Gamma_i}\lambda_{ca}\hat{c}_a + c_a\hat{\lambda}_{ca}\,\mathrm{d}s + \frac{1}{\Psi_0}\int_{\Gamma_o} 2\left(c-\bar{c}\right)\hat{c}_a\,\mathrm{d}s = 0,\ \forall\hat{c}_a \in \mathscr{H}(\Omega) \\
&\text{and } \forall\hat{\lambda}_{ca} \in \mathscr{H}^{\frac{1}{2}}(\Gamma_i)
\end{aligned}
\tag{6.33}
$$

$$
\begin{aligned}
&\text{find } \mathbf{u}_a \in \left(\mathscr{H}(\Omega)\right)^3,\ p_a \in \mathscr{L}^2(\Omega),\ \boldsymbol{\lambda}_{fa} \in \left(\mathscr{H}^{-\frac{1}{2}}(\Gamma_w)\right)^3 \text{ and } \lambda_{Qa} \in \mathbb{R} \text{ satisfying} \\
&\int_\Omega \rho\left[\left(\hat{\mathbf{u}}_a\cdot\nabla\mathbf{u}\right)\cdot\mathbf{u}_a + \left(\mathbf{u}\cdot\nabla\hat{\mathbf{u}}_a\right)\cdot\mathbf{u}_a\right] \\
&+ \nabla\mathbf{u}_a : \left[\eta\left(\nabla\hat{\mathbf{u}}_a + \nabla\hat{\mathbf{u}}_a^{\mathrm{T}}\right) - \hat{p}_a\mathbf{I}\right] - p_a\nabla\cdot\hat{\mathbf{u}}_a + \hat{\mathbf{u}}_a\cdot\nabla c c_a\,\mathrm{d}v \\
&+ \sum_{i=1}^{N_e}\int_{\Omega_i} -\tau_{gls}\nabla p_a\cdot\nabla\hat{p}_a + \left(\frac{\partial\tau_{supg}}{\partial\mathbf{u}}\cdot\hat{\mathbf{u}}_a\right)\left(\mathbf{u}\cdot\nabla c_a\right)\left(\mathbf{u}\cdot\nabla c - D\nabla^2 c\right) \\
&+ \tau_{supg}\left(\hat{\mathbf{u}}_a\cdot\nabla c_a\right)\left(\mathbf{u}\cdot\nabla c - D\nabla^2 c\right) + \tau_{supg}\left(\mathbf{u}\cdot\nabla c_a\right)\left(\hat{\mathbf{u}}_a\cdot\nabla c\right)\mathrm{d}v \\
&+ \int_{\Gamma_i}\hat{\lambda}_{Qa}\mathbf{u}_a\cdot\mathbf{n} + \lambda_{Qa}\hat{\mathbf{u}}_a\cdot\mathbf{n}\,\mathrm{d}s + \int_{\Gamma_w}\boldsymbol{\lambda}_{fa}\cdot\hat{\mathbf{u}}_a + \left(\hat{\boldsymbol{\lambda}}_{fa} - \hat{p}_a\mathbf{n}\right)\cdot\mathbf{u}_a\,\mathrm{d}s = 0, \\
&\forall\hat{\mathbf{u}}_a \in \left(\mathscr{H}(\Omega)\right)^3,\ \forall\hat{p}_a \in \mathscr{L}^2(\Omega),\ \forall\hat{\boldsymbol{\lambda}}_{fa} \in \left(\mathscr{H}^{\frac{1}{2}}(\Gamma_w)\right)^3 \text{ and } \hat{\lambda}_{Qa} \in \mathbb{R}
\end{aligned}
\tag{6.34}
$$

$$\text{find } V_a \in \mathscr{H}(\Omega) \text{ satisfying} \int_{\Omega} \sigma \nabla V_a \cdot \nabla \hat{V}_a \,\mathrm{d}v + \int_{\Gamma_{de}} \alpha V_a \hat{V}_a \,\mathrm{d}s \\ + \int_{\Gamma_w} \mu_{eo} \left[\nabla \hat{V}_a - \left(\mathbf{n} \cdot \nabla \hat{V}_a \right) \mathbf{n} \right] \cdot \boldsymbol{\lambda}_{fa} \,\mathrm{d}s = 0, \ \forall \hat{V}_a \in \mathscr{H}(\Omega) \tag{6.35}$$

find $\gamma_{fa} \in \mathscr{H}(\Gamma_{de})$ satisfying

$$\int_{\Gamma_{de}} r^2 \nabla_s \gamma_{fa} \cdot \nabla_s \hat{\gamma}_{fa} + \gamma_{fa} \hat{\gamma}_{fa} + (V - V_0) V_a \frac{\partial \alpha}{\partial \gamma_{fp}} \frac{\partial \gamma_{fp}}{\partial \gamma_f} \hat{\gamma}_{fa} \,\mathrm{d}s = 0, \ \forall \hat{\gamma}_{fa} \in \mathscr{H}(\Gamma_{de}) \tag{6.36}$$

where c_a, $\mathbf{u}_a$, p_a, V_a and γ_{fa} are the adjoint variables of the corresponding state variables, respectively; $\boldsymbol{\lambda}_{fa}$ and λ_{Qa} are the adjoint of the Lagrangian multiplier $\boldsymbol{\lambda}_f$ and λ_Q, respectively; λ_{ca} is the adjoint of the Lagrangian multiplier λ_c; $\hat{c}_a$, $\hat{\mathbf{u}}_a$, $\hat{p}_a$, $\hat{V}_a$, $\hat{\gamma}_{fa}$, $\hat{\boldsymbol{\lambda}}_{fa}$ and $\hat{\lambda}_{ca}$ are the test functions of the corresponding adjoint variables, respectively. The first-order adjoint derivative of the variational problem in Eq. 6.28 is

$$\delta \Psi = \int_{\Gamma_d} -\gamma_{fa} \delta \gamma \,\mathrm{d}s \tag{6.37}$$

where $\delta \Psi$ and $\delta \gamma$ are respectively the first-order variational of the design objective and design variable. In Eq. 6.37, γ_{fa} is derived by sequentially solving the Eqs. 6.29–6.36. Then, the adjoint derivative can be used to determine the evolution of the design variable.

After the adjoint analysis, the iterative procedure includes the following steps can be used to solve the variational problem: (a) the PDE constraints are solved with the current design variable; (b) the adjoint equations are solved based on the solution of the PDE constraints; (c) the adjoint derivative of the design objective is computed; (d) the design variable is updated using the method of moving asymptotes (MMA) [48]; (e) the convergent criterion are checked to stop the iterative loop, and the procedure will return to (a), if the convergent criterion are not satisfied. The convergent criterion is specified to be that the change of the objective values in five consecutive iterations satisfies $\frac{1}{5} \sum_{i=0}^{4} |J_{k-i} - J_{k-i-1}| / |J_k| \leq 1 \times 10^{-3}$ in the k-th iteration or the maximal iteration number 160 is reached, where J_k is the objective value in the k-th iteration.

During the iterative procedure, the threshold parameter ξ in Eq. 6.27 is set to be 0.5; the initial value of the projection parameter β is set to be 1 and it is doubled every 40 iterations until the preset maximal value 16 is reached; the finite element solving of the partial differential equations and corresponding adjoint equations are implemented in the finite element software COMSOL Multiphysics (http://www.comsol.com), where linear cubic elements are used for all the variables. On the relevant programming setting, one can refer to [5, 49] for the more details.

6.2.3 Results and Discussion

In this section, the inverse design of the electrode patterns is investigated for electroosmotic micromixers. In the following, the liquid of the microflow is the electrolyte with density $\rho = 1 \times 10^3\,\mathrm{kg/m^3}$, dynamic viscosity $\eta = 1 \times 10^{-3}\,\mathrm{Pa\,s}$, dielectric constant $\varepsilon_r = 80.2$, conductivity $\sigma = 0.12\,(\Omega\,\mathrm{m})^{-1}$, and Zeta potential $\zeta_0 = -0.1\,\mathrm{V}$. Considering the fabrication, different materials may be used for the top/bottom wall and lateral walls, then one only needs to change the value of the Zeta potential at the corresponding wall. The characteristic size of the cross section of the computational domain sketched in Fig. 6.20 is set to be $l = 400\,\mu\mathrm{m}$, the length of the microchannel covered by the electrodes is $n_l \cdot l$ with n_l representing the fold number, and the length of the inlet and outlet parts of the design domain is equal to l. The Reynolds number and Péclet number of the microflow are respectively calculated to be $Re = \rho U l/\eta$ and $Pe = Ul/D$, where U is the average velocity at the inlet. The computational domain is discretized by $20 \times 20 \times 20$ brick elements per cubic space with edge size equals to the characteristic size. In the computational domain, the design domain is set to be the top and bottom surfaces, where anode with specified potential is localized on the top surface and the cathode is localized on the bottom surface.

6.2.3.1 Effect of Electric Potential

The electrode patterns are inversely determined in the straight microchannel sketched in Fig. 6.20 to demonstrate the corresponding mixing mechanism and performance of the corresponding electroosmotic micromixers. By solving the variational problem in Eq. 6.28, the electrode patterns are derived as shown in Fig. 6.21a–e with different potentials specified on the anode, where the microchannel covered with electrodes is set to be the length 2.4 mm with $n_l = 6$, the Reynolds number and Péclet number of the microflow are respectively $Re = 10$ and $Pe = 1 \times 10^3$. The mixing measurement of the micromixers with derived electrode patterns is plotted in Fig. 6.21f, from which one can conclude that higher electrical potential corresponds to better mixing performance. This is because that higher electrical potential generates stronger electric field, and stronger electrical field induces more efficient disturbance and chaotic convection in the microflow. When the electrical potential is 4 and 5 V (Fig. 6.21d, e), the achieved mixing measurement is 0.0280 and 0.0072, which are less than the threshold level of complete mixing defined as 0.050 in [1]. Therefore, the derived electrode patterns for the electrical potentials 4 and 5 V achieve the complete mixing performance.

The distribution of the streamlines, projected velocity vectors and concentration has been plotted in Fig. 6.22 for the electroosmotic micromixers with derived electrode patterns imposed with 5 V electrical potential. From Fig. 6.22, it is concluded that the inversely designed electrode patterns for the electroosmotic micromixer generates the tangential velocity at the wall of the micromixer, this tangential velocity can effectively enhance the chaotic convectionof the microflow by inducing the sec-

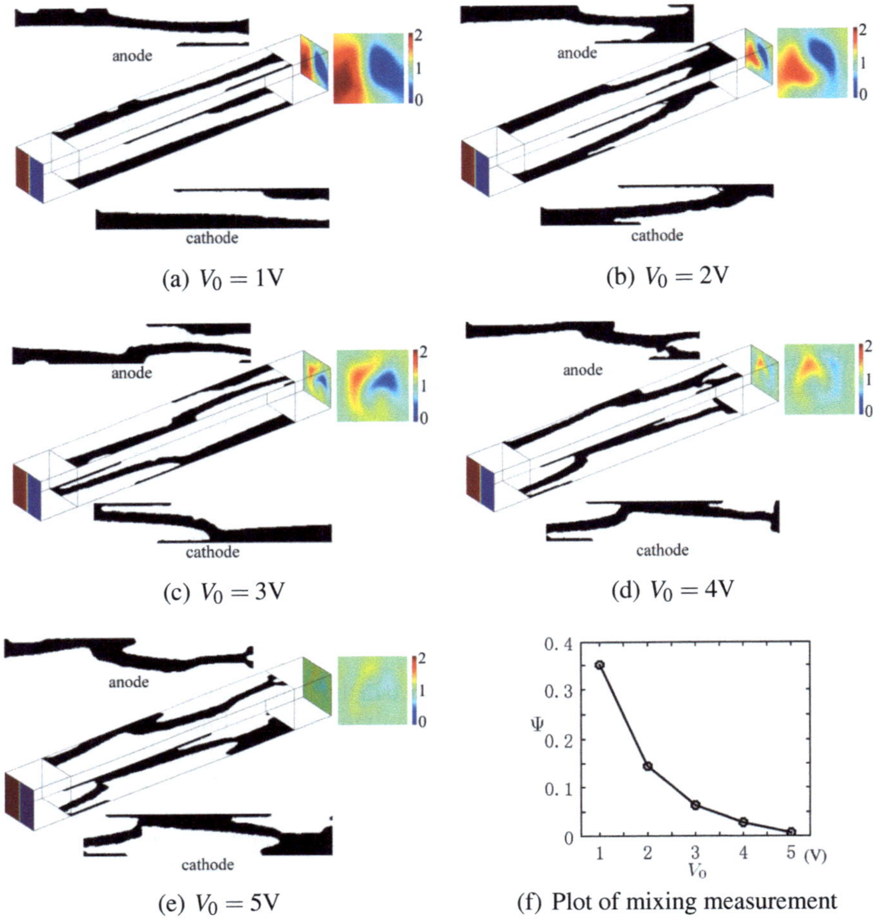

Fig. 6.21 **a–e** Inversely designed electrode patterns of electroosmotic micromixers with different electrical potentials; **f** plot of the mixing measurement corresponding to the electroosmotic micromixers with the derived electrode patterns shown in (**a**)–(**e**)

ondary flow in the cross section, and the enhanced convection further achieves the complete mixing performance in the mixing distance of 2.4 mm.

6.2.3.2 Effect of Mixing Distance

For different mixing distances, the inverse design of the electrode patterns is implemented for the electroosmotic micromixers. By setting the fold number of the microchannel length to be different values, the electrode patterns are derived as shown in Fig. 6.23a–f, where the Reynolds number, Péclet number and electrical potential are respectively $Re = 10$, $Pe = 1 \times 10^3$ and $V_0 = 5$ V. The mixing mea-

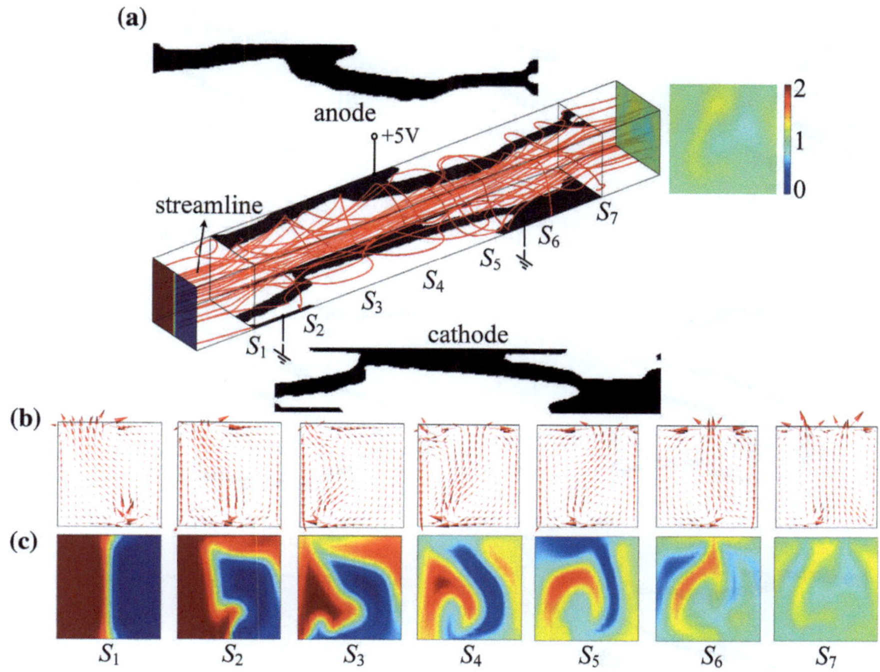

Fig. 6.22 **a** Streamlines in the electroosmotic micromixer with the inversely designed electrode patterns and 5 V electrical potential; **b** distribution of the projected velocity vectors in the cross sections of the electroosmotic micromixer with inversely designed electrode pattern; **c** concentration distribution in the cross sections of the electroosmotic micromixer with inversely designed electrode pattern

surement values corresponding to the length of the electroosmotic micromixers with the derived electrode patterns are plotted in Fig. 6.23g. In Fig. 6.23g, the mixing measurement of the electroosmotic micromixers with mixing length 2 mm ($n_l = 5$) and 2.4 mm ($n_l = 6$) is less than 0.050, which means the complete mixing performance.

6.2.3.3 Effect of Péclet Number and Reynolds Number

To demonstrate the effect of the Péclet number and Reynolds number, the inverse design of electrode patterns are implemented for the electroosmotic micromixers with different values of Péclet number and Reynolds number. The Péclet number is used to measure the relative importance of the convection of the flow and diffusion for the mixing progress. The results shown in Fig. 6.24 are the derived electrode patterns for different Péclet number and the plot of the corresponding mixing measurement, where the Reynolds number is kept to be $Re = 10$. Figure 6.24 shows that lower Péclet numbers result in better mixing performance. This is because lower Pélet numbers correspond to higher values of the diffusive constant under the precondition of consistent Reynolds number. Therefore, the value of the mixing

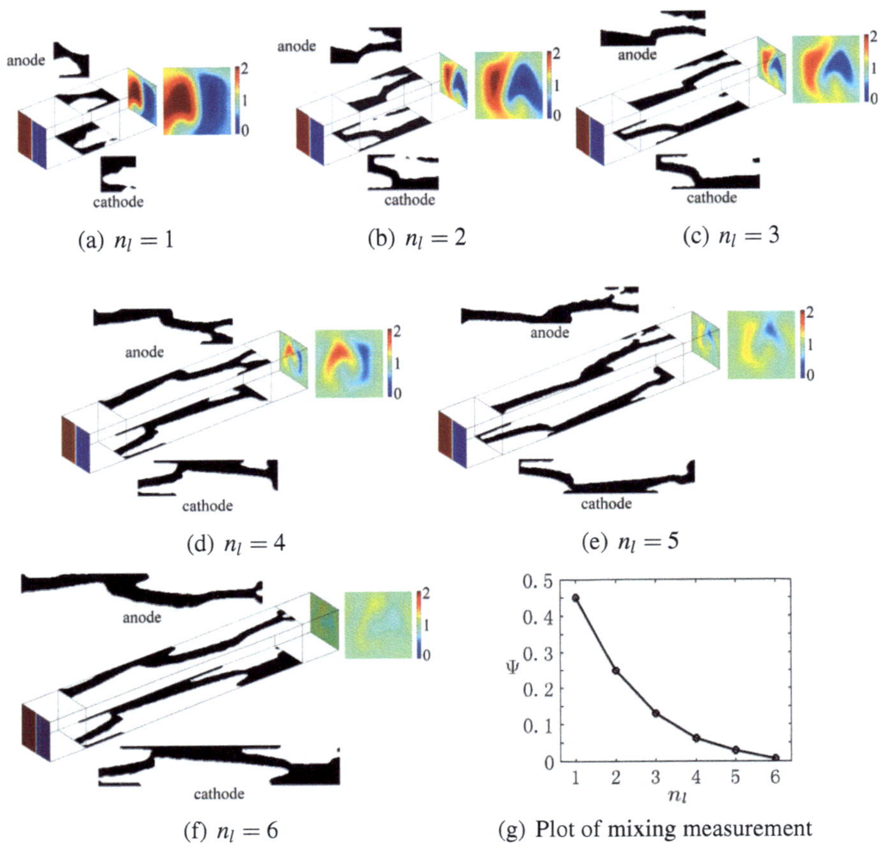

Fig. 6.23 **a–f** Inversely designed electrode patterns of electroosmotic micromixers with different mixing distances; **g** plot of the mixing measurement corresponding to the electroosmotic micromixers with the derived electrode patterns shown in (**a**)–(**f**)

measurement increases, i.e. the mixing performance becomes better as the Pélet number decreases. The Reynolds number is used to present the relative importance of the convection and viscosity of the fluidic flow. Because convection is an important aspect for the micromixing, the Reynolds number influences the electrode patterns of the electroosmotic micromixer. As shown in Fig. 6.25, the electrode patterns of the micromixers with different Reynolds numbers are inversely designed with plotting the corresponding mixing measurement values, where the Péclet number is kept to be $Pe = 1 \times 10^3$. Figure 6.25 shows that different Reynolds numbers correspond to different electrode patterns; in the low Reynolds number flows, the mixing progress is influenced dominantly by the diffusion, and the convection of the flow can be strengthened by the derived electrodes patterns; the derived patterns corresponding to low Reynolds number have relatively better performance than those corresponding to relatively high Reynolds number, because higher Reynolds number represents higher fluidic velocity meaning short mixing time in electroosmotic micromixer with

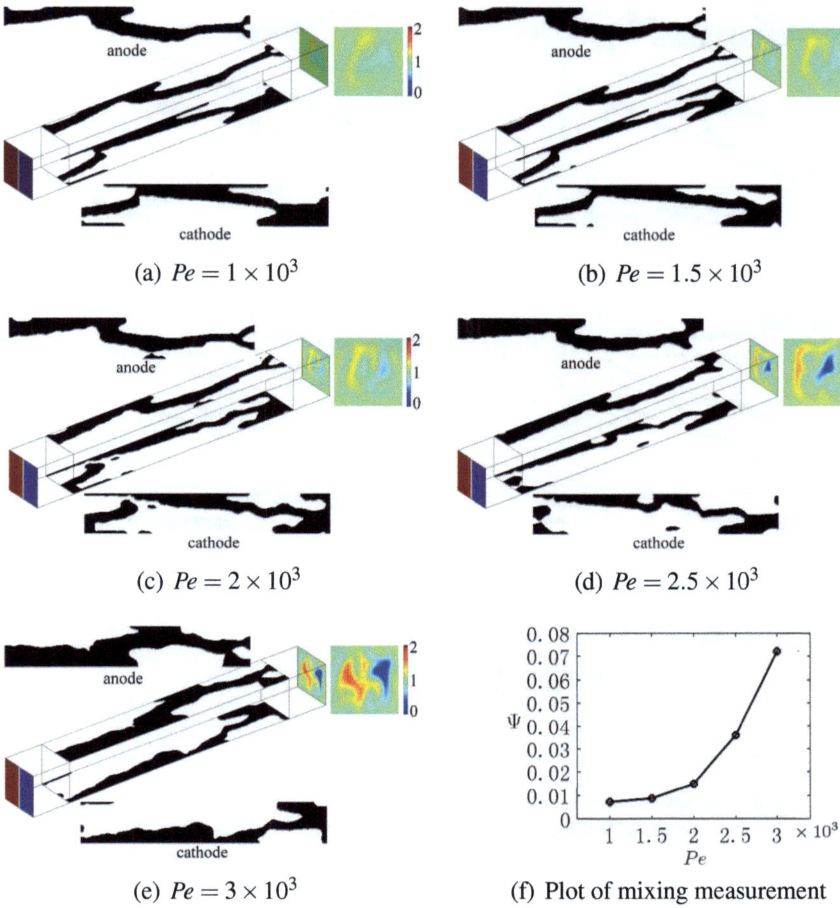

Fig. 6.24 **a–e** Inversely designed electrode patterns of electroosmotic micromixers with different Péclet number; **g** plot of the mixing measurement corresponding to the electroosmotic micromixers with the derived electrode patterns shown in (**a**)–(**e**)

fixed length, even when the electroosmotic micromixer has reasonable electrode patterns on the wall.

6.2.4 Summary

This section has presented an inverse design method for the electrode patterns of electroosmosis, which is a widely used actuation mechanism in microfluidics. The inverse design method is implemented based on the interpolation of electric insulation and electric potential on the specified walls of microchannels, where the

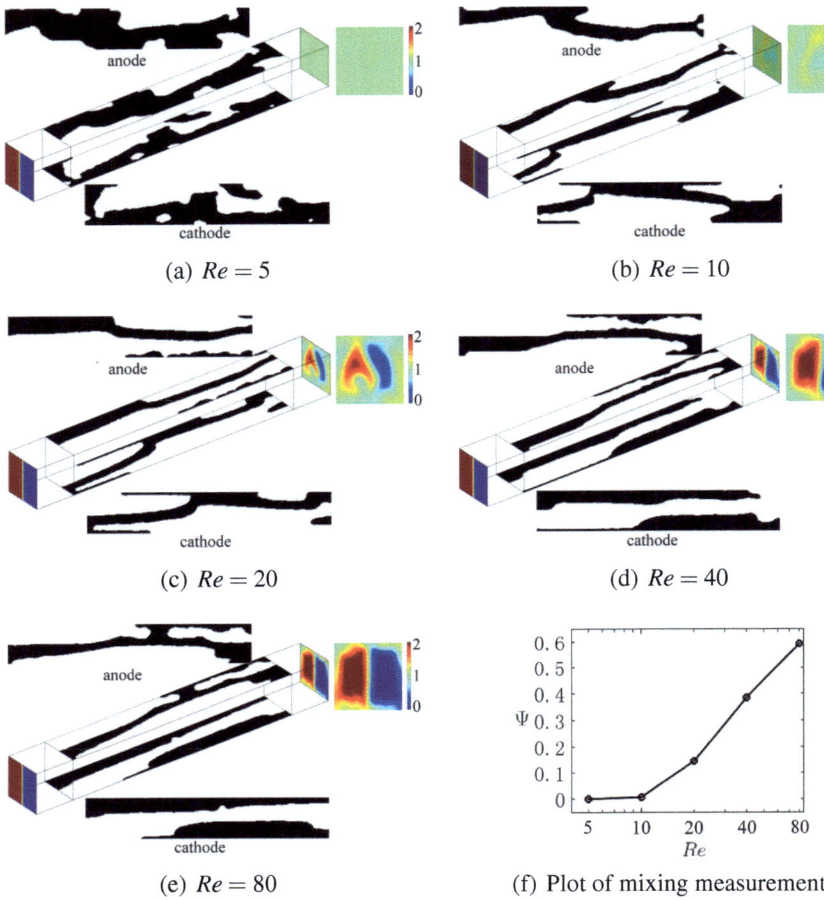

Fig. 6.25 **a–e** Inversely designed electrode patterns of electroosmotic micromixers with different Reynolds numbers; **g** plot of the mixing measurement corresponding to the electroosmotic micromixers with the derived electrode patterns shown in (**a**)–(**e**)

electrode patterns are determined to achieve the designer desired microfluidic performance. Electroosmotic micromixer has been demonstrated as the typical electroosmotic microdevice. By the proposed method, complete mixing has been achieved within the mixing distance 2.4 mm for the electroosmotic microflows with 5 V electrical potential, Reynolds number $Re \leq 10$ and Péclet number $Pe \leq 1.5 \times 10^3$. This inverse design method has the advantage on overcoming the limits of physical intuition-based design method. It can provide a symmetrical approach for the design of electroosmotic microfludics, with potential applications in chemicals and biochemistry due to the outstanding performance of the derived microdevices.

6.3 Inverse Design of Micro Venturi Diode

Micro Venturi diodes have been broadly used in microfluidics to rectify liquid flow. A key feature of these diodes is that they have no moving parts. This section treats the numerical optimization of such a micro Venturi diode for the case of steady Newtonian flow, with Reynolds numbers less than 100, which typically applies to microfluidic devices. Key design factors are discussed, such as the ratio of length to width of the design domain, the values of lower and upper bounds for the constraint of the volume of the diodes, and the Reynolds number, all of which influence the outcome of the diode effect. Curved micro Venturi diodes are found by the optimization. The method presented provides a systematic way to design Venturi diodes for steady Newtonian flows.

6.3.1 *Diodicity of Micro Venturi Diode*

A schematic drawing of a micro Venturi diode is shown in Fig. 6.26, where the pressure drop in the forward direction is lower than that in the reverse direction. The diffuser action is the fluid flow in the forward direction and the nozzle action in the reverse. The diodicity parameter measures the diode effect of a micro Venturi diode. A lower value of diodicity means a better diode effect. The diodicity of a micro Venturi diode can be described from the momentum or kinetic energy viewpoints [50].

From the momentum point of view, the diodicity is expressed as the ratio of the forward pressure drop to the reverse pressure drop, under the assumption that the flux in two opposite directions are equal

$$D_p = \Delta p_f / \Delta p_r \tag{6.38}$$

where Δp_f and Δp_r are the pressure drops of the forward and reverse flows, respectively, and the pressure drop is calculated using the averaged value of pressure on the inlet and outlet boundaries

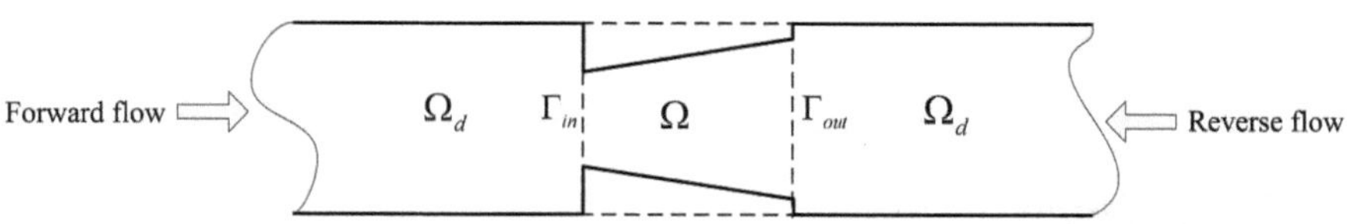

Fig. 6.26 Schematic of the micro Venturi diode. Γ_{in} and Γ_{out} are the inlet and outlet of the micro Venturi diode for the forward flow. Ω is the Venturi diode region, and Ω_d are ducts connected to the micro Venturi diode

$$\Delta p = \frac{\left|\int_{\Gamma_{in}} p\mathbf{n} \cdot d\mathbf{s}\right|}{\int_{\Gamma_{in}} 1\, ds} - \frac{\left|\int_{\Gamma_{out}} p\mathbf{n} \cdot d\mathbf{s}\right|}{\int_{\Gamma_{out}} 1\, ds} \tag{6.39}$$

where Γ_{in} and Γ_{out} are the inlet and outlet boundaries. From the kinetic energy point of view, the diodicity is expressed as the ratio of the consumption of the forward pressure work to the reverse pressure work, under the assumption that the flux in two opposite directions is equal. Hence, the diodicity can be alternatively defined as

$$D_w = \Delta W_{pf} / \Delta W_{pr} \tag{6.40}$$

where ΔW_{pf} and ΔW_{pr} are the consumption of the pressure work in a micro Venturi diode for the forward and reverse flows, respectively, and the consumption of the pressure work is expressed as

$$\Delta W_p = -\int_{\Gamma_{in}} p\mathbf{u} \cdot \mathbf{n} ds - \int_{\Gamma_{out}} p\mathbf{u} \cdot \mathbf{n} ds \tag{6.41}$$

where p is the pressure, $\mathbf{u}$ is the flow velocity, and $\mathbf{n}$ is the unit outward normal vector on the corresponding boundary. As a design may have non-paralleling inlet and outlet ports, it becomes somewhat complicated to utilize vector-based momentum to obtain a scalar measurement of valve diodicity. Whereas, the kinetic energy has no such difficulty [50]. At the same time, the description of diodicity has the following transformation

$$\Delta W_p = -\int_{\Gamma_{in}} p\mathbf{u} \cdot \mathbf{n} ds - \int_{\Gamma_{out}} p\mathbf{u} \cdot \mathbf{n} ds = -\int_{\Omega} \nabla \cdot (p\mathbf{u}) d\Omega \tag{6.42}$$

in the case where the boundary conditions are under no-slip conditions $\mathbf{u} = 0$ applied on all external walls of domain Ω, except for the inlet Γ_{in} and outlet Γ_{out} (Fig. 6.26). This transformation can improve the numerical accuracy of sensitivity in optimization procedures. Therefore, the description of diodicity from the kinetic energy point of view was adopted in this work.

6.3.2 Modeling

In this work, the gradient type optimization method based on the finite element solution of incompressible Navier-Stokes equation was used to improve the performance of micro Venturi diodes. The expression of diodicity in Eq. 6.40 could be used as the objective function directly. However, it is known that the numerically optimized results may manifest artificial numerical oscillation for ill-posed optimization models. One typical remedy is to add a regularization term into the optimization objective to recover the stability of an optimization model. Generally, the regularization term can be chosen as a positive definite function which is related to the design variables. In this work, the weighted consumption of the reverse pressure work was chosen as the regularization term, and thus the objective function is defined as

$$J(\mathbf{u}_f, \mathbf{u}_r, p_f, p_r; \gamma) = \Delta W_{pf} / \Delta W_{pr} + \delta \Delta W_{pr} / \Delta W_{pr0} \tag{6.43}$$

where $\mathbf{u}_f$, $\mathbf{u}_r$, p_f and p_r are the velocity and pressure of the forward and reverse flows, respectively, ΔW_{pf} and ΔW_{pr} are the consumption of the forward and reverse pressure work, ΔW_{pr0} is the initial consumption of reverse pressure work, and δ is a constant used to tuning the weight of $\Delta W_{pr} / \Delta W_{pr0}$. According to Eqs. 6.40–6.42, ΔW_{pf} and ΔW_{pr} are expressed as

$$\Delta W_{pf} = -\int_{\Omega} \nabla \cdot (p_f \mathbf{u}_f) d\Omega\ ,\ \ \Delta W_{pr} = -\int_{\Omega} \nabla \cdot (p_r \mathbf{u}_r) d\Omega \tag{6.44}$$

Therefore, the optimization model for designing a micro Venturi diode can be constructed as

$$\begin{aligned}
&\text{Find } \gamma \in [0, 1] \text{ to minimize } J(\mathbf{u}_f, \mathbf{u}_r, p_f, p_r; \gamma) = \Delta W_{pf} / \Delta W_{pr} + \delta \Delta W_{pr} / \Delta W_{pr0} \\
&\text{Subject to} \\
&\begin{cases}
\rho(\mathbf{u}_f \cdot \nabla)\mathbf{u}_f - \eta \Delta \mathbf{u}_f + \nabla p_f = -\alpha(\gamma)\mathbf{u}_f, \ \ -\nabla \cdot \mathbf{u}_f = 0, \ \forall \mathbf{x} \in \Omega \\
\rho(\mathbf{u}_r \cdot \nabla)\mathbf{u}_r - \eta \Delta \mathbf{u}_r + \nabla p_r = -\alpha(\gamma)\mathbf{u}_r, \ \ -\nabla \cdot \mathbf{u}_r = 0, \ \forall \mathbf{x} \in \Omega \\
\rho(\mathbf{u}_f \cdot \nabla)\mathbf{u}_f - \eta \Delta \mathbf{u}_f + \nabla p_f = \mathbf{0}, \ \ -\nabla \cdot \mathbf{u}_f = 0, \ \forall \mathbf{x} \in \Omega_d \\
\rho(\mathbf{u}_r \cdot \nabla)\mathbf{u}_r - \eta \Delta \mathbf{u}_r + \nabla p_r = \mathbf{0}, \ \ -\nabla \cdot \mathbf{u}_r = 0, \ \forall \mathbf{x} \in \Omega_d \\
\underline{V}_r \le \int_{\Omega} \gamma \, d\Omega \Big/ \int_{\Omega} 1 \, d\Omega \le \overline{V}_r
\end{cases}
\end{aligned} \tag{6.45}$$

where γ is the optimization design variable and $\alpha(\gamma)$ is the interpolation function of design variable to control the distribution of the fluidic velocity and pressure fields based on the modified layout of Venturi diode during an optimization procedure; $\underline{V}_r$ and $\overline{V}_r$ are the lower and upper bounds for the constraint of volume fraction; η and ρ are the dynamic viscosity and density of the liquid; Ω_d is the domain of the ducts connected with the micro Venturi diode. The optimization method used in this section is the same as the method proposed in [3–5]. The interpolation function $\alpha(\gamma)$ was chosen as

$$\alpha(\gamma) = \alpha_{\min} + (\alpha_{\max} - \alpha_{\min})\, q\, (1 - \gamma) / (q + \gamma) \tag{6.46}$$

where α_{min}, α_{max} and q are scalar parameters to tune the magnitude and convexity of the interpolation function $\alpha(\gamma)$, respectively. The initial design domain of the micro Venturi diode was chosen to be the domain enclosed by the dash lines shown in Fig. 6.26.

After the construction of the optimization model, the boundary conditions of the Navier-Stokes equations were chosen as follows. As mentioned in Sect. 6.3.1, the assumption of the equal flux on the forward and reverse directions was used to measure the diodicity of the micro Venturi diode. When designing a micro Venturi diode in micro channel with the initial design domain, as shown in Fig. 6.26, the parabolic velocity distribution with the same maximal value U_{max} was loaded on the inlets of the ducts for the two opposite flows. To avoid the entrance effect taken by the parabolic velocity boundary condition, the length of the ducts needs sufficiently long [51]. Based on numerical test, the optimized result will be uninfluenced by the entrance effect when the length of the duct is 3 to 12 times the width of duct when Reynolds number varies between 10 and 100. The boundary condition on the outlet was set to be

$$\left[-p\mathbf{I} + \eta\left(\nabla\mathbf{u} + \nabla\mathbf{u}^{\mathrm{T}}\right)\right]\mathbf{n} = \mathbf{0} \tag{6.47}$$

6.3.3 Analyzing

The analysis for optimization model in Eq. 6.45 can be calculated using the adjoint method [52–54]. For solving the sensitivity of an objective in a bounded domain Ω with Lipschitz continuous boundary $\partial\Omega$, the augmented functional $\hat{\Phi}$ with the finite element discretization form is defined as [53]

$$\hat{\Phi} = \Phi - \boldsymbol{\lambda}\cdot\mathbf{R}\big(\mathbf{U}(\gamma), \gamma\big) \tag{6.48}$$

where Φ is the discretized objective function, $\boldsymbol{\lambda}$ is the Lagrange multiplier, $\mathbf{U}$ is the combined vector of velocity $\mathbf{u}$ and pressure p, and $\mathbf{R}\big(\mathbf{U}(\gamma), \gamma\big) = \mathbf{0}$ is the discretized Navier-Stokes equations in residual form. Differentiation of the above functional with respect to the design variable γ yields

$$\frac{\partial\hat{\Phi}}{\partial\gamma} = \frac{\partial\Phi}{\partial\mathbf{U}}\frac{\partial\mathbf{U}}{\partial\gamma} + \frac{\partial\Phi}{\partial\gamma} - \boldsymbol{\lambda}\cdot\left[\frac{\partial\mathbf{R}(\mathbf{U},\gamma)}{\partial\mathbf{U}}\frac{\partial\mathbf{U}}{\partial\gamma} + \frac{\partial\mathbf{R}(\mathbf{U},\gamma)}{\partial\gamma}\right] \tag{6.49}$$

Equation 6.49 can be rearranged as

$$\frac{\partial\hat{\Phi}}{\partial\gamma} = \frac{\partial\Phi}{\partial\gamma} + \left[\frac{\partial\Phi}{\partial\mathbf{U}} - \boldsymbol{\lambda}\cdot\frac{\partial\mathbf{R}(\mathbf{U},\gamma)}{\partial\mathbf{U}}\right]\frac{\partial\mathbf{U}}{\partial\gamma} - \boldsymbol{\lambda}\cdot\frac{\partial\mathbf{R}(\mathbf{U},\gamma)}{\partial\gamma} \tag{6.50}$$

and terms related with $\partial\mathbf{U}/\partial\gamma$ can be eliminated by solving the adjoint equation

$$\frac{\partial\Phi}{\partial\mathbf{U}} - \boldsymbol{\lambda}\cdot\frac{\partial\mathbf{R}(\mathbf{U},\gamma)}{\partial\mathbf{U}} = \mathbf{0} \tag{6.51}$$

For detailed numerical implementation of the adjoint sensitivity analysis, one can refer to [5, 55]. Once the Lagrange multipliers are obtained, the sensitivity expression is reduced to

$$\frac{\partial \hat{\Phi}}{\partial \gamma} = \frac{\partial \Phi}{\partial \gamma} - \boldsymbol{\lambda} \cdot \frac{\partial \mathbf{R}(\mathbf{U}, \gamma)}{\partial \gamma} \tag{6.52}$$

For sensitivity analysis of the optimization model in Eq. 6.45, the adjoint Eq. 6.51 needs to be solved for the Lagrange multiplier $\boldsymbol{\lambda}_f$ and $\boldsymbol{\lambda}_r$, respectively. Therefore, the adjoint sensitivity of the consumption of the forward and reverse pressure work can be obtained as

$$\begin{aligned} \frac{\partial \Delta \widehat{W}_{pf}}{\partial \gamma} &= \frac{\partial \Delta W_{pf}}{\partial \gamma} - \boldsymbol{\lambda}_f \cdot \frac{\partial \mathbf{R}_{\Omega}(\mathbf{U}_f, \gamma)}{\partial \gamma} \\ \frac{\partial \Delta \widehat{W}_{pr}}{\partial \gamma} &= \frac{\partial \Delta W_{pr}}{\partial \gamma} - \boldsymbol{\lambda}_r \cdot \frac{\partial \mathbf{R}_{\Omega}(\mathbf{U}_r, \gamma)}{\partial \gamma} \end{aligned} \tag{6.53}$$

and the adjoint sensitivity of the optimization model in Eq. 6.45 with the finite element discretization form can be computed as

$$\frac{\partial \hat{J}}{\partial \gamma} = \frac{\partial \Delta \widehat{W}_{pf} / \partial \gamma}{\Delta W_{pr}} - \frac{\Delta W_{pf} \cdot \partial \Delta \widehat{W}_{pr} / \partial \gamma}{\Delta {W_{pr}}^2} + \delta \frac{\partial \Delta \widehat{W}_{pr} / \partial \gamma}{\Delta W_{pr0}} \tag{6.54}$$

6.3.4 Results and Discussion

Several micro Venturi diodes are designed based on different setups of design parameters and initial design domains. In the optimization procedure, the Navier-Stokes equations were solved to obtain the distribution of velocity and pressure fields for the forward and reverse flows; the adjoint analysis is done by solving two linear adjoint equations; the design variable is updated using MMA [48] based on sensitivity vector and design constraints; the stop criteria of the iterative optimization was specified by the maximum change of design variables between two consecutive iterations.

6.3.4.1 Effect of Optimization Parameters

For the optimization model in Eq. 6.45, two parameters α_{max} and q, which are related to the interpolation function of the design variable γ in Eq. 6.46, need to be chosen before optimization. In principle, α_{max} should be infinite in order to approximate the solid material. However, α_{max} has to be finite in order to maintain the stability of the numerical solution of the Navier Stokes equations. The parameter q was used for tuning convex property of the interpolation function. A clear black-white type optimal geometry can be obtained for a relatively larger value of q. Currently, there is no unique rule in choosing the most suitable values of α_{max} and q for a general optimization problem, because of the strong nonlinearity of the optimization model.

Therefore, a parametric sweeping for the diodicity optimization was used to choose the design parameters α_{max} and q. The straight ducts Ω_d and the initial design domain Ω based on the design variables γ had the dimensions $2400 \times 200\,\mu$m and $200 \times 200\,\mu$m, respectively (Fig. 6.26). The design domain Ω is discretized with 80×80 structured quadrilateral elements. The liquid flowing through the design domains was chosen to be water with physical parameters as $997\,\text{kg/m}^3$ for the density and 8.55×10^{-4} Pa s for the dynamic viscosity. The value for α_{max} was swept from 1.0×10^6 to 4.0×10^8, and for q from 1.0×10^{-1} to 1.0×10^0, respectively. The maximal value U_{max} of the parabolic velocity distribution at the inlet for the forward and reverse flows was 5×10^{-2} m/s. The lower and upper bounds of the volume constraint were $\underline{V}_r = 0.49$ and $\overline{V}_r = 0.50$ such that the optimized result had similar values of channel volume. The constant δ, used to tune the relative importance of the consumption of the reverse pressure work, was chosen to be 1. The stop criterion of the optimization procedure was that the maximal change of the design variable γ was less than 5.0×10^{-3}. Figure 6.27 shows the results obtained by solving the optimization model in Eq. 6.45 based on different choices of the parameters α_{max} and q.

Two criteria were used in this work to choose the most reasonable values of the design parameters: (1) the optimization result should have a lower value of diodicity; (2) the optimization result should have clear black-white type boundary so that it can be fabricated using available materials for microfluidics, and the result with tiny gray region should be excluded (Fig. 6.28). For the optimized micro Venturi diodes, the curves in Fig. 6.27 show that there are two points ($\alpha_{max} = 1.1 \times 10^8$, $q = 1.0 \times 10^0$) and ($\alpha_{max} = 3.3 \times 10^8$, $q = 1.0 \times 10^{-1}$) at which the optimized results have relatively

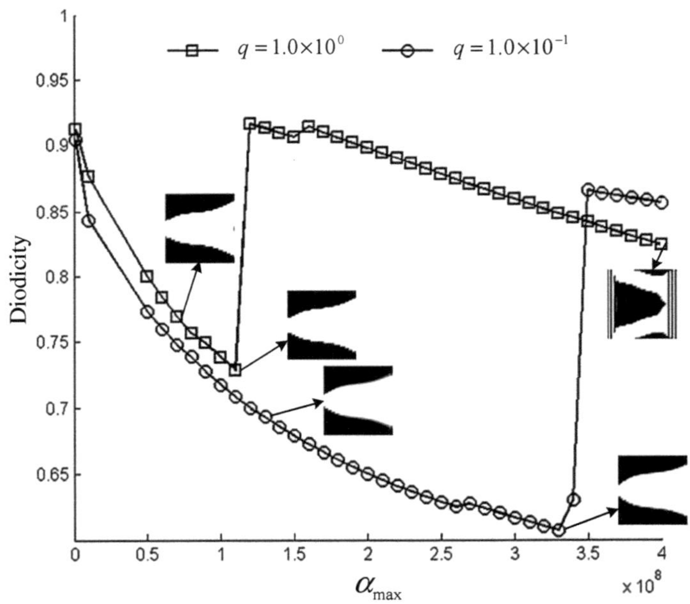

Fig. 6.27 Values of diodicity and optimized micro Venturi diode obtained by parametric sweeping of design parameters α_{max} and q

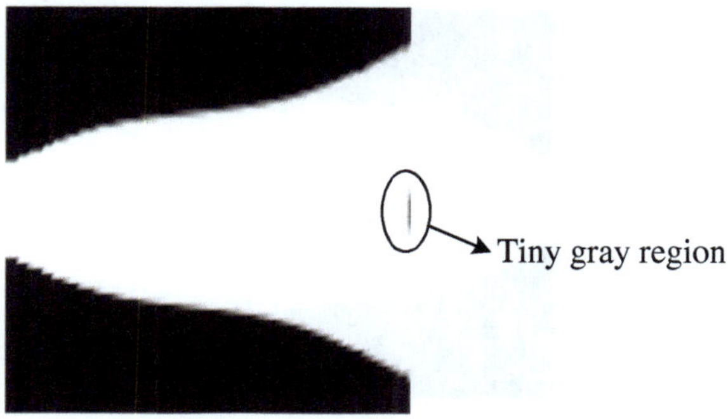

Fig. 6.28 An optimized result with a tiny *gray region* for the design parameters $\alpha_{max} = 7.0 \times 10^7$, $q = 1.0 \times 10^0$

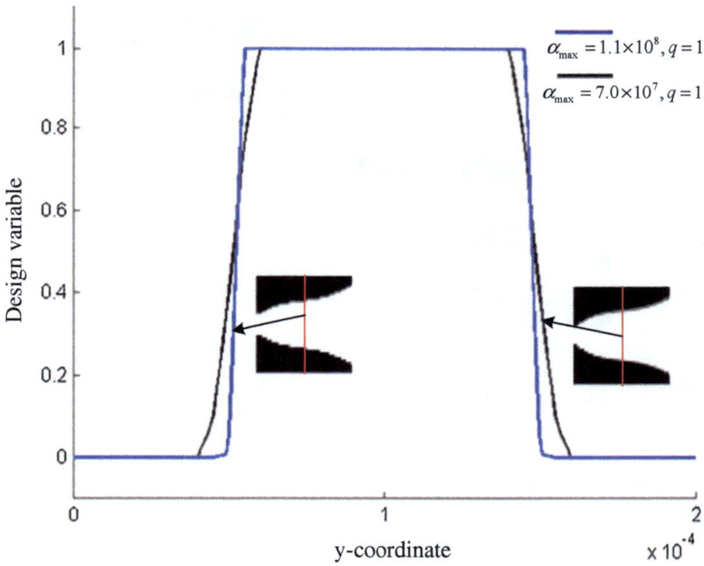

Fig. 6.29 The value of design variable γ on a specified cross section. The optimized result with *gray color* means relatively smooth change of the design variable (*the black line*) when compared with the result with relatively sharp *black-white* interface (*the blue line*)

lower diodicity, 0.73 and 0.61, respectively. However, the optimized result which has the lowest diodicity has a smeared solid and fluid interface. Figure 6.29 shows the comparison of the design variable γ for the above two optimized results. In short, the parameters ($\alpha_{max} = 1.1 \times 10^8, q = 1.0 \times 10^0$) were chosen for obtaining the optimal micro Venturi diode which is shown in Fig. 6.30.

Fig. 6.30 The optimal micro Venturi diode

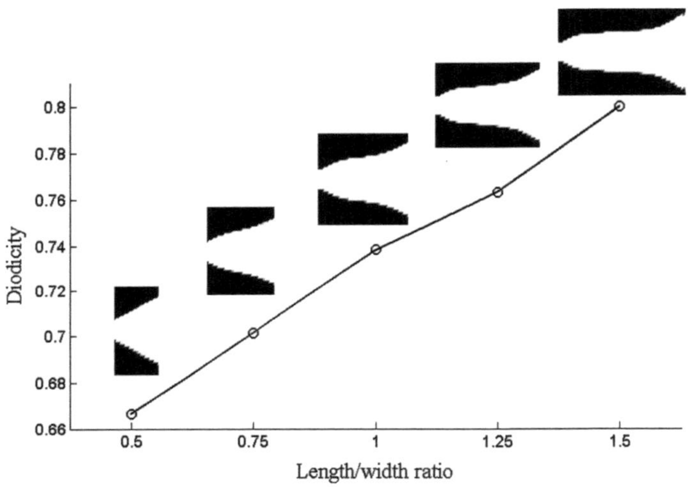

Fig. 6.31 The values of diodicity for the optimized diodes in the design domains with different ratio of length to width

6.3.4.2 Effect of Ratio of Length to Width

A series of micro Venturi diodes were designed for the initial design domains shown in Fig. 6.26, where the ratio of length to width of the design domain varied from 0.5 to 1.5 while the width was fixed as 200 μm. For reasonable comparison of the diodicity, the upper bound of the volume constraint was set to be 0.5 for all numerical examples in this section. The parameters α_{max} and q were chosen to be the optimal one that were obtained by parametric sweeping as described in Sect. 6.3.4.1. Figure 6.31 shows the value of diodicity and the optimized micro Venturi diodes. The results illustrate that a micro Venturi diode designed in a design domain with lower ratio of length to width has a better diode effect.

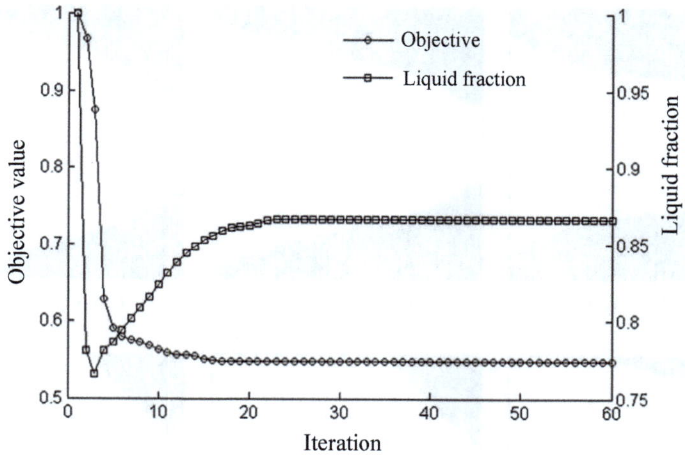

Fig. 6.32 The history of optimization for the micro Venturi diode, where $\underline{V}_r = 0.00$ and $\overline{V}_r = 1.00$. The initial value of objective function is scaled as 1

6.3.4.3 Effect of Volume Fraction

It is known that a suitable constraint in an optimization model can regularize an optimized result. For the optimization model in Eq. 6.45, the volume constraint was used to regularize the optimal micro Venturi diodes. Whereas, the suitable values of lower and upper bounds of volume constraint need to be specified by the designers. Trivially, one can always set $\underline{V}_r$ to 0 and $\overline{V}_r$ to 1, so that the volume constraint has no limitation to the design variable, or it is an inactive constraint. Based on the numerical tests, an upper bound of the liquid fraction does exist for the micro Venturi diodes. Figure 6.32 shows the optimization history for the value of the objective function and volume fraction of the Venturi diode, where design parameters were chosen to be the optimal ones used in Sect. 6.3.4.1, except for $\underline{V}_r = 0.00$ and $\overline{V}_r = 1.00$. It shows that the volume fraction of the Venturi diode converges to 0.87 which is lower than $\overline{V}_r = 1.00$. In order to check how the value of the upper bound influences the optimized results, the value of $\overline{V}_r$ was chosen to be 0.50, 0.60, 0.70, 0.80, 0.90 and 1.00, and the corresponding optimal micro Venturi diodes are listed in Fig. 6.33. The corresponding values of diodicity and convergent volume fraction are plotted in Fig. 6.34. It shows that the diodicity decreases with increasing upper bound $\overline{V}_r$, and the diodicity reaches the minima when the volume fraction converges to 0.87. When varying the maximal value of the inlet velocity in the range 2.5×10^{-3} to 2.5×10^{-1} m/s, the optimized micro Venturi diodes shows an upper bound of volume fraction which varies between 0.87 and 0.89 (Fig. 6.35). When increasing the ratio of length to width of the design domain, the volume of the optimized micro Venturi diode has upper bounds also (Fig. 6.36). The above examples show that the volume constraint can be excluded in the optimization of micro Venturi diodes.

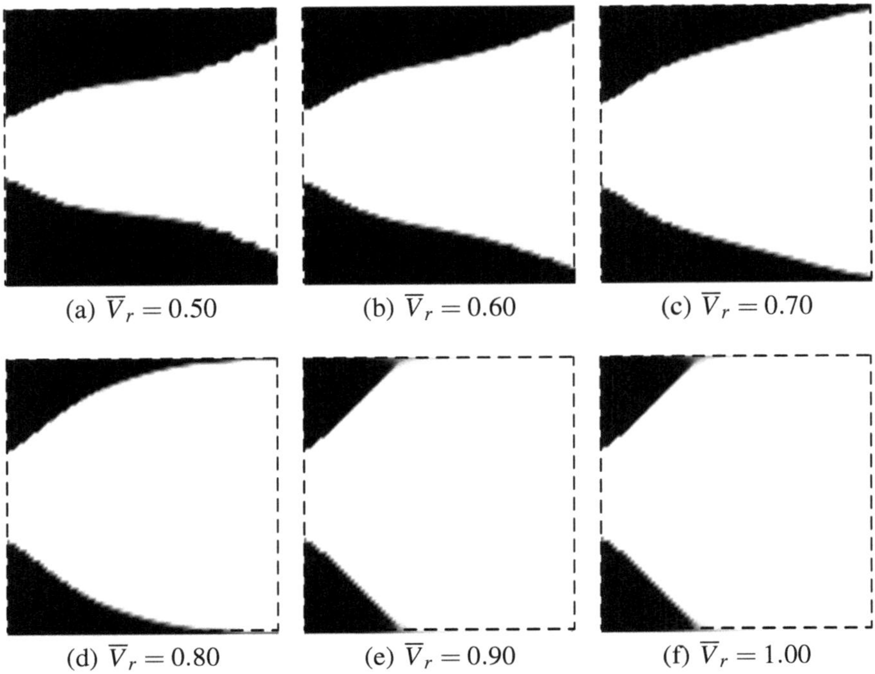

Fig. 6.33 Optimized micro Venturi diodes. The *upper* bound of volume constraint varies from 0.50 to 1.00

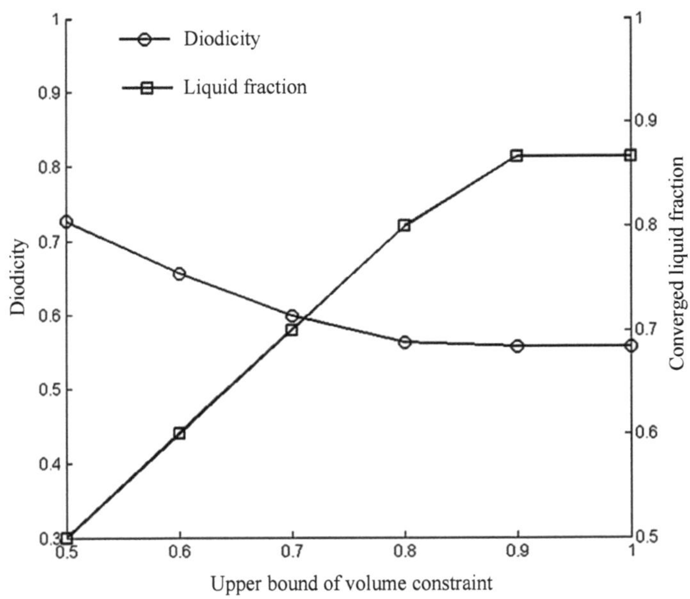

Fig. 6.34 The diodicity and converged liquid fraction for the optimized results in Fig. 6.33

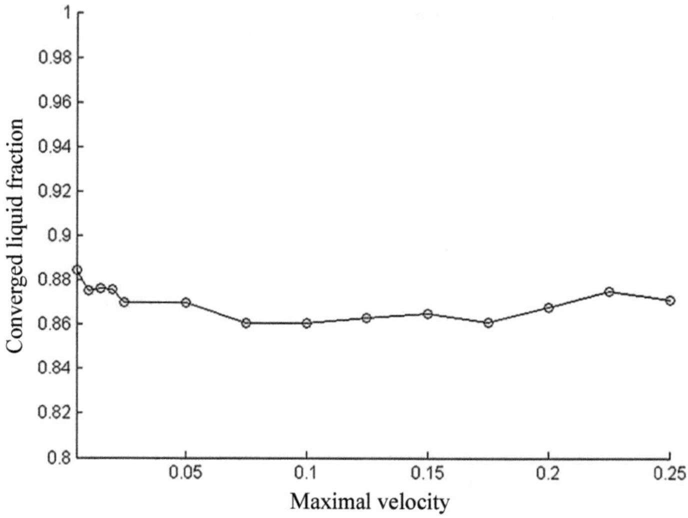

Fig. 6.35 The converged liquid fraction obtained by varying the maximal value of the inlet velocity

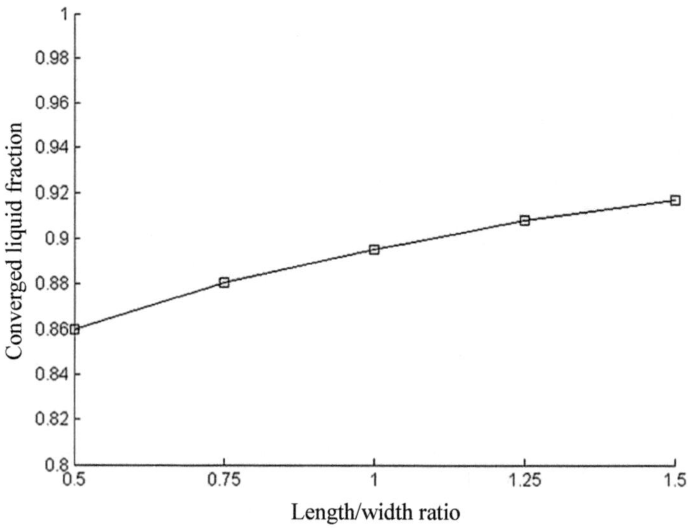

Fig. 6.36 The converged liquid fraction by varying the ratio of length to width for the initial design domain

6.3.4.4 Effect of Reynolds Number

In microfluidics, it is known that fluidic dynamic property depends on the Reynolds number

$$\mathrm{Re} = \rho U_{max} L \big/ \eta \tag{6.55}$$

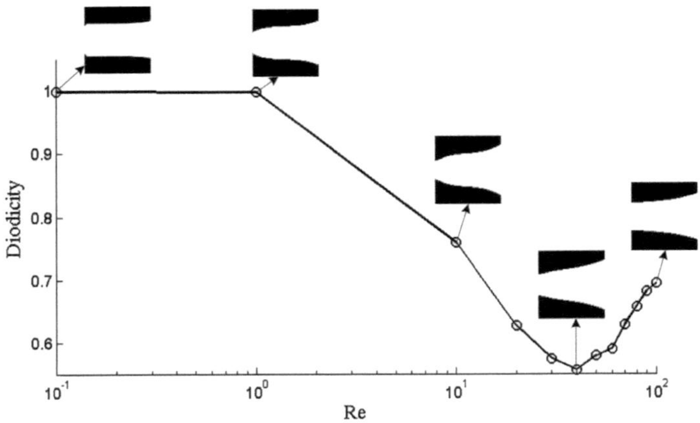

Fig. 6.37 The optimized diodes and the corresponding values of diodicity for different Reynolds numbers. The *upper* bound of volume constraint $\overline{V}_r = 0.50$ and the ratio of length to width of the design domain is 1.00

where L is the characteristic size of the design domain. Therefore, it is not surprising that the optimal micro Venturi diode depends on the Reynolds number. As discussed in Sects. 6.3.4.2 and 6.3.4.3, the optimal value of the diodicity depends on the ratio of length to width for the initial design domain and the upper bound of the volume constraint. In order to discuss how the optimized result is influenced by the Reynolds number, the ratio of length to width for the initial design domain was chosen as 1.0 and the upper bound of the volume constraint was fixed to $\overline{V}_r = 0.50$. The other parameters are chosen to be the optimal values which are used in Sect. 6.3.4.1. The micro Venturi diode is optimized for the flows with the Reynolds numbers between 1×10^{-1} to 1×10^2. The results listed in Fig. 6.37 show that micro Venturi diode has almost no diode effect for the Reynolds number lower than $\mathrm{Re} < 1 \times 10^{-1}$. This effect is verified numerically for the case that the ratio of length to width for the initial design domain equal to 0.5, and the upper bound of volume constraint was chosen to be $\overline{V}_r = 1.00$. At the same time, the results in Fig. 6.37 show that the optimal micro Venturi diode does vary with the Reynolds number. It is necessary to mention that the relationship between the optimal diodicity and the Reynolds number obtained in Fig. 6.37 is only meaningful for the current setup of optimization model and design parameters. A micro Venturi diode usually works in the steady flows with different Reynolds numbers. A more realistic optimization objective is to consider the diode effect in several Reynolds numbers cases. Therefore, the objective of the optimization model in Eq. 6.45 can be modified as

$$J = \sum_{i=1}^{N} \varpi_i (\Delta W_{pfi} / \Delta W_{pri} + \delta \Delta W_{pri} / \Delta W_{pr0i}) \tag{6.56}$$

where ϖ_i is the weight for each single objective and the value of ϖ_i is chosen to be $1/N$ in this work, and N is the number of the Reynolds number values considered simultaneously; ΔW_{pfi} and ΔW_{pri} are the consumption of the pressure work

Fig. 6.38 The optimized micro Venturi diodes that the objective function considers multiple Reynolds numbers

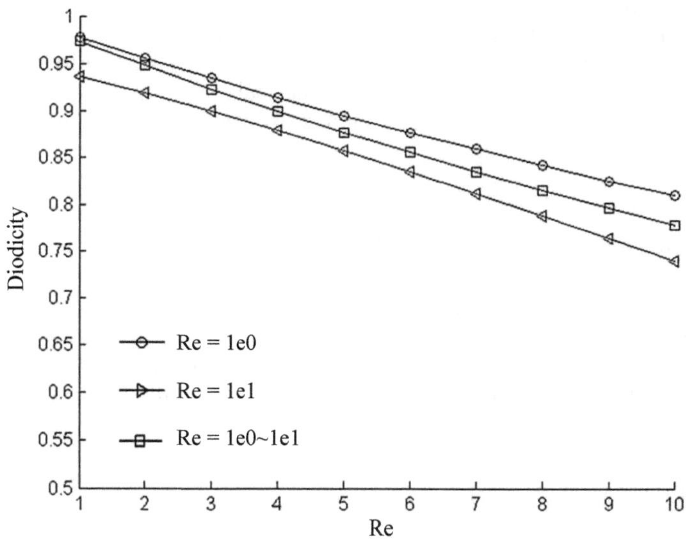

Fig. 6.39 The comparison of diodicity of the optimized micro Venturi diode for the objective function including multiple Reynolds numbers

to the i-th Reynolds number, ΔW_{pr0i} is the initial consumption of the reverse pressure work for the i-th Reynolds number case. Using the same parametric choice in Sect. 6.3.4.1, the optimized micro Venturi diode is shown in Fig. 6.38 where the Reynolds numbers considered in the optimization objective are integers in the range 1×10^0 to 1×10^1, and $N = 10$ for the objective in Eq. 6.56. Figure 6.39 shows how the value of the diodicity varies with the Reynolds numbers for the optimization model which considers single and multiple Reynolds numbers, respectively. The value of diodicity is compromised for the objective considering multiple Reynolds numbers when comparing with the objective considering single Reynolds number.

Whereas, it is concluded that the diodicity strongly depends on the Reynolds number for the optimized micro Venturi diode in this work.

6.3.4.5 Curved Micro Venturi Diode

In most cases, a micro Venturi diode was designed inside a straight channel. For the requirement of connecting complicated channel networks, one may need a curved micro Venturi diode. Therefore, the design of micro Venturi diode with a curved design domain is presented in this section. The initial design domain is a sector with outer radius 300 μm and inner radius 5 μm (Fig. 6.40). The initial design domain is discretized using structured quadrilateral meshes. The central angle β of the sector was chosen between 30 and 150°. The other optimization parameters were the same as those used in Sect. 6.3.4.1. The optimal designs corresponding to different central angles are listed in Fig. 6.41 and the corresponding diodicity is plotted in Fig. 6.42. Figure 6.42 shows that the curved micro Venturi diode has better diodicity than the one designed in a straight channel. These numerical examples show that the curved micro Venturi diode provides a flexible choice to design a microfluidic channel with a diode effect.

6.3.4.6 Optimization of 3D Micro Venturi Diode

The above micro Venturi diodes were designed using the Navier-Stokes equations in the two dimensional case. For a real microfluidic device, the dimension along the depth has the limited size. Therefore, a three dimensional optimization model will be more realistic than the two dimensional one. Considering the fabrication procedure

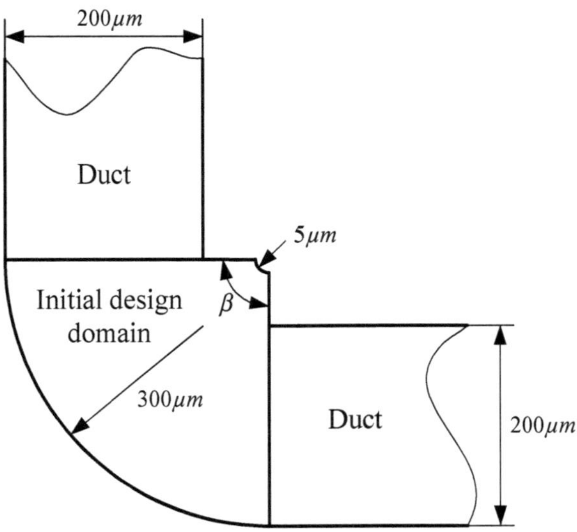

Fig. 6.40 The initial design domain for the curved micro Venturi diode. The radius is 300 μm, and a sector with radius 5 μm is removed from the center in order to discretize the domain with structured quadrangular mesh. The length of duct is scaled

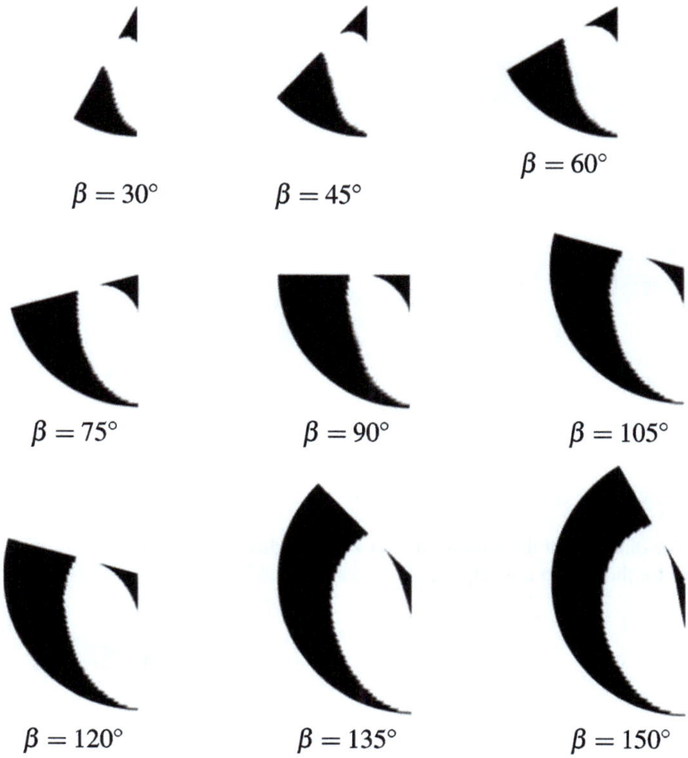

Fig. 6.41 The optimized micro Venturi diodes corresponding to different central angles β of a curved initial design domain

of a microfluidic device, a fluidic channel usually has a rectangular cross section for a specified layer. Therefore, this fabrication constraint needs to be included into the optimization model to design three-dimensional microfluidic devices. For a three-dimensional initial design domain, one can use the extrusion method to generate a three dimensional finite element mesh which is extruded from a two dimensional mesh. In the case that the extruding direction is along the z direction, the design variables can be grouped, where a group of design variables have the same coordinates in x and y directions. In addition, the number of fabrication layers can be specified by choosing a different number of groups along the z direction. When calculating the sensitivity vector, one can set all the design variables to be independent of each other. After the sensitivity analysis, the component of sensitivity which belongs to the same group is replaced by the averaged value of original sensitivity. In this section, a three dimensional micro Venturi diode with merely one fabrication layer along the z direction is presented. To save on the computational costs, the initial value for the 3D example was chosen to be the result obtained from the 2D optimization. After 3D optimization, the optimized result in Fig. 6.43 is similar to the two dimensional case. As the computational cost of Navier-Stokes equations in the three dimensional

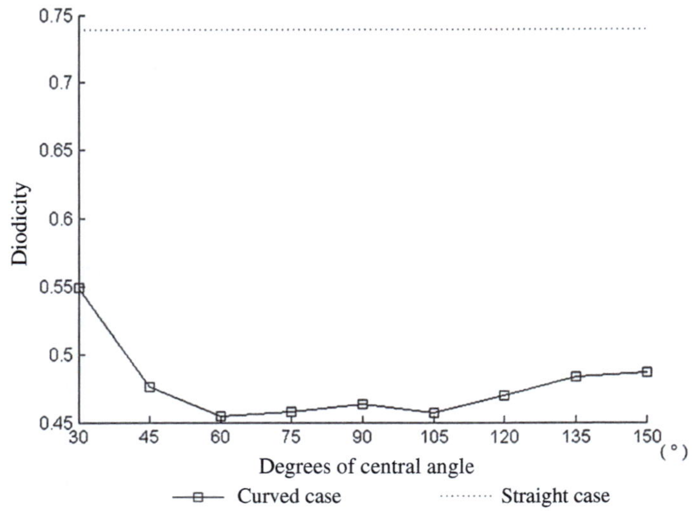

Fig. 6.42 The diodicity of the optimized micro Venturi diodes shown in Fig. 6.41. The *dash line* is the diodicity for the design in straight case shown in Fig. 6.30a

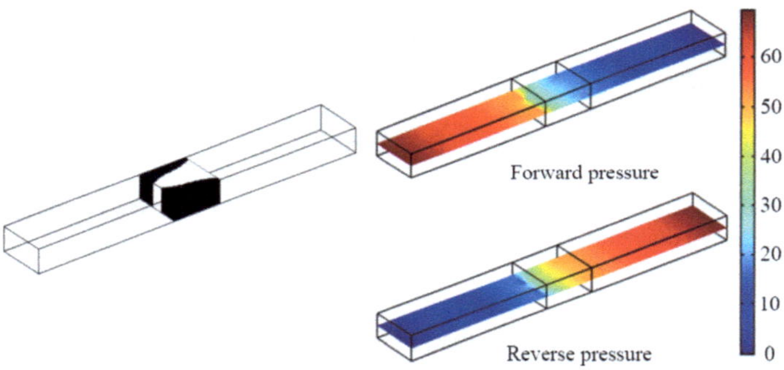

Fig. 6.43 The pressure distribution of the forward and reverse flow for the optimized 3D micro Venturi diode where the Reynolds number is 10

case is significantly more expensive than the two dimensional case, one can firstly design a micro Venturi diode using the two dimensional optimization method, and then extrude the designed two dimensional Venturi diode into a three dimensional domain. This three dimensional layout can be used as an initial design to restart the optimization in the three dimensional case. This strategy saves the computational cost in the case that there is merely a limited number of channel layers along the z direction.

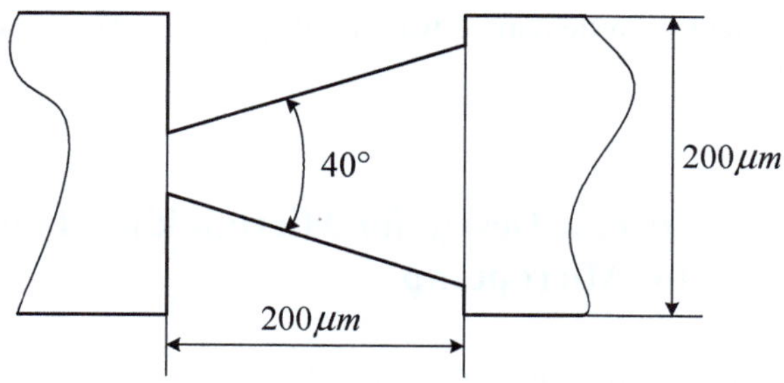

Fig. 6.44 The micro Venturi diode optimized by [59] for Re = 1×10^2

6.3.4.7 Comparison of Diodicity of Optimized Micro Venturi Diodes

There are a couple of papers which focus on the design of micro Venturi diodes [56–59]. Due to the lack of precise data about the optimized shape of micro Venturi diode in the above literatures, or the optimized micro Venturi diode being mainly for flows with Reynolds number higher than 1×10^2, the micro Venturi diodes with straight boundary was used to compare the diode effect with the one optimized in this section. [59] optimized the Venturi diode (Fig. 6.44) where the optimal diverging angle is 40°, where the corresponding diodicity is 0.86. The designed micro Venturi diode has a value of diodicity 0.70 for the same Reynolds number 100 and the channel volume as [59]'s design in. The performance of diodicity was improved up to 19%. [58] proposed a Venturi diode by adding 3-fin structure which has the best diodicity at a Reynolds number around 70, and improvement of the diodicity is about 16% compared with the conventional micro Venturi diode in [59]. The designed diode in this work improves the performance of diodicity up to 24% compared with the conventional diodes with a Reynolds number of 70.

6.3.5 Summary

Based on the gradient type optimization method and the finite element discretization of incompressible Navier-Stokes equations, this section presents a systematic way to design micro Venturi diodes. Through a couple of numerical examples, we have numerically shown that the optimized micro Venturi diode will only change the shape of Venturi diode's boundaries. Several rules are discussed to obtain a reasonable design of a micro Venturi diode: (1) the lower ratio of length to width of an initial design domain will result in a design with better diodicity; (2) the volume constraint in the optimization model can be an active constraint so that choosing a reasonable value of the upper bound of the constraint is important to obtain a design with a better diodicity; (3) the Reynolds number of the flow influences the optimal micro Venturi diode, and numerical examples illustrate that a micro Venturi diode has almost no diode effect when the Reynolds number is lower than 0.1; (4) the curved micro Venturi diodes may have better diode effect than straight ones. All the above rules provide a systematical way to design micro Venturi diodes for the case of steady flows.

6.4 Inverse Design for Microfluidic Channel for Micropump

In this section, the output net flux for valveless micropumps is improved by the effective topology optimization method. The considered micropump is actuated by the vibration of membranes and can generate a net flow along the optimized rigid fluidic channel. The detailed layout of the micropump channel is optimized with respect to a number of factors which influence the output net flux of liquid, such as the number of actuation elements, the positions of actuation elements, the driving phase differences and the driving frequency. Accordingly, the layouts of several novel valveless micropumps are presented.

6.4.1 *Methodology*

A valveless micropump is usually driven by bimorph actuators. Figure 6.45 shows the layout of a valveless micropump, which includes piezoelectric plates, membranes, pumping chambers, connection channels and a main channel. The piezoelectric plate and membrane can work together as a bimorph actuator by applying waveform voltage. Therefore, the pumping chambers generate suction and injection flows successively along the sides of the main channel. The measure of the performance is the net flux pumped by the micropump with a specified actuation. Therefore, the optimization of performance can be achieved by maximizing the net flux, which can be expressed as

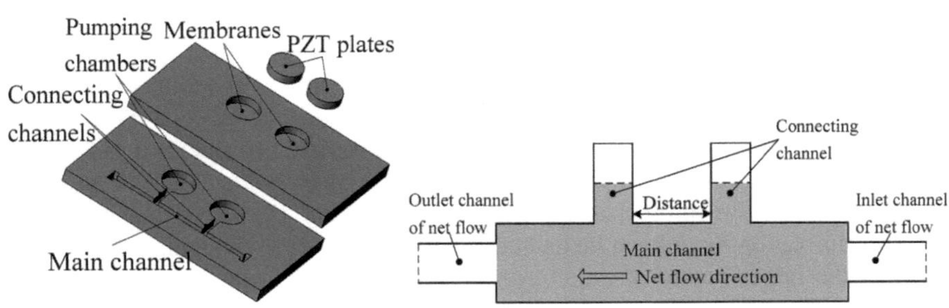

Fig. 6.45 Schematic of the optimized valveless micropump. The micropump is driven by bimorph actuators made up of membranes and piezoelectric plates. The vibration of membranes produces a directed fluidic net flux between the two ends of the main channel

$$Q_{net} = \int_0^T \int_{\Gamma_{o2}} \mathbf{u} \cdot \mathbf{n} \, \mathrm{d}\Gamma \mathrm{d}t - \int_0^T \int_{\Gamma_{o1}} \mathbf{u} \cdot \mathbf{n} \, \mathrm{d}\Gamma \mathrm{d}t \tag{6.57}$$

where Γ_{o1} and Γ_{o2} are the two ends of the main channel; $(0, T)$ is the considered time interval; and $\mathbf{n}$ is the outward normal vector. In the topology optimization of unsteady flows, the incompressible Navier-Stokes equations are subject to an artificial friction force which is proportional to the fluidic velocity $\mathbf{u}$. Based on the Darcy law, the friction force is expressed as $\mathbf{f} = -\alpha \mathbf{u}$, where α is the inverse of the local permeability of an artificial porous medium. Therefore, the incompressible Navier-stokes equations are modified as

$$\begin{aligned} -\nabla \cdot \left[-p\mathbf{I} + \eta \left(\nabla \mathbf{u} + \nabla \mathbf{u}^{\mathrm{T}}\right)\right] + \rho \left(\mathbf{u} \cdot \nabla\right) \mathbf{u} = \mathbf{f}, \text{ in } \Omega \\ -\nabla \cdot \mathbf{u} = 0, \text{ in } \Omega \end{aligned} \tag{6.58}$$

where ρ is the fluid density; $\mathbf{u}$ is the fluid velocity; p is the fluid pressure; and η is the dynamic viscosity. The artificial material property α is interpolated by the design variable γ as

$$\alpha(\gamma) = \alpha_{\min} + (\alpha_{\max} - \alpha_{\min})\, q\,(1-\gamma)\,/\,(q+\gamma) \tag{6.59}$$

where $\alpha_{\min}$ and $\alpha_{\max}$ are the minimal and maximal values of α, respectively. $\alpha_{\min}$ is set to 0. In order to obtain a perfect impermeable solid wall, $\alpha_{\max}$ should be infinity. However, for numerical reasons, $\alpha_{\max}$ has to be chosen as a finite but large enough value so as to ensure the numerical stability and to approximate a solid with negligible permeability. In order to maintain an energy-efficient design, the dissipation power is constrained to be no more than a given value. Based on the modified Navier-Stokes Equation 6.58, the energy dissipation is defined as [3, 49, 60]

$$\Phi = \int_0^T \int_\Omega \frac{1}{2}\eta \left(\nabla \mathbf{u} + \nabla \mathbf{u}^{\mathrm{T}}\right) : \left(\nabla \mathbf{u} + \nabla \mathbf{u}^{\mathrm{T}}\right) + \alpha \mathbf{u}^2 \, \mathrm{d}\Omega \mathrm{d}t \tag{6.60}$$

where Ω is the domain occupied by the flow. In addition, a constraint which is related with the volume of the main channel may also be necessary in order to regularize the optimization problem. Then the optimized design of a valveless micropump can be obtained by solving the following optimization problem as

$$\text{Find } \gamma \in [0, 1] \text{ to minimize } -\frac{Q_{net}}{Q_0}, \text{ subject to}$$

$$\begin{cases} \rho \dfrac{\partial \mathbf{u}}{\partial t} - \eta \nabla \cdot \left(\nabla \mathbf{u} + \nabla \mathbf{u}^{\mathrm{T}}\right) + \rho \left(\mathbf{u} \cdot \nabla\right) \mathbf{u} + \nabla p = -\alpha \mathbf{u}, \text{ in } \Omega \\ -\nabla \cdot \mathbf{u} = 0, \text{ in } \Omega \\ \Phi \le \theta_\Phi \Phi_0 \\ \displaystyle\int_\Omega \gamma \, \mathrm{d}\Omega \le \theta_V V_0 \end{cases} \tag{6.61}$$

where Q_0 and Φ_0 are user specified values of net flux and energy dissipation; θ_Φ ($0 \leq \theta_\Phi \leq 1$) is the upper bound of the normalized energy dissipation; V_0 is the volume of the initial design domain, and θ_V ($0 \leq \theta_V \leq 1$) is the upper bound of the volume fraction. The optimization problem in Eq. 6.61 includes one objective and multiple constraints. Therefore, the method of moving asymptotes (MMA) [48], which has been shown to work for optimization tasks involving a large number of design variables and multiple design constraints, is used to implement the topology optimization of valveless micropumps. When MMA is used in conjunction with an optimization task, the "user program" needs to provide the values of the objective function, the residuals of constraints, and the first-order derivatives of the objective function and constraints, all for a given state of the topology. The sensitivity analysis is done by the continuous adjoint method [49]. The commercial finite element software COMSOL [9] is used to implement optimization procedure. The detailed implementation for the case of the topology optimization of a fluidic channel has been presented by the work in [49, 60].

6.4.2 Results and Discussion

The performance of the micropump is influenced by the number of actuation elements, the positions of actuation elements, the driving phase differences and the driving frequency. Therefore, any modification of the above factors will influence the optimized layout of the main channel. For all the presented numerical examples, the density and dynamic viscosity of fluid are set to as 1×10^3 kg/m^3 and 1×10^{-3} Pa, respectively. Q_0 and Φ_0 are set as the output net flux and dissipation power corresponding to the initial layout of fluidic channels. The choices of the other optimization parameters are shown in Table 6.5.

6.4.2.1 Number of Actuation Elements

The performance of the valveless micropump depends on the number of actuation elements. In order to investigate how the output net flux is influenced, the main channels of the valveless micropumps with one, two or three actuation elements are optimized by solving the optimization problem in Eq. 6.61. Figure 6.46 shows the initial design domains. The actuation velocity posed on the inlet of the connecting channels are chosen as $\mathbf{u}_{in1} = \mathbf{u}_{in2} = \mathbf{u}_{in3} = -4 \times 10^{-3} s\,(1 - s)\sin(\pi t)\,\mathbf{n}$, where s is the boundary parametric coordinates of the inlet duct. The optimized main channels are shown in Fig. 6.47. One can conclude that the output net flux is large for microp-

Table 6.5 The setting of the parameters in the optimization problem

α_{max}	α_{min}	q	θ_{Phi}	θ_V
5×10^5	0	1	0.3	0.5

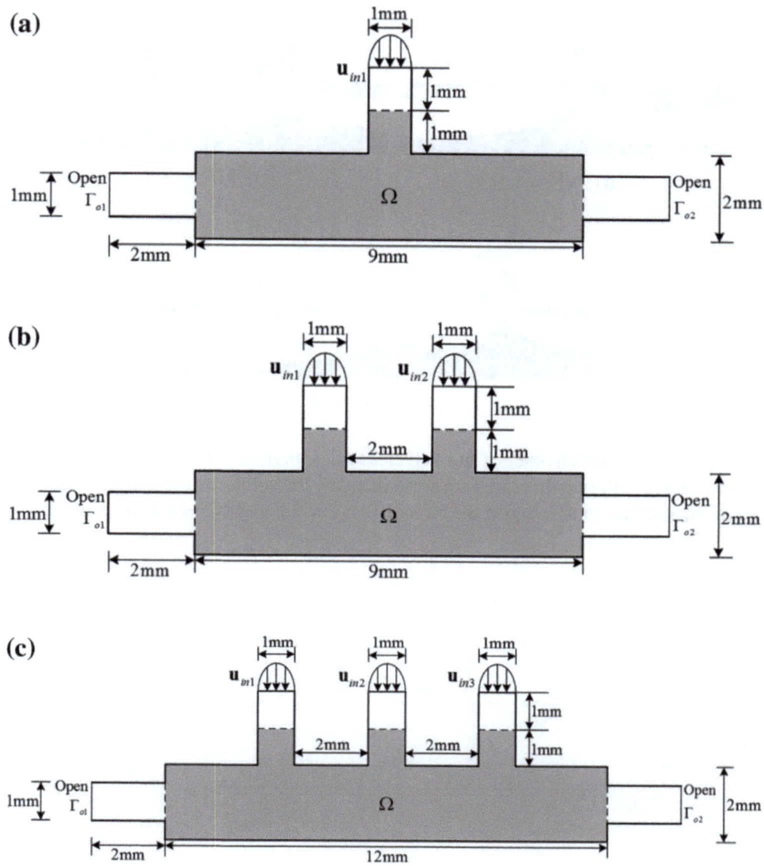

Fig. 6.46 Initial design domains of the valveless micropumps with one, two or three actuation elements

umps with multiple actuation films. In the following numerical examples, the value of the net flux (NF) for a micropump actuated by two or three actuation elements are all referred to the case with one actuation element as presented in Fig. 6.47a.

6.4.2.2 Positions and Phase Differences of Actuation Elements

When the valveless micropump has two or more actuation elements, the performance of the micropump will be affected by their relative positions and the phase differences of the actuation pulses. The design of the main channel with two actuation elements is investigated firstly.

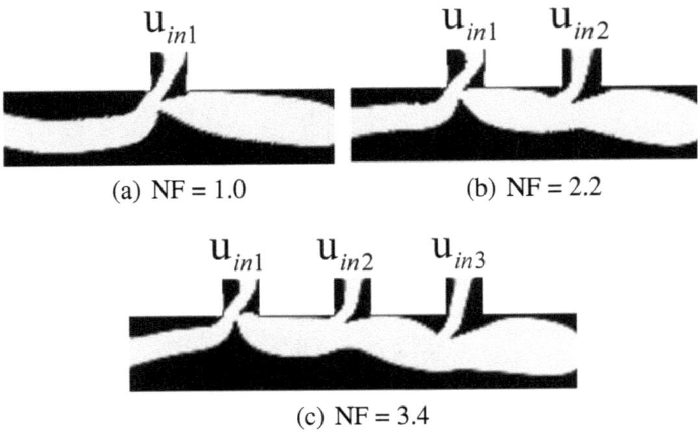

Fig. 6.47 Optimized main channel of the valveless micropumps with one, two or three actuation elements. The values of the net flux (NF) are normalized by the value corresponding to (**a**). Therefore, the values of the net flux in **a**, **b** and **c** are NF = 1.0, NF = 2.2 and NF = 3.4, respectively

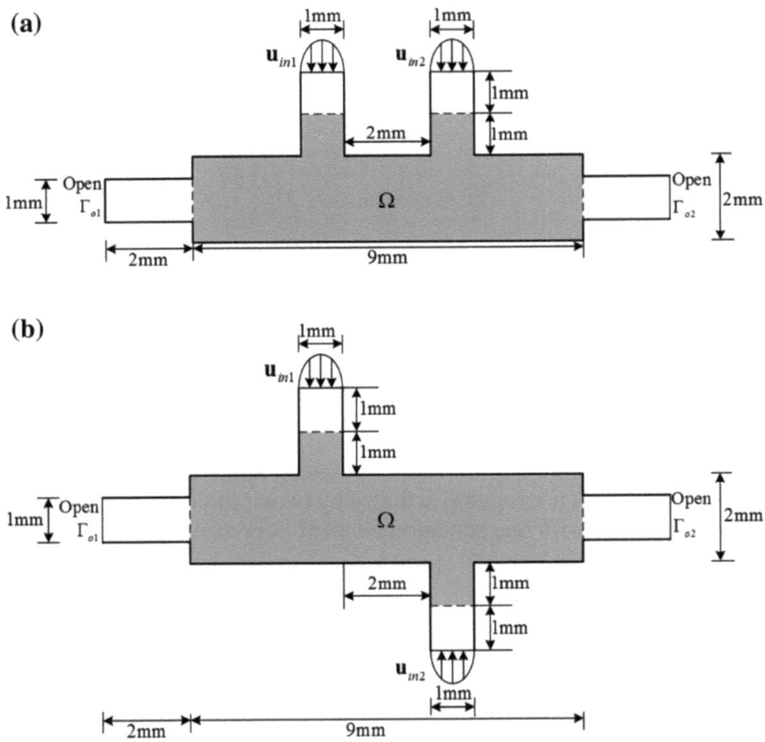

Fig. 6.48 Initial design domains of valveless micropumps with two actuation elements located at the same or both sides of the main channel

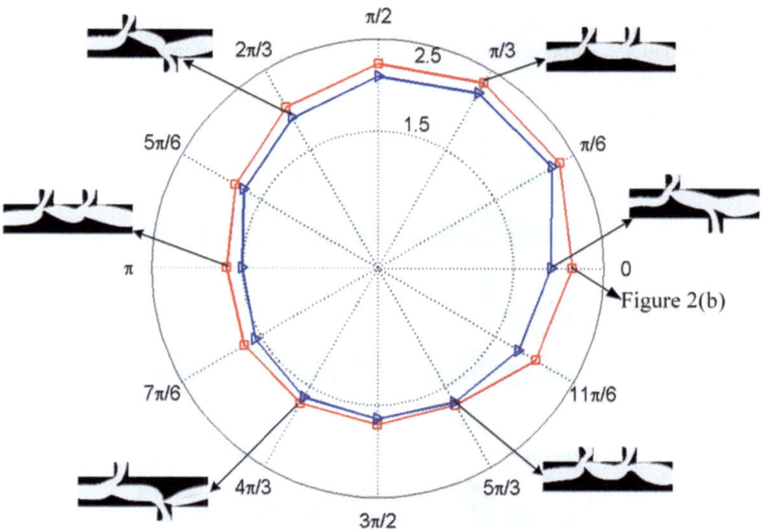

Fig. 6.49 Optimized micropumps with two elements at the same side or both sides of the main channel corresponding to varying phase differences. The values of NF are plotted in polar coordinate system

The initial design domains are shown in Fig. 6.48, where the two actuation elements are located at the same side or both sides of the main channel. The flows actuated by the membranes in the connecting channels are $\mathbf{u}_{in1} = -4 \times 10^{-3} s\,(1-s)\sin(\pi t)\,\mathbf{n}$ and $\mathbf{u}_{in2} = -4 \times 10^{-3} s\,(1-s)\sin(\pi t + \Delta\phi_1)\,\mathbf{n}$, where $\Delta\phi_1$ is the phase difference. For different phase differences, the optimized micropumps are shown in Fig. 6.49. The results show that the optimized micropumps with actuation elements at the same side of the main channel have higher value of pumping net flux, and the optimal phase difference for the optimized main channel with two actuation elements is $\pi/3$.

For the case with three actuation elements set at the same side of the main channel, the effect of phase difference is investigated next. The initial design domain is the same as shown in Fig. 6.46c. The velocity conditions posed on the inlets of the connecting channels is $\mathbf{u}_{in1} = -4 \times 10^{-3} s\,(1-s)\sin(\pi t)\,\mathbf{n}$, $\mathbf{u}_{in2} = -4 \times 10^{-3} s\,(1-s)\sin(\pi t + \Delta\phi_1)\,\mathbf{n}$, and $\mathbf{u}_{in3} = -4 \times 10^{-3} s\,(1-s)\sin(\pi t + \Delta\phi_2)\,\mathbf{n}$ respectively, where $\Delta\phi_1$ and $\Delta\phi_2$ is the phase differences. Because the computational cost is very high for the three actuation case, merely four groups of the phase differences are considered. By solving the topology optimization problem in Eq. 6.61, the optimized results are shown in Table 6.6. Based on these results, one can conclude that three actuation elements can achieve large pumping net flux than two, as long as the phase differences are chosen reasonably.

For two actuation elements, the distance between them affects performance of the micropump and the layout of the main channel. Based on the above conclusion, the two actuation elements are set at the same side of the main channel, and the phase difference is fixed to be π.

Table 6.6 Optimized micropumps actuated by three elements. $\Delta\phi_1$ is the phase difference between the first and second actuation element; and the $\Delta\phi_2$ is the phase difference between the first and third actuation element

$\Delta\phi_1$	$\Delta\phi_2$	Optimal	Times
0	π		1.4
π	0		1.9
π	π		3.1
0	0	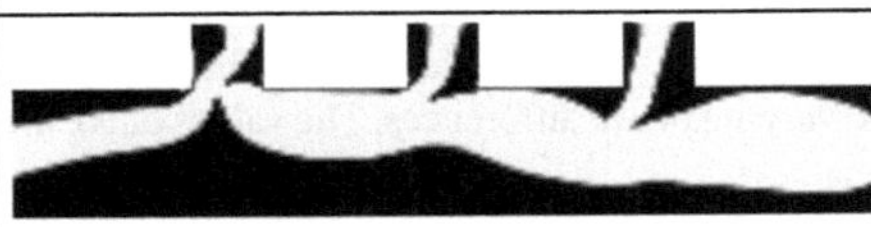	3.4

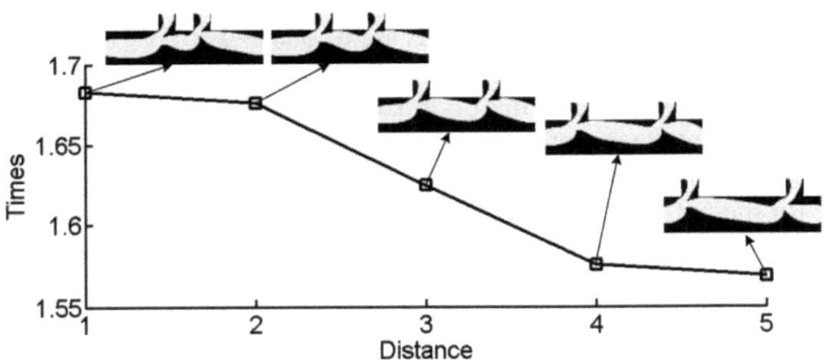

Fig. 6.50 Optimized micropumps with different distances between two actuation elements

By setting different distances between the two actuation elements, the optimized designs of the main channel and the corresponding values of NF (referred to the case in Fig. 6.47a) are shown in Fig. 6.50. The results show that a shorter distance between the two actuation elements generates larger value of the output net flux.

6.4.2.3 Driving Frequency of Actuation Elements

The influence of the driving frequency on the net flux is discussed here. For a valveless micropump with two actuation elements, the velocity conditions posed on the inlets

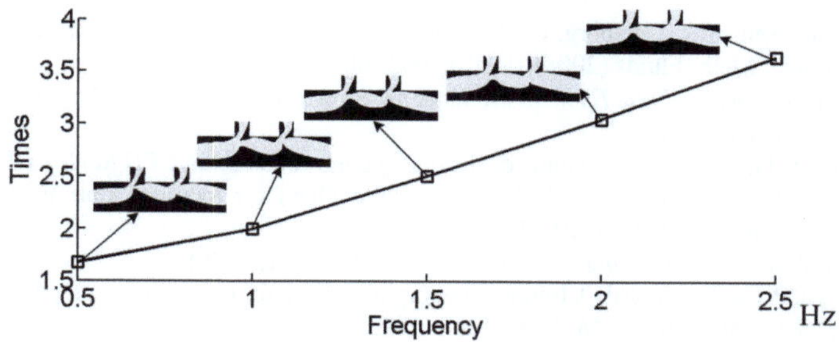

Fig. 6.51 Optimized micropumps actuated by different frequencies

of the connecting channels are $\mathbf{u}_{in1} = -4 \times 10^{-3} s(1-s)\sin(\omega t)\,\mathbf{n}$ and $\mathbf{u}_{in2} = -4 \times 10^{-3} s(1-s)\sin(\omega t + \Delta\phi_1)\,\mathbf{n}$, where $\Delta\phi_1$ is fixed to be π. For different values of ω, the optimization problem is solved. The optimized designs of the main channel and the corresponding values of NF (referred to the case in Fig. 6.47a) are shown in Fig. 6.51. The results indicate that a valveless micropump with a higher actuation frequency generates larger value of output net flux.

6.4.3 Summary

This section has presented the optimization of valveless micropumps driven by the vibration of actuation membranes, based on a topology optimization method that was developed for unsteady incompressible liquid flows. A number of factors which influence the output net flux, such as the number of actuation elements, positions of the actuation elements, the driving phase differences and the driving frequency, are considered. Based on numerical verification, we find that more actuation elements with suitable driving phases can achieve higher output net flux than pumps based on a single actuation element. At the same time, a higher driving frequency for the actuation element will improve the performance of the output net flux further.

References

1. D. Li, *Encyclopedia of Microfluidics and Nanofluidics* (Springer, 2007)
2. P. Danckwerts, The difinition and measurement of some characteristics of mixtures. Appl. Sci. Res. **A 3**, 279–296 (1952)
3. T. Borrvall, J. Petersson, Topology optimization of fluid in Stokes flow. Int. J. Numer. Meth. Fluids **41**, 77–107 (2003)
4. A. Gersborg-Hansen, O. Sigmund, R.B. Haber, Topology optimization of channel flow problems. Struct. Multidiscip. Optim. **29**, 1–12 (2005)
5. L.H. Olesen, F. Okkels, H. Bruus, A high-level programming-language implementation of topology optimization applied to steady-state Navier-Stokes flow. Int. J. Numer. Meth. Eng. **65**, 975–1001 (2006)
6. M. Hinze, R. Pinnau, M. Ulbrich, S. Ulbrich, *Optimization with PDE Constraints* (Springer, 2009)

7. C.S. Andreasen, A.R. Gersborg, O. Sigmund, Topology optimization of microfluidic mixers. Int. J. Numer. Meth. Fluids (2008). doi:10.1002/fld
8. J. Donea, A. Hureta, *Finite Element Methods for Flow Problems* (Wiley, West Sussex, 2003)
9. http://www.comsol.com
10. A. Jameson, Optimum aerodynamic design using control theory, in *CFD Review* (Wiley, 1995)
11. O. Sigmund, Morphology-based black and white filters for topology optimization. Struct. Multidiscip. Optim. **33**, 401–424 (2007)
12. J. Nocedal, S.J. Wright, *Numerical Optimization* (Springer, 1998)
13. R.H. Byrd, P. Lu, J. Nocedal, A limited memory algorithm for bound constrained optimization. SIAM J. Sci. Stat. Comput. **16**, 1190–1208 (1995)
14. C. Zhu, R.H. Byrd, J. Nocedal, L-BFGS-B: Algorithm 778: L-BFGS-B. FORTRAN routines for large scale bound constrained optimization. ACM Trans. Math. Softw. **23**, 550–560 (1997)
15. A.P. Sudarsan, V.M. Ugaz, Multivortex micromixing. PNAS **103**, 7228–7233 (2006)
16. A.D. Stroock, S.K.W. Dertinger, A. Ajdari, I. Mezić, H.A. Stone, G.M. Whitesides, Chaotic mixer for microchannels. Science **295**, 647 (2002)
17. M.S. Williams, K.J. Longmuirb, P. Yager, A practical guide to the staggered herringbone mixer. Lab Chip **8**, 1121–1129 (2008)
18. Y. Du, Z. Zhang, C.H. Yim, M. Lin, X. Cao, A simplified design of the staggered herringbone micromixer for practical applications. Biomicrofluidics (2010). doi:10.1063/1.3427240
19. S.P. Kee, A. Gavriilidis, Design and characterisation of the staggered herringbone mixer. Chem. Eng. J. **142**, 109–121 (2008)
20. Z. Xu, C. Li, D. Vadilloc, X. Ruanb, X. Fu, Numerical simulation on fluid mixing by effects of geometry in staggered oriented ridges micromixers. Sens. Actuators B: Chem. **153**, 284–292 (2011)
21. Y.Z. Liu, B.J. Kim, H.J. Sung, Two-fluid mixing in a microchannel. Int. J. Heat Fluid Flow **25**, 986–995 (2004)
22. C.A. Cortes-Quiroza, A. Azarbadegana, M. Zangeneha, A. Goto, Analysis and multi-criteria design optimization of geometric characteristics of grooved micromixer. Chem. Eng. J. **160**, 852–864 (2010)
23. R. Choudhary, T. Bhakat, R.K. Singh, A. Ghubade, S. Mandal, A. Ghosh, A. Rammohan, A. Sharma, S. Bhattacharya, Bilayer staggered herringbone micro-mixers with symmetric and asymmetric geometries. Microfluid. Nanofluid. **10**, 271–286 (2011)
24. M.A. Ansari, K.Y. Kim, Evaluation of surrogate models for optimization of herringbone groove micromixer. J. Mech. Sci. Technol. **22**, 387–396 (2008)
25. N.S. Lynn, D.S. Dandy, Geometrical optimization of helical flow in grooved micromixers. Lab Chip **7**, 580–587 (2007)
26. C.A. Cortes-Quiroz, M. Zangeneh, A. Goto, On multi-objective optimization of geometry of staggered herringbone micromixer. Microfluid. Nanofluid. **7**, 29–43 (2009)
27. H. Song, X.Z. Yin, D.J. Bennett, Optimization analysis of the staggered herringbone micromixer based on the slip-driven method. Chem. Eng. Res. Des. **86**, 883–891 (2008)
28. M.A. Ansari, K.Y. Kim, Shape optimization of a micromixer with staggered herringbone groove. Chem. Eng. Sci. **62**, 6687–6695 (2007)
29. S. Hossain, A. Husain, K.Y. Kim, Shape optimization of a micromixer with staggered-herringbone grooves patterned on opposite walls. Chem. Eng. Sci. **162**, 730–737 (2010)
30. H.Y. Lee, J. Voldman, Optimizing micromixer design for enhancing dielectrophoretic micro-concentrator performance. Anal. Chem. **79**, 1833–1839 (2007)
31. D.R. Mott, P.B. Howell Jr., J.P. Golden, C.R. Kaplan, F.S. Ligler, E.S. Oran, Toolbox for the design of optimized microfluidic components. Lab Chip **6**, 540–549 (2006)
32. T.G. Kang, M.K. Singh, T.H. Kwon, P.D. Anderson, Chaotic mixing using periodic and aperiodic sequences of mixing protocols in a micromixer. Microfluid. Nanofluid. **4**, 589–599 (2008)
33. R.H. Liu, M.A. Stremler, K.V. Sharp, M.G. Olsen, J.G. Santiago, R.J. Adrian, H. Aref, D.J. Beebe, Passive mixing in a three-dimensional serpentine microchannel. J. Microelectromech. Syst. **9**, 190–197 (2000)

34. J.M. Park, T.H. Kwon, Numerical characterization of three-dimensional serpentine micromixers. AIChE J. **54**, 1999–2008 (2008)
35. J.M. Park, D.S. Kim, T.G. Kang, T.H. Kwon, Improved serpentine laminating micromixer with enhanced local advection. Microfluid. Nanofluid. **4**, 513–523 (2008)
36. K. Malecha, L.J. Golonka, J. Baldyga, M. Jasińska, P. Sobieszuk, Serpentine microfluidic mixer made in LTCC. Sens. Actuators B **143**, 400–413 (2009)
37. P.E. Neerincx, R.P.J. Denteneer, S. Peelen, H.E.H. Meijer, Compact mixing using multiple splitting, stretching, and recombining flows. Macromol. Mater. Eng. **296**, 349–361 (2011)
38. T.G. Kang, M.K. Singh, P.D. Anderson, H.E.H. Meijer, A chaotic serpentine mixer efficient in the creeping flow regime: from design concept to optimization. Microfluid. Nanofluid. **7**, 783–794 (2009)
39. C.K. Chung, T.R. Shih, T.C. Chen, B.H. Wu, Mixing behavior of the rhombic micromixers over a wide Reynolds number range using Taguchi method and 3D numerical simulations. Biomed. Microdevices **10**, 739–748 (2008)
40. C.K. Chung, T.R. Shih, Effect of geometry on fluid mixing of the rhombic micromixers. Microfluid. Nanofluid. **4**, 419–425 (2008)
41. E. Cummings, S. Griffiths, R. Nilson, P. Paul, Conditions for similitude between the fluid velocity and the electric field in electroosmotic flow. Anal. Chem. **72**, 2526–2532 (2000)
42. L. Formaggia, J.F. Gerbeau, F. Nobile, A. Quarteront, Numerical treatment of defective boundary conditions for the Navier-Stokes equations. SIAM J. Numer. Anal. **40**(1), 376–401 (2002)
43. L. Formaggia, A. Veneziani, C. Vergara, A new approach to numerical solution of defective boundary value problems in incompressible fluid dynamics. SIAM J. Numer. Anal. **46**(6), 2769–2794 (2008)
44. B. Lazarov, O. Sigmund, Filters in topology optimization based on Helmholtz type differential equations. Int. J. Numer. Methods Eng. **86**, 765–81 (2011)
45. J. Guest, J. Prevost, T. Belytschko, Achieving minimum length scale in topology optimization using nodal design variables and projection functions. Int. J. Numer. Methods Eng. **61**, 238–254 (2004)
46. S. Hossain, M.A. Ansari, K.Y. Kim, Evaluation of the mixing performance of three passive micromixers. Chem. Eng. J. **150**, 492–501 (2009)
47. Y. Deng, Z. Liu, P. Zhang, Y. Liu, Q. Gao, Y. Wu, A flexible layout design method for passive micromixers. Biomed. Microdevices **14**, 929–945 (2012)
48. K. Svanberg, The method of moving asymptotes: a new method for structural optimization. Int. J. Numer. Meth. Eng. **24**, 359–373 (1987)
49. Y. Deng, Z. Liu, P. Zhang, Y. Liu, Y. Wu, Topology optimization of unsteady incompressible Navier-Stokes flows. J. Comput. Phys. **230**, 6688–6708 (2011)
50. R.L. Bardell, The diodicity mechanism of Tesla-type no-moving-parts valves. Ph.D. Thesis, University of Washington, 2000
51. J. Koo, C. Kleinstreuer, Liquid flow in microchannels: experimental observations and computational analyses of microfluidics effects. J. Micromech. Microeng. **13**, 568–579 (2003)
52. K. Dems, Z. Mroz, Variational approach to sensitivity analysis in thermoelasticity. J. Therm. Stress **10**, 283–306 (1987)
53. P. Michaleris, D.A. Tortorelli, C.A. Vidal, Tangent operators and design sensitivity formulations for transient nonlinear coupled problems with applications to elasto-plasticity. Int. J. Numer. Meth. Eng. **37**, 2471–2499 (2005)
54. D.A. Tortorelli, G. Subramani, S.C.Y. Lu, R.B. Haber, Sensitivity analysis for coupled thermoelastic systems. Inter. J. Solids Struc. **27**, 1477–1497 (1991)
55. T. Slawig, PDE-constrained control using FEMLAB-control of the Navier-Stokes equations. Numer. Algor. **42**, 107–126 (2006)
56. C.T. Wang, P.C. Chang, C.C. Huang, Optimization of diffuser valves. J. Mar. Sci. Tech. **16**, 134–138 (2008)
57. A. Olsson, G. Stemme, E. Stemme, Numerical and experimental studies of flatwalled diffuser elements for valve-less micropumps. Sens. Actuators A **84**, 165–175 (2000)

58. K.S. Yang, I.Y. Chen, C.C. Wang, J.C. Shyu, Novel no-moving-part valves for microfluidic devices. Microsyst. Technol. **24**, 359–373 (2010)
59. Y.C. Wang, J.C. Hsu, P.C. Kuo, Loss characteristics and flow rectification property of diffuser valves for micropump applications. Int. J. Heat Mass Transfer **52**, 328–336 (2009)
60. F. Okkels, H. Bruus, Design of micro-fluidic bio-reactors using topology optimization, in *ECCOMAS CFD* (2006)

MIX
Papier aus verantwortungsvollen Quellen
Paper from responsible sources
FSC® C105338

If you have any concerns about our products,
you can contact us on
ProductSafety@springernature.com

In case Publisher is established outside the EU,
the EU authorized representative is:
Springer Nature Customer Service Center GmbH
Europaplatz 3, 69115 Heidelberg, Germany

Printed by Libri Plureos GmbH
in Hamburg, Germany